SOLUTE MOVEMENT IN
THE SOIL-ROOT SYSTEM

STUDIES IN ECOLOGY

GENERAL EDITORS

D. J. ANDERSON BSc, PhD
Department of Botany
University of New South Wales
Sydney

P. GREIG-SMITH MA, ScD
School of Plant Biology
University College of North Wales
Bangor

FRANK A. PITELKA PhD
Department of Zoology
University of California, Berkeley

STUDIES IN ECOLOGY VOLUME 4

SOLUTE MOVEMENT IN THE SOIL-ROOT SYSTEM

P.H. NYE

MA, BSc
Reader in Soil Science and
Fellow of St Cross College
University of Oxford

P.B. TINKER

MA, PhD
Professor of Agricultural Botany
University of Leeds

Now Head of Department
of Soils and Plant Nutrition,
Rothamsted Experimental Station

BLACKWELL SCIENTIFIC PUBLICATIONS

OXFORD LONDON EDINBURGH MELBOURNE

First published 1977

British Library Cataloguing in Publication Data

Nye, Peter Hague
Solute movement in the soil-root system.
—(Studies in ecology; vol. 4).
1. Soil moisture 2. Plant-soil relationships
3. Roots (Botany)
I. Title II. Tinker. Philip Bernard III. Series
631.4'1   S594

ISBN 0-632-09730-2

Distributed in the United States of America by
the University of California Press

Typeset by Enset Ltd, Midsomer Norton, Bath
and
Printed in Great Britain at
The Alden Press, Oxford
and bound at
Kemp Hall Bindery, Oxford

CONTENTS

PREFACE

In this book we describe in detail how plant nutrients and other solutes move in the soil in response to plant uptake. The plants we consider may grow in isolation, or as a crop, a mixture of crops or a natural community. The way their roots interact with the soil is not so fully understood as the way their shoots respond to the atmosphere, because the root–soil system is both complex and too inaccessible to measure easily. But our aim is to understand processes in the root zone so fully that we can model them realistically, and predict the effects of variations in natural conditions or in our own practices. Although this aim is not achieved, we think the approach is correct, and has great potential. As the ecologist, A. R. Clapham (1969) has commented: 'We must edge our way in the general direction of the goal. The important point is to learn more than we know at present about comparative magnitudes. What components of the system can be safely neglected in the early rounds of model making?'

At present, the world's rich countries have a vast experience of the effects of the major nutrient elements on important crops, and a sketchier one of those of the other elements. The poor countries, after long delay, are rapidly acquiring like experience. But experience is confined to existing or past conditions, and new varieties of crops and new cultural practices bring with them the need to reassess former conclusions with repeated field trials. For an economically important crop the need can be met, e.g. for maize in Iowa, rubber in Malaysia, or sugar beet in East Anglia; but more often resources do not match the range of crops or vegetation, or diversity of soil, climate and treatment. To take an example: there are more than twenty kinds of vegetables grown commercially in the United Kingdom. Each is available in several varieties. They are grown on soils that differ as much as the sands of Woburn from the peats of the Fens; and under annual rainfalls ranging from 1500 mm in the West to 500 mm in the East, with considerable seasonal variation. The use of herbicides has often removed the need for planting in rows and new patterns are being introduced. The cost of nitrogen and phosphate has varied four-fold within a year. How in these circumstances can reliable advice be given without fundamental insight into the likely influence of these variables?

In recent years an increasingly bewildering variety of insecticides, fungicides, herbicides, stimulants and repressants, radionuclides, heavy metals, toxins and all manner of wastes are being deliberately or inadvertently added to the soil; and we must know their effects on plants, the animals that eat them and the quality of the drainage water. For example, the recommended rates of application of pesticides are continually being altered. How are the consequences to be assessed? All conditions cannot possibly be covered by field trials and modelling may be the only way of obtaining an answer.

These phenomena are dynamic: soil solutes move, and plants grow; yet, as we show in the introductory chapter the intimate connection between the two has only recently been understood. We think therefore that an account of solute transport processes is timely. Hitherto, it has been difficult to link all the separate steps involved in the movement of solutes through the soil and their uptake by extending roots, because the mathematics was too difficult or tedious; and hence simplifying assumptions had to be made; for example, that rate of uptake of solute is directly proportional to its concentration. Fluctuations of external variables, such as rainfall were difficult to include in any detail. The computer has removed these obstacles and become an essential tool in assembling the various pieces of the system. The scientist studying a single process can therefore be fairly confident that if he can describe it quantitatively, either mechanistically or stochastically, it can readily be incorporated in a larger system.

The effort to set down the steps in a system, and to quantify them, helps us to discover the most serious gaps in our understanding, to formulate precise questions and to devise experiments under controlled conditions to answer them. In describing these steps we have relied on experiments, remembering Sir Fredrick Bawden's remark 'The trouble about computer models is that you feed in your own prejudices, and they come back at you' (BBC January 5th, 1972). Most of the mechanisms described have been worked out for the major nutrient elements, simply because they were first in the field. But they are equally relevant, with modifications, to other solutes whether beneficial or harmful.

Even a detailed understanding of linked-transport processes will not remove the need for field experiments and observations. Soils are usually too heterogeneous and plant behaviour too complex for us to achieve the accuracy possible with physico-chemical systems—a point emphasized by Passioura (1973). But a good model or theory will suggest questions, and combinations of treatments or effects that had been overlooked or undervalued. It will help to decide whether new trials are worthwhile. Without it, almost any empirical approach is worth a shot, and it is alarmingly evident that the power of the computer for correlating data assures an abundant supply of raw material. Even so, essential measurements are often omitted for lack of a coherent model. We have been unable to make use of many experimental results achieved at great expense, simply because a measurement which could readily have been made, such as root length, has been omitted.

In addition it does not seem to be sufficiently appreciated that statistical techniques such as multiple regression do not generate causal relations between variables. In our subject these are rarely simple, e.g. equation 7.12, and can only be deduced from insight into the mechanisms involved.

Most models of natural ecosystems or crop production are coarse grained, the object being to establish a framework and fill in the details later. Our approach is different in that we analyse the working of small scale, often simplified systems first before combining them into more complicated ones. The first five chapters describe basic processes of solute movement in the root zone. Chapter 1 outlines the history of ideas on the subject and concludes with a simple account of the continuity equation, which underlies most quantitative treatments. Chapter 2 deals with water movement in sufficient depth to show how it should be introduced into models of solute movement. Chapter 3 describes how solutes are distributed between the solid, liquid and gas phases of the soil; and Chapter 4 their local movements, particularly by diffusion. Chapter 5 concentrates on plant roots and the knowledge we have gained from solution culture about the uptake rate of solutes. In the next three chapters these processes are combined in increasingly complex systems. Chapter 6 examines the complicated changes occurring in the soil around single roots of intact plants. Chapter 7 describes the interactions within the root zone of a single whole plant; and this treatment is extended in Chapter 8 to include a crop or plant community.

We have not attempted to deal with anaerobic soils by these methods because the problems they raise are very complicated and to treat them adequately—if that is possible—would have unduly prolonged the appearance of this book.

We assume readers will know some soil science and rather less plant physiology. Rather than attempt an exhaustive review we have mentioned work that has contributed most to our own understanding. However, where points are contentious we review the evidence more thoroughly.

We have used the International System of Units and its abbreviations. Since so many of the results we quote use the centimetre, e.g. diffusion coefficients are nearly always reported in $cm^2 s^{-1}$, we normally use this as the most convenient unit of length.

ACKNOWLEDGMENTS

Many of the experimental results recorded here have been obtained by research staff and students, supported by the Agricultural Research Council, and the Ministry of Agriculture, Fisheries and Food. We should like to acknowledge the far-sighted policies of these bodies which has encouraged work of the type described in this book.

We thank our students and colleagues for the keen stimulation they have provided; Jennifer Rowles for her help with references; Phyllis Nye for indexing; and our publishers for their patience and flexibility.

We are grateful for permission to make use of the material contained in the following sources for the figures and tables listed.
Journal of Applied Ecology, Figures 2.6; 2.8; 2.10; 6.18; 7.13; 8.2. Tables 2.1; 8.6; *Soil Science*, Figures 2.4; 2.5; 2.9; 3.4; 4.8; 4.10; 5.5; 6.11; 7.9; 8.4; *British Journal of Applied Physics* (Institute of Physics), Figures 2.3; 4.6; 4.7; *Canadian Journal of Soil Science*, Figure 2.1; *Nature*, Table 5.6; *Plant Physiology*, Figures 5.8; 5.9; *Journal of Agricultural Science (Camb.)*, Figures 3.14; 6.9; 8.12; 8.13; 8.14; 8.17. Table 8.8; *Plant and Soil*, Figures 3.8; 3.11; 6.2; 6.3; 6.5; 6.6; 6.7; 6.8; 6.12; 6.13; 6.14; 6.15; 6.16; 6.20; 7.16; 7.18; 7.19; 7.20; 7.21; 7.23; 7.24; 7.25; 8.16. Tables 6.6; 6.12; 8.4; *Annual Report of the ARC Letcombe Laboratory*, Tables 6.7; 7.5; *Oecologia Plantarum*, Table 8.7; (1970, 5, 328—publ. Gauthier–Villars); *Journal of Soil Science*, Figures 3.1; 3.2; 3.3; 3.5; 3.15; 4.1; 4.4; *Bulletin of the International Society of Soil Science*, Figure 3.6; *Proceedings of the Fertilizer Society*, Figure 3.12; *North-West Science*, Figure 3.13; *Annals of Applied Biology*, Figure 8.5; *Weed Research*, Figures 3.17; 3.18; 8.11; *Journal of General Physiology*, Figure 4.3; *Journal of the Science of Food and Agriculture*, Figures 4.5; 8.6; *Physiologia Plantarum*, Figures 5.2; 5.12; *Pesticide Science*, Figure 6.2; *Rothamsted Experimental Station Annual Report for 1973*, Figure 7.5. Table 7.4; *Chemistry and Industry*, Figure 7.7, Table 8.1; *New Phytologist*, Tables 6.10; 7.1; *Journal of Forestry*, Table 6.9; *Forest Science*, Table 6.5; *Royal Institute of Chemistry Reviews*, Table 3.4; *Soil Science Society of America Proceedings* By permission of Soil Science Society of America. Figures 2.2 (1943, 8, 117); 2.7 (1970, 34, 384); 3.7 (1969, 33, 207); 3.10 (1947, 12, 120); 7.8 (1970, 34, 424); 7.10 (1964, 28, 603); 8.7 (1957, 21, 23); 8.8 (1969, 33, 831); 8.10 (1964, 28, 472). Table 7.8 (1970, 34, 424); *Advances in*

Agronomy, Figures 5.4 (1973, *25*, 181); 7.3 and 7.17 (1970, *22*, 162 and 178); *Agronomy Journal*, Figures 5.11 (1975, *67*, 334); 7.11 (1971, *63*, 771). Tables 7.3 (1970, *62*, 9); 7.7 (1969, *61*, 810); 8.3 (1974, *66*, 401); *Proceedings of the American Society for Horticultural Science*, Table 7.6 (1963, *83*, 682); *Journal of Experimental Botany*, Table 5.7 (1959, *10*, 308) By permission of Oxford University Press); *Planta*, Figure 5.10.
Books
Society of Experimental Biology Symposium No. 19. 1965, Figure 2.10.
National Agricultural Advisory Service Open Conference: Residual value of applied nutrients. *Ministry of Agriculture, Fisheries and Food, Tech. Bull. 20*, Figure 3.9.
International Atomic Energy Agency, Technical Report No. 48, Figure 4.2.
Epstein, E.: *The Mineral Nutrition of Plants*. Wiley, New York, Figure 5.2.
Carslaw, M. S. & Jaeger, J. C., *Conduction of Heat in Solids*. Clarendon Press, Oxford, Figure 6.4.
Sanders, F. E. T., *DPhil. thesis*, Oxford, Figure 6.10.
Spectrum No. 126. Central Office of Information, Crown copyright, Figure 6.21.
United States Department of Agriculture Technical Bulletin 923. Bureau of Plant Industry, Figure 7.2.
Carson, F. W. (ed.), *The Plant Root and its Environment*. University of Virginia Press, Figures 7.4; 7.12.
Troughton, A., *The Underground Organs of Herbage Grasses. Bulletin 44*, Imperial Bureau of Plant Genetics, Figures 5.1; 7.6.
Van Schilfgaarde, J. H. (ed.), *Drainage for Agriculture*, American Society of Agronomy 1974, Figure 8.9.
Weaver, J. E., *Root Development of Field Crops*. Used with permission of McGraw-Hill Book Co., New York, Figure 7.1.
Esau, K., *Anatomy of Seed Plants*, Wiley, New York, Figure 5.1.
Jennings, D. H. & Lee, D. L. (ed), *Symbiosis*. Society of Experimental Biology Symposium *29*, Cambridge University Press, Figure 6.24.
Whittington, J. H. (ed.), *Root Growth*. 15 Easter School, University of Nottingham. Butterworths, London, Figures 8.3.; 8.15.
International Potash Institute. Proceedings 10th Colloquium, Abidjan, 1973, Table 8.5.
Rorison, I. (ed.), *Ecological Aspects of the Mineral Nutrition of Plants*, British Ecological Society Symposium 1968, Figure 7.13.
Sparling, G., *PhD thesis*, Leeds, U.K. 1976, Table 5.1.
Goring, C. A. I. & Hamaker, J. W. (eds.), *Organic Chemicals in the Soil Environment*. Marcel Dekker, New York, Tables 3.2; 3.3.

LIST OF MAIN SYMBOLS

Symbol	Definition	Units
A_L	leaf area	cm^2
A_R	root surface area	cm^2
C	concentration of diffusible solute in soil	mol cm^{-3} of soil
C_g	concentration of solute in gas phase	mol cm^{-3} of gas
C_1	concentration of solute in liquid phase	mol cm^{-3} of liquid
C_s	concentration of solute in solid phase	mole cm^{-3} of solid
D	diffusion coefficient of solute in soil	cm^2 s^{-1}
D^*	dispersion coefficient of solute in soil	cm^2 s^{-1}
D_g, D_1, D_s	diffusion coefficient of solute in gas, liquid, solid phase	cm^2 s^{-1}
D_1^*	longitudinal dispersion coefficient of solute (equation 4.22)	cm^2 s^{-1}
E	net assimilation rate	g cm^{-2} d^{-1}
F	flux of solute	mol cm^{-2} s^{-1}
G	Gibbs free energy	J
I	inflow (net absorption rate by unit length root axis)	mol cm^{-1} s^{-1}
I_w	absorption rate of water by unit length root axis	cm^3 cm^{-1} s^{-1}
K	hydraulic conductivity (equation 2.2)	cm s^{-1} or cm^2 s^{-1} bar^{-1}
L	root length	cm
L_A	root length under unit area of land surface	cm cm^{-2}
L_P	root length per plant	cm
L_V	root density (length per unit volume of soil)	cm cm^{-3}
LAR	leaf area ratio (A_L/W)	cm^2 g^{-1}
M	molecular weight	
M_t, M_∞	amounts absorbed at time t, ∞	mol
N	equivalent fraction	
P	external pressure	bar (1 bar $= 10^5$ Pa $= 10^5$ Nm^{-2})

R_W	relative growth rate $(dW/dt)/W$	d^{-1} or s^{-1}
S	unit absorption rate (absorption rate per unit weight of fresh root)	$mol\ g^{-1}\ s^{-1}$
U, U_R, U_S	solute content of plant, root, shoot	mol
V	volume	cm^3
V	uptake rate per unit fresh weight (Chapter 5)	$mol\ g^{-1}\ s^{-1}$
W, W_R, W_S	dry weight of plant, root, shoot	g or mg
W_{RF}	root fresh weight	g
X	concentration of solute in dry plant material	$mol\ g^{-1}$
a	radius of root axis	cm
a_h	radius of root hair	μm
b	solute buffer power of soil (dC/dC_1)	
f_g, f_1, f_s	diffusion impedance factor in gas, liquid, solid phase	
k	rate constant	s^{-1} or d^{-1} or yr^{-1}
q	water (uptake rate per unit crop area)	$cm^3\ cm^{-2}\ s^{-1}$
r	radial distance	cm
t	time	s or d or yr
v	water flux	$cm^3\ cm^{-2}\ s^{-1}$
v_g, v_1, v_s	volume fraction of gas, liquid, solid phase	
z	ionic charge	
α	root absorbing power (F/C_{1a})	$cm\ s^{-1}$
$\overline{\alpha a}$	root demand coefficient	$cm^2\ s^{-1}$
β	gas solubility coefficient (C_1/C_g)	
γ	flux parameter (equation 7.5)	
δ	unstirred layer thickness	μm
θ	soil moisture fraction by volume	$cm^3\ cm^{-3}$
λ	packing factor (equation 4.22)	
μ	chemical potential	$J\ mol^{-1}$
μ_e	electrochemical potential	$J\ mol^{-1}$
ϕ	water potential (Chapter 2)	bar (10^5 Pa)
ψ	water matric potential (Chapter 2)	bar (10^5 Pa)

Subscripts

a	at root surface	i	initial value	s	solid phase
g	gas phase	l	liquid phase	t	at time t
h	root hair			w	water

Note: if a symbol is used in another sense this is noted in the text.

1

INTRODUCTION

It is now widely accepted that under given growth conditions uptake of a solute by roots is related to its concentration in the soil solution and the extent to which this is buffered by the soil. Though these apparently simple ideas were advanced more than a century and a half ago, it is only recently that they have been defined clearly enough to form a basis for detailed understanding of the effect of solutes on plants grown in soil. They have, in particular, been obscured by specific effects of roots with their associated organisms: for roots not only vary widely in their response to solute concentration, but alter in their proximity the soil properties we measure in the bulk of the soil. Thus it is only in the past 20 years or so that we have come within reach of the objective clearly set us by Liebig in 1840 when he wrote: 'A rational system of agriculture must be based on an exact acquaintance with the means of nutrition of vegetables, and with the influence of soils and action of manure upon them'.

THE ORIGINS OF CURRENT IDEAS

1.1 Foundations

The history of ideas about soil and plant relations has been well described by Sir John Russell (1937) for the period up to the beginning of this century. Even though Jethro Tull in 1731 could not decide whether nitre, water, air, fire or earth was the food of plants, John Woodward (1699), 30 years earlier, seems to have grasped the essential idea of a flow of solutes to roots induced by the transpiration stream when, having grown spearmint in a primitive 'nutrient' culture solution, he wrote: 'It has been shown that there is a considerable quantity of (terrestial) matter contained in rain, spring and river water; that the greatest part of the fluid mass that ascends up into plants does not settle there but passes through their pores and exhales up into the atmosphere: that a great part of the terrestial matter, mixed with the water, passes up into the plant along with it, and that the plant is more or less augmented in proportion as the water contains a greater or lesser quantity of that matter; from all of which we may reasonably infer, that earth, and not water, is the matter that constitutes vegetables'. Then, in 1804, de Saussure showed that the root was selective and could absorb the salts in water at different rates, excluding some so that they became more concentrated in the external solution. de Saussure isolated soil solution by compacting the soil; but as long ago as 1866 Schloesing collected the solution in the soil pores by displacing it with water coloured with carmine, a method that is still as good as any available. The idea that the soil solution was the main source of plant nutrients was further advanced by Whitney & Cameron (1903).

A fascinating history of solute–solid adsorption studies has been chronicled by Forrester & Giles (1971). The capacity of the soil to adsorb substances from solution—so rendering barnyard liquor and other coloured waters clear and odourless—had long been known; but a clear statement of the buffering role of the soil lies in the following quotation from Gazzeri (1823): 'Loam and especially clay take possession of soluble matters which are entrusted to the soil, and retain them in order to give them by degrees to plants, conformably with their needs'. And Lambruschini (1830) wrote that 'fertilizing liquids and the constituents of a well-prepared soil enter into a peculiar combination, by virtue of a special affinity. This combination is not weak enough to allow any very easy loss of the fertilizing constituents, and to permit plants to consume them too rapidly, and yet the combination is not so strong, but that the vital action of growing plants can gradually overcome it.' Such ideas as these were the forerunners of experiments made between 1845 and 1852 by Huxtable, Thompson, and particularly Way (1850, 1852) that established the effect of 'base exchange'.

Meanwhile Daubeny (1846) was considering what fraction of the total soil phosphorus was available to plants, and decided the active fraction was that dissolved in water impregnated with carbon dioxide. As Larsen (1967) has pointed out, Daubeny recognized that 'if the results of soil phosphorus studies are to be meaningful, the soil must be kept as chemically intact as possible. . . . His principle was widely violated in the following century, and only in recent times has it been reinstated'.

These ideas about the soil solution and its buffering had to contend with others which considered that the feeding power of plants should be attributed to the total acidity of sap in their roots (Dyer 1894); and that carbon dioxide excreted by the roots forms carbonic acid which can make available a fraction of the nutrients in 'readily available' minerals (Czapek 1896).

It is remarkable that there were no really new developments in this position for the first forty years of this century except in understanding of the exchange chemistry and structure of soil clays. The agricultural chemists and the plant physiologists scarcely seemed to communicate with each other. The former concentrated on finding a suitable extractant for 'available' nutrients in the soil, the first example being Dyer's 1% citric acid (1894), to be followed by numerous others, all based on a search for the 'available' fraction. The main method of investigation of soil nutrient questions during this period—and indeed since—was field experiments, designed and interpreted with the aid of the powerful statistical methods developed by R. A. Fisher (1925). In practical terms this approach has been extremely successful, and forms the basis of modern fertilizer practice. However, for scientific purposes, it was over-emphasized, since it led agricultural chemists to be satisfied with correlations and regressions between fertilizer responses and chemical extracts, and in-hibited the search for more fundamental and detailed explanations of their results.

Meanwhile, plant physiologists, eschewing such a complex medium as soil in favour of nutrient culture solutions, established the 'essential' nutrient elements, and determined the main metabolic factors affecting salt absorp-tion. Unfortunately, their experiments were usually made at concentrations very much higher than that of the soil solution, so that the quantitative as-pects of their work are all too often irrelevant to soil conditions. During this period, Californian workers (Hoagland 1944) were exceptional in attempting to bridge the gap between soil and nutrient-solution cultured plants—possibly because salt concentrations in their soils are unusually high.

Much of the early history of agricultural chemistry was dominated by the search for the role of humus in plant nutrition, and the source of plant nitrogen—a story well told by Sir John Russell (1937). It was one of the major triumphs of the last quarter of the 19th century that bacteriologists discovered the main outlines of the nitrogen cycle and particularly the sources of ammonium and nitrate in the soil. During the present century the incredible diversity of the whole soil population of microbes, fungi and animals has been recognized, and a new branch of the subject—soil ecology—has come into being. It was also recognized that there was a particularly active population in the root–soil interfacial region, to which, according to Russell (1973 p. 241), the term 'rhizophere' was first given in 1904 by Hiltner. The special properties of the rhizophere have subsequently received much attention, and we discuss this work in Chapter 6.

From the standpoint of soil solutes, it has proved very difficult to relate the numbers and types of organisms to the rates of mineralization of organic matter or microbial metabolism. We now have a wide experience of the rates of mineralization of litter and humus, and also specific pesticides, gained in a variety of environments, but know little about the details of the mech-anisms involved. We have in general therefore to be content with empirical rate functions.

1.2 The beginning of the modern period (*c.* 1940–1960)

In 1939 Jenny & Overstreet amplified the concept of 'contact exchange' which had been suggested by Devaux (1916). He noted that roots, like the soil, had cation exchange properties, and thought that since they were in such close contact they could readily exchange cations. Jenny & Overstreet (1939a) held that the nutrient cations were absorbed in this way in exchange for hydrogen ions produced at its surface by the root. They further designated 'contact exchange' as an overlapping of the volumes within which the individual absorbed ions oscillated at their adsorption sites, and added 'the ions do not enter the soil solution *per se*, but, in the moment of contact they jump directly from one particle to another.' So stated, contact exchange, while it may exist, is quite inadequate to account for the number of cations absorbed by plants. However, Jenny & Overstreet (1939b) further claimed that cations could migrate along clay surfaces, and could pass from one clay surface to another if they were in close contact with one another. Hence the roots could draw on a much greater number of cations than they could by immediate contact (Albrecht *et al.* 1942). Here we can recognize the beginnings of our current ideas about the zones around the root from which nutrients are diffusing. However, as Russell (1950 p. 444) realized: 'The process of contact exchange does not in fact differ fundamentally from uptake from the solution, for in either case the ion can only be transferred through the water film surrounding the root, and the composition of the soil solution is controlled by the solid phase of the soil as well as by the uptake of nutrients by the crop.'

It proved difficult to demonstrate that plants would grow as well in dilute nutrient solutions, similar to typical soil solutions, as they did in the same solution buffered by a suspension of clay or ion exchange resin. However, when precautions were taken to ensure that the solution was well stirred and the supply of nutrient ions maintained as they were depleted by uptake, Lagerwerff (1960) was able to conclude that 'It seems clear that uptake rates of cations from solution, sand, and sand and resin are the same provided the soil solution is renewed sufficiently often'; and Olsen & Peech (1960) concluded that their results of rubidium uptake from clay suspensions 'fully support the classical soil solution theory of mineral nutrition of plants proposed long ago by Cameron'. This equivalence has not been conclusively shown with soil suspensions, probably because of microbial effects and also the influence of hormones derived from the soil.

There is another difficulty about the extended theory of contact exchange. It has for long been generally accepted that plant roots make the soil more acid, Jenny & Overstreet (1939a) suggesting that they exchange hydrogen ions. Other sources of hydrogen ion have been thought to be carbon dioxide and organic acids excreted by the roots (Schander 1941). However, as we discuss more fully in Chapter 6, in general, roots in soil take up more anions than cations, and consequently have to liberate bicarbonate ion if a high potential difference is not to develop between the root and the soil. The soil near them thus as a rule becomes more alkaline.

If the concentration of an element in a plant is only a minute fraction (10^{-3}–10^{-5}) of its concentration in the soil, then it is possible that the layer of mucilage, a few μm thick, often observed on the root surface, may contact or envelop sufficient soil to solubilize enough of the element required. Passioura (1966) concludes that finely-divided manganese and iron oxides (Jones 1957; Jenny 1960), and molybdenum adsorbed on ferric hydroxide probably provide elements in this way.

The next notable advance was made independently by Schofield (1955) and Woodruff (1955). They attempted to define more precisely the idea of the 'availability' of an ion, and suggested that a measure of this was the work needed to withdraw it from an equipotential pool in the soil. This work was to be measured as the chemical potential of the ion referred to the calcium ion, which is usually the dominant ion in the soil solution. Schofield, writing of phosphate, drew a very clear distinction between the intensity of its supply, to be measured by the chemical potential of $Ca(H_2PO_4)_2$, and the quantity in the exchangeable pool of phosphate that he considered maintained it. He used the analogy of a well: 'It is the depth to water which determines the work needed to get the water to the top. Similarly (the chemical potential of $Ca(H_2PO_4)_2$) is a measure of the "depth" of the "level" of the pool of soil phosphate. As water is taken out of the well it is replaced by lateral movement of ground water, and as phosphate is taken from the soil solution it is replaced by desorption. In neither case is replacement quite complete; withdrawals generally cause some lowering of the "level", depending on the "capacity" of the system'.

These thoughts led to a great deal of valuable work on the relation between changes in chemical potential of a nutrient ion in solution and related changes in the amount adsorbed on the solid, which we describe in Chapter 3.

Nevertheless, it has not been shown that nutrient uptake is controlled by the work the root has to do to absorb a nutrient: and indeed it seems unlikely that it should be, since the energy available as a result of root respiration greatly exceeds that required over the same period to accumulate ions such as potassium, which are at a higher concentration in plant cell vacuoles than in the soil solution (Nye 1968). In addition Wild (1964) and Wild *et al.* (1969) have shown that the uptake of phosphorus and of potassium is more closely related to their concentrations in solution than to their chemical potentials referred to the calcium ion potential.

Meanwhile, plant physiologists concentrated their attention on salt absorption by portions of plant tissue, rather than whole plants. Radio-isotopes, now readily available, greatly assisted them. Much work was done with roots excised from their shoots, barley seedlings being particularly suitable (see Epstein 1972). In addition, slices of storage tissue, such as carrot discs; the nutrient absorbing leaves of aquatic plants like *Elodea*; single and multicellular algae; cell organelles, e.g. mitochondria; and ectotrophic mycorrhizal tissue were all studied (see Sutcliffe 1962). For the purpose of our theme many of these experiments are important because they were made at lower solution concentrations than hitherto, and we discuss them more fully in Chapter 5. They distinguished between passive and active uptake, and revealed that the initial rates of active uptake were usually related to the solution concentration by a diminishing returns form of curve, though the underlying mechanisms are still keenly debated. They also revealed quantitatively the effects of competition between ions on their rates of uptake. Consequently, it became clear that to interpret the effect of applying a salt to a soil on the composition of a plant, e.g. the effect of adding potassium chloride on the magnesium composition, one had to know both the change produced in the composition of the soil solution, and also the consequent effect on the plant's absorption mechanism.

During this period the importance of knowing the rate of uptake of solutes per unit of root in intact growing plants was not widely appreciated, and in the discussion of this question in Chapter 5 there are few references to work before 1960.

The approach of soil scientists to the problem of nutrient uptake from the soil had so far been essentially static. It considered that as roots ramified through the soil they took up nutrients at particularly active zones just behind their root tips, tapping the available nutrients or rendering them available as they progressed. The question how the individual ions reached the surface of the root does not seem to have been raised, and even the role of transpiration in sucking the soil solution to the root surface seems to have been neglected. The first signs of a more dynamic view emerged from a seminal paper by Bray (1954) aptly entitled 'A nutrient mobility concept of soil–plant relationships'. He wrote: 'The mobility of nutrients in soils is one of the most important single factors in soil fertility relationships. The term "mobility" as used here, means the overall process whereby nutrient ions reach sorbing root surfaces, thereby making possible their sorption into the plant. Thus the term involves the solution or exchange of the nutrient as well as its movement to the root surfaces. A correlative process, just as important, is the growth of the roots and the extension of the (working) root surfaces into areas where the nutrients occur. These two processes, complementing each other, largely (determine) the soil fertility requirements of a plant'. Bray also recognized that individual roots would be more likely to compete with each other for mobile nutrients such as nitrate, than for relatively immobile ones such as phosphate.

Tepe & Leidenfrost (1958) also expressed the new dynamic view clearly when they wrote: 'To obtain a realistic estimate of the availability of plant nutrients in soils due consideration should be given to the degree of mobility of ions and their proximity to an absorbing surface (root hair), and to the fact that the uptake of nutrients on the one hand and the growth of plant roots on the other prevents the plant–nutrient system from ever reaching equilibrium'.

These ideas soon bore fruit and the subject was put on a quantitative basis. The 'mobility' of ions was resolved into two processes: mass flow of the soil solution to the root induced by transpiration; and diffusion of ions to the root induced by lowering of their concentration by uptake at its surface. Barber (1962) calculated the proportion of nutrients in a plant that could be supplied by mass flow of the soil solution induced by the transpiration stream; and Walker & Barber (1961) also demonstrated by radioautographs zones of nutrient depletion and also of accumulation around roots. Bouldin (1961) and Olsen *et al.* (1962) applied the mathematics of diffusion theory to ions diffusing to the root. It is interesting to note that the first theoretically sound treatment of water movement to roots was made at about the same time, by Philip (1957b).

Measurements of ionic diffusion in pure clay systems and ion exchange resins had been made for many years previously, but it was Porter *et al.* (1960), and Schofield & Graham-Bryce (1960) who showed how the diffusion coefficients of ions could be measured in soil.

This book is largely an account of the quantitative work over the past decade that sprang from these beginnings.

1.3 Wider perspectives

So far we have given an account of the development of ideas about the localized movement of ions in the immediate neighbourhood of the absorbing surfaces of roots. To apply these ideas to problems of crop production and ecology we need a model of uptake by whole plants or communities of plants. Thus we have to show how behaviour of solutes at the root–soil interface may be integrated over whole root systems; and to do this we have to understand in detail the grosser movements of solutes within the profile.

A broad-scale approach to this question has been adopted in innumerable studies, ever since the main plant nutrients derived from the soil were clearly recognized in the middle of the last century. In 1868 Augustus Voelcker reported analysing the water draining from variously fertilized plots of Broadbalk Field, Rothamsted, anticipating that his investigation was likely to open up 'quite a mine of theoretical enquiry' (Russell 1966 p. 126); and the relative ease of leaching of various nutrients from the topsoil was early appreciated, e.g. Dyer (1901). Nutrient balance sheets for cropped soils have been drawn up, and their importance in assessing long-term changes in fertility realized—a subject thoroughly reviewed by Cooke (1967, 1969). Foresters, particularly in Germany and France, measured the nutrient uptake of mature trees, and published numerous papers between 1876 and 1893. Rennie (1955) has reviewed this and later work, and discussed the longterm effects of timber production on the nutrient balance in the soil.

Ecologists were slower to study the wider aspects of the nutrient balance in the ecosystem. Sampling of mixed communities is laborious, and natural vegetation does not provide the commercial incentive of a crop. Where the natural vegetation is relied on to restore fertility of worn out soils, as in the practice of shifting cultivation, considerable information does exist (Nye & Greenland 1960). Available information on nutrient contents and cycles in a range of natural vegetation from tundra to tropical rain forest has been assembled by Rodin & Bazilevich (1965). Much more data on natural communities will shortly be published as a result of work on Primary Productivity co-ordinated under the International Biological Programme.

In the exploitation of the soil by plant roots there is a marked difference between a root system developing from seed in a fertilized soil, and progressively exploiting deeper layers of the soil—characteristic of an annual crop—and the established root system of perennial plants, such as occurs in many natural communities, e.g. a humid woodland. Nutrient levels in the soil solution in the latter are usually much lower; the overall rate of uptake will depend on rates of mineralization of humus, release by weathering, and leaching from living vegetation and litter by rain; and finally, competition between roots is intense. Factors that may aid this competition such as mycorrhizal hyphae, or root surface enzymes are likely to be particularly important in explaining the success of a particular species. Since the action of these factors is little understood, our understanding of such systems is correspondingly much less complete.

The possibility of a more exact prediction of movement of solutes in the whole root zone arose when the ideas of ion exchange chromatography were applied to leaching of ions through columns of soil by Ribble & Davis (1955). A soil profile can in fact be viewed as a large, irregularly packed, erratically eluted, multilayer, ion exchange column. Needless to say, no exact predictions of its behaviour are possible; but as long as this is recognized, the principles of exchange chromatography provide a sound theoretical starting point for models that can be increasingly refined as the complexities are understood.

We aim in Chapter 8 to show how the detailed understanding we are now gaining of solute movement in small volumes of soil can give greater insight into these larger and complex systems.

1.4   The continuity equation

Although our preliminary survey of solute movement has ranged from small to large scale processes, their quantitative treatment, which is developed in subsequent chapters, is usually based on some form of the 'continuity' equation. According to the problem this may be formulated in Cartesian, cylindrical or spherical co-ordinates; and be solved to satisfy various boundary conditions, e.g. that there is a specified concentration of solute at the boundary between a root surface and the adjacent soil. It is therefore convenient to explain this simple concept here, and to present together for future reference a variety of forms in which it appears.

1.4.1 Solute transfer

We may illustrate the principle involved by considering a movement of a solute through soil (e.g. by diffusion) in the direction of the x axis. Figure 1.1 shows two imaginary planes of unit cross-section normal to the axis, and distance δx apart. The volume enclosed is $\delta x \times 1 = \delta x$. Within this volume we have:

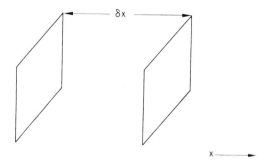

FIG. 1.1

$$\begin{Bmatrix} \text{rate of gain in solute} \\ \text{within volume } \delta x \end{Bmatrix} = \begin{Bmatrix} \text{rate of entry across} \\ \text{the plane at } x \end{Bmatrix} - \begin{Bmatrix} \text{rate of exit across the} \\ \text{plane at } x + \delta x \end{Bmatrix}$$

i.e. $\left(\delta x \, \dfrac{\partial C}{\partial t} \right)_x \simeq \left(F_x - F_{x + \delta x} \right)_t \simeq \left(-\dfrac{\partial F}{\partial x} \, \delta x \right)_t$ (1.1)

where $F_x, F_{x + \delta x} = $ flux of solute at $x, x + \delta x$
  $C = $ amount of solute per unit volume of soil.
As $\delta x \to 0$

$$\left(\frac{\partial C}{\partial t} \right)_x = \left(-\frac{\partial F}{\partial x} \right)_t. \tag{1.2}$$

Equation 1.2 is the continuity equation in one dimension.

If movement of solute is by diffusion alone, by Fick's first law (p. 69)

$$F = -D\frac{dC}{dx} \tag{1.3}$$

where D is the coefficient of diffusion,

$$\therefore \frac{\partial C}{\partial t} = \frac{\partial}{\partial x}\left(D\frac{dC}{dx}\right) \tag{1.4}$$

an equation known as Fick's second law.

If the solvent (water) is also moving, solute is carried by mass flow, or convection as it is sometimes termed.

$$F = -D^*\frac{dC}{dx} + vC_1 \tag{1.5}$$

where D^*, the dispersion coefficient, differs from the diffusion coefficient because the water movement itself causes some dispersion of the solute molecules (see Chapter 4). v is the water flux in the direction x, and C_1 is the concentration of solute in the soil solution.

For combined mass flow and diffusion the continuity equation, obtained by substituting F in equation 1.5 into equation 1.2, becomes

$$\frac{\partial C}{\partial t} = \frac{\partial}{\partial x}\left(D^*\frac{dC}{dx}\right) - \frac{\partial(vC_1)}{\partial x}. \tag{1.6}$$

Notice v occurs within the differential unless there is a steady flow of water.

It is evident that the flux of solute often depends to a great extent upon the movement of water. We consider the essential aspects of water movement in Chapter 2.

If the solute is volatile, we have to consider its movement through the soil air, usually by diffusion, though convection may also contribute in some instances. These complications are considered as they arise. We do not treat the movement of gases or problems of soil aeration in any depth.

1.4.2 Solute reaction

The level of solute may also change as a result of processes occurring within the volume δx. To allow for the rate of such reactions we add a term $f(C)_{x,t}$, often called a 'source' term, to equation 1.2 to give

$$\left(\frac{\partial C}{\partial t}\right)_x = \left(-\frac{\partial F}{\partial x}\right)_t + f(C)_{x,t}. \tag{1.7}$$

Examples of processes that are specified by this rate function are.
(a) Slow release of ions from non-exchangeable to exchangeable form.

(b) Release or fixation of solutes by soil organic matter transformations.
(c) Degradation of organic solutes by purely chemical, or microbial action.
We do not attempt to cover all this ground in detail, though we aim to mention any quantitative generalizations that can be made.

A function of this form is also particularly useful for expressing the rate of uptake by a system of roots in a small finite volume of soil, and we develop this approach for single plants in Chapter 7, and for plant communities in Chapter 8.

1.4.3 Generalization of the continuity equation

Cartesian co-ordinates

When movement may occur in three dimensions, equation 1.7 becomes

$$\left(\frac{\partial C}{\partial t}\right)_{x,y,z} = -\left(\frac{\partial F_x}{\partial x} + \frac{\partial F_y}{\partial y} + \frac{\partial F_z}{\partial z}\right)_t + f(C)_{x,y,z,t}. \tag{1.8}$$

Cylindrical co-ordinates

If movement occurs in a direction normal to a cylinder, e.g. to a root, equation 1.8 may be expressed as

$$\frac{\partial C}{\partial t} = -\frac{1}{r}\frac{\partial}{\partial r}(rF_r) + f(C)_{r,t} \tag{1.9}$$

where r is the radial distance from the axis of the cylinder.

Spherical co-ordinates

For some problems, such as movement into a soil aggregate, movement may be considered normal to the surface of a sphere and equation 1.8 may be expressed as

$$\frac{\partial C}{\partial t} = -\frac{1}{r^2}\frac{\partial}{\partial r}(r^2 F_r) + f(C)_{r,t} \tag{1.10}$$

where r is the distance from the centre of the sphere.

In simple cases, these equations may be solved analytically. Most of these solutions will be found in the books of Carslaw & Jaeger (1959) and Crank (1975). More often, numerical computer methods must be used, and many examples of these will be encountered in subsequent chapters. Crank (1975, Chapter 8) gives an introduction to them, and multi-dimensional problems are treated more fully by Mitchell (1969, Chapter 2).

2

MOVEMENT OF WATER THROUGH SOILS TO PLANTS

Water is the main agent moving solutes in the soil profile, and an important one in conveying them to root surfaces and thence to the shoot. Since it moves from points of high to low water potentials this concept will be considered first.

2.1 Water potential

The equilibrium state of water in soil is that in which all the water has the same partial molar Gibbs free energy, $\bar{G}$. This is also known as the chemical potential, μ. Various physical mechanisms affect the potential energy, and it is necessary to identify these, since they do not all have similar consequences for soil water movement (see for discussion ISSS, 1974). A general statement of the components of the partial molar free energy of water in the soil is:

$$d\bar{G} = (\bar{V})_{T,\theta,n}\, dP - (\bar{S})_{P,\theta,n}\, dT$$
$$+ \left(\frac{\partial \bar{G}}{\partial \theta}\right)_{T,P,n} d\theta + \left(\frac{\partial \bar{G}}{\partial n}\right)_{T,P,\theta} dn + Mgh \qquad (2.1)$$

where $\bar{G}$ is the partial molar Gibbs energy, $\bar{V}$ the molar volume, P the external pressure, $\bar{S}$ the molar entropy, T the temperature, n the mol fraction of solute, θ the volumetric water content, M the molar mass of water, g the acceleration due to gravity and h the height above some specified reference level. The pressure may be the hydrostatic pressure of a column of water in saturated soil, an increase of atmospheric pressure above standard reference level, or the overburden pressure of the soil where part of the load is carried by the soil water. The second term may cause some theoretical difficulty, since in the derivation of this equation (Bolt & Frissel 1960; Slatyer 1967) it is postulated that $\bar{S}$ is at constant pressure. In fact, the pressure within the soil water will vary with temperature, which alters the surface tension. Wilkinson & Klute (1962) investigated the variation of soil water potential (defined below) with temperature, but the explanation of the results in terms of change in surface tension was only successful for very sandy soils. No

12

full and exact treatment of the temperature effect on soil water potential has been published, and the theory of transfer of water due to a temperature gradient is consequently not entirely satisfactory (see p. 19). The third term includes all matric suction effects in the soil pores, which arise from interactions of the water with the solid material of the soil. It thus includes both the effects of surface tension and of surface adsorption, and avoids any discussion of the relative effects of these. The fourth term deals with the osmotic component of the free energy, which results from the presence of freely-diffusible solutes in the soil solution. Considerable care is needed in the use of this term for problems of soil water flow, since a solute gradient may drive both a flow of water and a flow of solute. This point is of some interest in relation to dispersion of fertilizer and to mass flow accumulations around plant roots. The last term allows for the effects of gravity, which are particularly significant in drainage problems.

The term 'water potential' ϕ is now in common use (Slatyer & Taylor 1960), and is defined as $(\mu_w - \mu_w{}^0)/\bar{V}$ where μ_w^0 refers to the chemical potential of a reference pool of pure water at the same temperature as the water being considered. The dimensions of pressure are the same as those of energy per unit volume, hence water potential is also expressed in pressure units (Table 2.1). Since the molar volume of water, in millilitres (ml), is effectively the same as the molecular weight in grammes, the water potential is thus the free energy per gramme. The usual energy unit is now J kg^{-1}, whereas the pressure unit is the bar, centimetres (cm) of water head, or pascal (N m^{-2}). (1 bar $\equiv 10^3$ cm water head $\equiv 10^5$ Pascals $\equiv 100$ J kg^{-1}).

2.2   Transfer of water in soil

2.2.1   Basic theory

Soil water moves in response to a difference in water potential. By Darcy's Law,

$$v = - K \frac{d\phi}{dx} \tag{2.2}$$

where K is the hydraulic conductivity in cm s^{-1} if ϕ is expressed in cm head of water, and v is the flux in cm^3 cm^{-2} s^{-1}. In a saturated soil K is a constant so long as the structure is stable, since the conformation of the flow path remains the same. The source of the potential difference is usually a difference in gravitational, pressure or overburden potential (Rose 1966; ISSS 1974; Baver, Gardner & Gardner 1972). In unsaturated soils K varies with the water content of the soil since this determines the cross-section for transfer, the mean effective pore radius and the effective path length. In unsaturated soil, it is possible to have transfer with a virtually constant θ and K, for example in the transmission zone behind the wetting front during infiltration where the potential difference is mainly due to gravity (Fig. 2.2). However, these are rather special cases, and it is usual for the driving force to be a matric potential difference. Since matric potential, ψ, is related to water content by the so called 'soil moisture characteristic', the driving force can therefore be expressed as a difference in water content, though it should always be remembered that this is not physically correct. The real criterion of equilibrium is uniform water potential, not water content, and the potential is the driving force for water transfer. For example, if water redistributes itself within a homogeneous soil volume until equilibrium is reached, there will be virtually no remaining potential gradient, but there may well be considerable gradients in water content, due to hysteresis. If we wish to transform the variable in equation 2.2, we use the slope of the moisture characteristic $\left(\dfrac{d\psi}{d\theta}\right)$, and get

$$v = - K\frac{d\psi}{d\theta}\left(\frac{d\theta}{dx}\right) = - D_{\mathrm{w}}\frac{d\theta}{dx} \qquad (2.3)$$

where the new proportionality factor between flux and moisture gradient is called the water diffusivity, D_{w}. It must be clearly understood that this does not mean that water flow in porous bodies is physically equivalent to diffusion; it is a bulk flow under a macroscopic pressure difference, whereas diffusion results from random thermal motion.

The major justification for the use of diffusivity lies in the greatly simplified mathematical treatment of transient flow situations which it allows. Unsaturated flow may occasionally occur between two regions of different but constant water content, so that the water content varies with the distance between them, but not with time (steady-state flow). However, it is much more usual for it to vary with both distance and time (transient-state flow). To such a flow through a thin slice of soil we apply the 'continuity equation' (see p. 9), and get

$$\frac{\partial\theta}{\partial t} = \frac{\partial}{\partial x}\left(K\frac{d\psi}{dx}\right) = \frac{\partial}{\partial x}\left(D_{\mathrm{w}}\frac{d\theta}{dx}\right). \qquad (2.4)$$

Hence $\dfrac{\partial\theta}{\partial t}$ may be stated either in terms of matric potential or water content.

If we use the latter form, the equation has only three variables, and is formally identical to Ficks 2nd Law in diffusion theory. Solutions of the latter can consequently be used for water transfer problems. The major difficulty is that both K and D_{w} are strongly dependent on θ, i.e. the diffusivity is analogous to a concentration-dependent diffusion coefficient, which greatly complicates the solutions, and may render them impossible except by approximations or by numerical methods (for discussion see Philip 1973; Nielsen, Kirkham & van Wijk 1961).

2.2.2   Effects of pore size distribution

Both D_w and K depend ultimately on the pore size distribution in the soil, and on the way in which these pores interconnect. Their magnitude is affected by degree of compaction, structural changes and disruption of channels during handling, and the exact method of wetting. There is a regrettable lack of methods for defining and expressing the complex geometry of the soil. The moisture characteristic gives a crude idea of the pore radius–pore volume relationship, but a clearer visualization may be gained from consideration of the hysteresis loop of the characteristic (Childs 1969). By this, the total volume of pore space in a soil may be considered to consist of 'domains', each of which is defined by the fact that it fills and empties of water over small intervals of water tension. A domain is therefore not a single physical group of pores, but all pores which empty and fill with water in a similar way, which is determined by their shape.

Various attempts have been made to consider both the pore radii and their continuity, in defining a relationship between the moisture characteristic and the unsaturated conductivity (Marshall 1958; Millington & Quirk

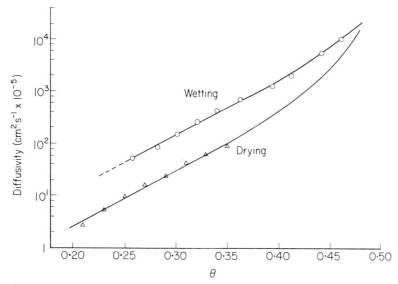

FIG. 2.1. Diffusivity as a function of moisture content for Grenville silt loam, during wetting and drying (after Staple 1965).

1960; Rogowski 1972), but none has been so successful that it has been adopted widely. In practice, hysteresis in the moisture characteristic causes considerable problems. The geometry of the water bodies in the soil will differ depending upon whether the soil is wetting or drying, and the conduc-

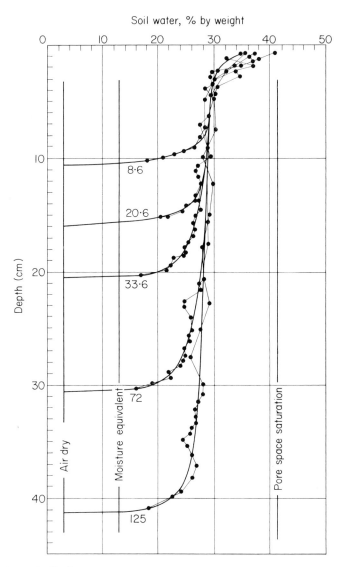

Fig. 2.2. Distribution of soil moisture in relation to depth of wetting into Yolo sandy loam. Infiltration times are indicated by numbers on curves, in minutes (after Bodman & Colman 1943).

tivity may, therefore, vary at a particular value of θ. Secondly, we recall that D_w is defined as $K\, d\psi/d\theta$ and the slope of the characteristic depends upon the wetting-drying history of the soil. Figure 2.1 shows the very large differences in D_w measured by Staple (1964) in a soil in the wetting and drying conditions.

2.2.3 Infiltration

This has some interest for our topic, in that it gives the maximum flow rate of water down a profile. The usual equation is

$$v = \tfrac{1}{2} st^{-\frac{1}{2}} + a \tag{2.5}$$

(Philip 1957a), where t is time and s and a are constants. A typical set of infiltration profiles, as found by Bodman & Colman (1943) are given in Fig. 2.2. After long times, when the top of the profile is wet, the infiltration rate is constant at the value a, which is numerically equal to the hydraulic conductivity of the soil at the particular water content in the 'transmission zone' (see Fig. 2.2). This varies widely from soil to soil; for example, Slatyer (1967) gives data of around $1\cdot0$ cm h^{-1} in sandy and well-aggregated soils, $0\cdot25$ cm h^{-1} in soils of high clay content, and less than $0\cdot1$ cm h^{-1} in soils of high swelling capacity. The velocity of flow within the pores of the sandy soil is thus of the order of 5 cm h^{-1}, with higher rates during the early stages of infiltration.

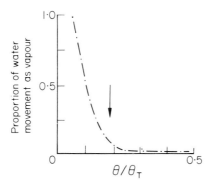

Fig. 2.3. Proportion of water movement in vapour phase in soil crumbs, in relation to water content (expressed as fraction of saturation). Arrow indicates 15 bar suction (after Rose 1963b).

2.2.4 Vapour phase transport

It has been accepted for a considerable time that water may move in the vapour phase in soils. The separation of the vapour and liquid fluxes is not easy, since it is likely that liquid 'necks' act as short circuits to a general vapour phase transport, by evaporation and condensation on the two exposed meniscuses. The total pathway is thus a complex series–parallel system, rather analogous to the solid phase—liquid phase system for ion diffusion in soils. Rose (1963a,b,c) has given results for several materials, showing the fraction of the total water flux in the vapour phase (Fig. 2.3) and the vapour and total water diffusivities at different water contents. For all six materials, the vapour and total diffusivities became essentially identical at relative humidities below 60%. The vapour phase diffusivity became negligible in relation to total diffusivity in two natural soils when the water content rose above 15% of saturation. This almost certainly means that there is no significant vapour phase movement at soil water potentials where flow to roots could be large (see p. 24 *et seq.*).

However, Cowan & Milthorpe (1968) have presented interesting calculations to show that vapour phase transport of water could bridge the gap around roots which shrink away from the surrounding soil when the water potential falls. Their calculations take into account both the transfer of vapour to the root, and the conduction of the latent heat of vaporization back to the soil. They suggest that vapour phase transport could supply water to roots, with a potential difference of 5 bars across a 40-μm gap, at the rate of 0·007 cm^3 cm^{-1} d^{-1}, which is a significant rate for a plant with a dense root system.

Considerable transfer of water as vapour can occur when there is a temperature gradient, rather than a water potential gradient, to drive it. The humidity, or concentration of water vapour in the air, is much more sensitive to small differences in temperature than in potential, and water will readily move from hot to cold regions, as vapour. This is considered below in more detail.

2.2.5   Water movement under temperature and solute gradients

Water may be transferred other than by gradients of water suction or water content, and these other driving forces may be of some importance in practice. Diurnal and seasonal temperature changes occur regularly in soils, and under bare soils these gradients may become quite intense. A change in temperature alters both the vapour pressure and the surface tension of water. The resulting vapour and liquid phase fluxes may vary almost independently. Gurr, Marshall & Hutton (1952) showed this neatly with a steady-state system in which the two ends of a sealed-soil column were maintained at different temperatures, the soil containing added sodium chloride. In the steady state the distribution of water and chloride was as shown in Fig. 2.4, due to a vapour phase flux of water from hot to cold, causing a liquid phase flux in the opposite direction which carried dissolved chloride to the hot end of the column. Cary & Mayland (1972) investigated

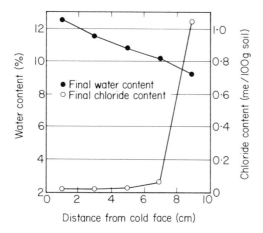

Fig. 2.4. Distribution of water and chloride in a column of loam soil, of 11·5 % initial water content, subjected to a temperature gradient for 18 days (after Gurr *et al.* 1952).

transfer of water and salt by thermal gradients in unsaturated soils near freezing point; they found that water moved from the warmer to colder areas as expected, but the simultaneous salt movement indicated that the main transfer mechanism in this case was by bulk flow in water films.

Quantitative theories have been put forward to deal with thermally driven fluxes of water and heat, by Philip & De Vries (1957) and Taylor & Cary (1964). These have been compared by Cassel, Nielson & Biggar (1969), who found that the Philip–De Vries equation predicted the flux reasonably well, whereas the Taylor–Cary equation underestimated it by a factor of 20 under one set of conditions. The latter theory, based on irreversible thermodynamics, thus seems less useful.

The essence of the Philip–De Vries theory is the separation of the water and thermal potential driving forces, and the absence of any interaction between the water and heat fluxes. It may be stated as

$$v = - D_\theta \frac{d\theta}{dx} - D_T \frac{dT}{dx}. \qquad (2.5)$$

D_θ is the water diffusivity appropriate to a flux caused by a water potential or water content difference, and D_T is the equivalent value for water flux driven by a temperature gradient. Both D_θ and D_T have a liquid and a vapour component, and Philip and de Vries gave equations for determining all four values from soil measurements. Selected data from Cassel *et al.* (1969) are given below as an indication of orders of magnitude of the various quantities.

Mean temp. °C	θ	D_θ cm² d⁻¹	D_T cm² d⁻¹ °C⁻¹
18	0·073	0·309	0·0127
16	0·076	0·219	0·0203

A gradient of solute concentration (i.e. osmotic pressure) will normally cause counter-fluxes of water and solute. If there is a degree of semipermeability to solute in the soil medium, the flow of water is enhanced and that of solute diminished. The flux of water can conveniently be expressed in terms of irreversible thermodynamics (Letey, Kemper & Noonan 1969, and quoted references).

$$v = -L_P \frac{dP}{dx} + L_{PD} \frac{d\pi}{dx} \tag{2.7}$$

where L_P and L_{PD} are the phenomenological coefficients and π, osmotic pressure. In a loam soil, they found that the former were respectively of the

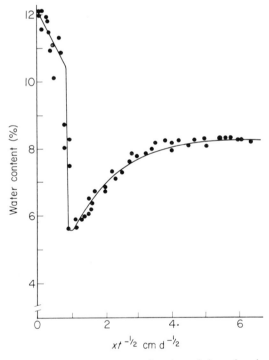

FIG. 2.5. Gravimetric water content as a function of the reduced variable $xt^{-\frac{1}{2}}$, with solid sodium chloride placed at $x = 0$. Dots give experimental measurements of Scotter & Raats (1970). Solid line gives results of theory (after Parlange 1973).

order 10^{-6} and 10^{-8} cm h^{-1} in steady-state experiments. The relative efficiency of matric and osmotic gradients can be expressed as L_{PD}/L_P (the osmotic efficiency coefficients) and the largest value measured, at -0.66 bar, was 0.157.

The semipermeability in soil normally arises from the exclusion of anions from the electrical double layers around clay particles (p. 44). In fine structured clay soils much of the water may be within double layers and, hence not available for salt diffusion. The semipermeability naturally increases as the water films in the soil become thinner, or the diffuse double layers thicker. The full explanation of water plus salt transfer is exceptionally complex, and cannot be dealt with here. Thus a theoretical treatment by Bresler (1973) lists convection, ionic diffusion, mechanical dispersion, anion exclusion and coupling phenomena in salt and water flows as effects which must all be considered. In natural soils there are the additional complications of structure formation and the wide range of pore sizes which this produces.

A rather interesting case of movement due to solute gradients has been considered by Parlange (1973), based on experiments by Scotter & Raats (1970). This arises when there is an exceptionally sharp solute gradient, as for example when a salt is placed on a moist soil surface. A discontinuous distribution of water results (Fig. 2.5), when plotted against the reduced variable $xt^{-\frac{1}{2}}$. It is suggested that the following processes operate to produce this result:

(a) Water vapour condenses on the surface of the salt and forms a saturated solution.

(b) This solution moves into the soil, but the salt concentration increases the surface tension, hence a steep water content gradient results.

(c) More water vapour, from the salt-free zone, condenses at the front of this saline and wet zone, with salt diffusing from the solid salt source into this freshly-condensed water.

(d) A sharply-defined dry zone forms immediately in front of the salt layer; Parlange postulates that this becomes so dry that only vapour transfer takes place across it, though it might be expected that this would depend upon the initial water content of the soil.

The process of a fertilizer salt diffusing into moist soil is, therefore, rather complex, and cannot always be treated as a simple matter of ionic co-diffusion at the existing water content of the soil.

2.3 Water use by plants

2.3.1 Soil–plant–atmosphere pathway

Plants draw by far the greater part of their total water requirements from the soil, but the size of that requirement is dictated by the climate, and the size and arrangements of the plants' transpiring surfaces. There are many close analogies between the transport of water and the transport of nutrient ions into a plant, but the climatically determined rate of water usage (the potential evapotranspiration for a crop) has no direct equivalent in the study of ion transport and uptake (Gardner *et al.* 1975).

The relationship between relative humidity and water potential is given by

$$\phi - \phi_0 = \frac{RT}{\bar{V}} \ln p/p_0 \qquad (2\cdot8)$$

where $\bar{V}$ is the partial molar volume of water, and p and p_0 are the partial pressures of water at potentials ϕ and ϕ_0 respectively. From this the usually accepted limit of plant availability of soil water, -15 bars, is reached at a relative humidity of 0·989. This is the basis of the usual assumption that soil air is always saturated with water vapour, and that there is normally a large difference in water potential between the atmosphere and the soil. Since the leaf water potential does not fall below that in the soil by more than a few bars, it is clear that most of this great potential difference exists across the boundary layers around the leaf and the stomatal cavities.

Continuing transpiration demands a steady supply of radiative or convective energy, otherwise latent heat loss will cool the leaf, the vapour pressure of water in it will drop and transpiration decrease. A short-term equilibrium is usually set up, in which the net energy input is balanced by the latent heat loss due to transpiration, plus the convective and radiative transfer of heat to the surrounding air. This statement presumes that the roots can supply this flow of water; if they cannot, the water potential of the leaf decreases until the stomata close, which reduces transpiration appreciably. The leaf temperature and heat loss then increase until a new equilibrium is attained. The capability of the root to supply water at the required rate in turn depends upon the water potential differences in the root and soil and the resistance to movement encountered by water during this transfer. This approach has been expressed in the 'Ohms Law analogy', in which the potential differences, resistances and water flows are regarded as equivalent to an electrical circuit (Cowan 1965) (Fig. 2.6).

The first difficulty with the model is that the leaf–atmosphere transfer rate is not proportional to the water potential difference, but to a vapour pressure difference, and these two are not linearly related. The 'Ohms Law' form cannot therefore possibly explain this part of the system properly. Secondly, the resistance to transfer of water from soil to shoot is variable and difficult to define. Part of it consists of the root permeability, which is known to vary with water potential and pressure (Slatyer 1967; Newman 1974). The other part is in the soil, and depends upon hydraulic conductivity, which is

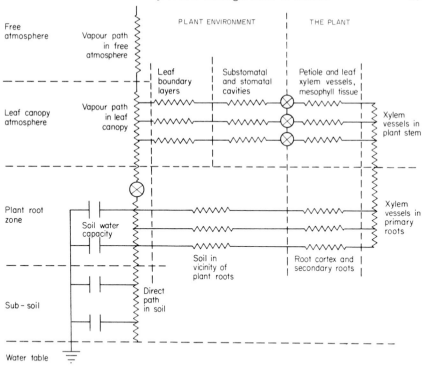

FIG. 2.6. Water pathways in soil–plant atmosphere continuum, drawn as an electrical circuit. ⊗ marks change of phase (after Cowan 1965).

extremely variable with water potential, and with the mean distance of move-ment of the water within the soil, which varies with time, root density and soil water potential. Any representation of this as a simple resistance must consequently be very approximate.

If we deal only with the soil–root–leaf part of the pathway, we can write the flow equation:

$$q = \frac{\phi_s - \phi_l}{R_p + R_s} \tag{2.9}$$

where q is the rate of transpiration, and subscripts l, s and p stand respectively for leaf, soil and plant. R_p in this case represents the resistance from the cortex surface to the leaf, rather than that from the cortex surface into the xylem, which has also been used. This distinction is important in view of the resistance to flow in xylem vessels, stem and leaves. It has often been shown that $R_s + R_p$ increases greatly as soil water content falls, and the conclusion has been drawn (Macklon & Weatherley 1965) that R_s is dominant, and that the increase is due to drying out of soil in the rhizosphere. This is examined further below.

2.3.2 Water movement in the rhizosphere

Gardner (1960) first treated water uptake by a root from soil as a water diffusivity problem, by the simple expedient of regarding the root as having a constant boundary potential. His solution was:

$$\psi_s - \psi_a = \frac{I_w}{4\pi K}\,(\ln 4D_w t/a^2 - 0.577) \qquad (2.10)$$

where ψ_a is the matric potential at the root surface, I_w the water uptake per unit root length per unit time, and a the root radius. D_w and K are both clearly variables if there is any change in θ with distance or time, but constant values may be used to get an order-of-magnitude solution. The treatment has been extended by Cowan (1965). According to this equation R_s is

$$\frac{\ln 4D_w t/a^2 - 0.577}{4\pi\,KL_A}$$

where L_A is root length under unit soil surface, which emphasizes its complexity and variability. Passioura & Cowan (1968) have stated appropriate equations for steady state and 'steady rate' approximate treatments—the former implying injections of water at some arbitrary radial distance outside the root, and the latter assuming a constant rate of removal of water from all points in a cylinder of soil around the root. They compared the results using these equations with a numerical solution by the Crank–Nicholson procedure, and showed that the agreement was good in most cases. The steady-rate

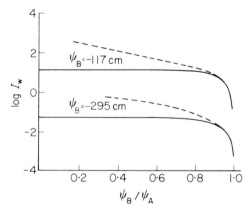

Fig. 2.7. Relation of log of the water inflow into a root in Columbia sandy loam to the ratio of the water potentials at the root surface (ψ_A) to that at an arbitrary distance B (ψ_B). Dotted line is for assumed constant K; full line shows effect of K varying with ψ, in which case I_w rapidly reaches maximum value ($B_B/A_A = 10$; I_w in cm³ cm⁻¹ d⁻¹) (after Whisler *et al.* 1970).

treatment yields the final approximate equation (for discussion see Tinker 1976)

$$\theta_b - \theta_a = \frac{I_w}{2\pi D_w} \ln \frac{b}{1 \cdot 65a} \qquad (2.11)$$

where θ_b is the water content at some distance b; this is usually considered to be the radius of the equivalent cylinder of soil from which the root draws its water (see p. 223 for discussion).

These methods all assume a constant D_w. If a dry zone developed, D_w would decrease, requiring a greater potential difference to maintain I_w. In such a situation, with a dry soil and a further dried zone around the root, the flow may no longer be sensitive to change in the water potential in the plant, and a maximum value of the water inflow can be calculated, which is analogous to the flow of ions to a sink of infinite strength, as discussed in later chapters. Two very similar methods of predicting this maximum rate of flow to roots were given in the same year (Lang & Gardner 1970; Whisler, Klute & Millington 1970). Essentially, it is as follows:

$$I_w = 2\pi r K \frac{d\psi}{dr} \qquad (2.12)$$

$$I_w \int_a^b \left(\frac{dr}{r}\right) = 2\pi \int_{\psi_a}^{\psi_b} K(\psi)\, d\psi \qquad (2.13)$$

where I_w is the annular flow rate of water into unit length of root, b is some specified but arbitrary distance from the root surface, and ψ_a is the water matric potential at the root surface. Lang and Gardner point out that

$$\frac{dI_w}{d\psi_a} = \frac{2\pi K_a}{\ln(a/b)} \qquad (2.14)$$

where K_a is the hydraulic conductivity at the root surface. Hence the effect of decreasing ψ_a becomes steadily less as K_a decreases. It has been found empirically that $K = A|\psi|^{-n}$ where A and n are constants, and the latter is in the range 2–5, therefore integrating equation 2.13

$$I_w = \frac{2\pi A}{\ln(b/a)(n-1)} \left(|\psi_b|^{(1-n)} - |\psi_a|^{(1-n)}\right). \qquad (2.15)$$

Hence as $|\psi_a| \to \infty$

$$\lim I_w = \frac{2\pi A}{\ln(a/b)(n-1)} |\psi_b|^{(1-n)} \qquad (2.16)$$

Figure 2.7 gives the log of the water flux against the ratio of the suction at the root surface and at the arbitrary distance b. Both papers suggest maximum values for the water flux at the root surface of the order of 1 cm³ cm⁻² d⁻¹, which is equivalent to about 0·1 cm³ cm⁻¹ d⁻¹ for I_w; this figure is often used in model calculations.

The equations discussed above predict a lowered water content around the root. This moisture gradient may also be calculated by more sophisticated procedures such as computer simulations in which variable D_w can be included (de Wit & van Keulen 1972; Dunham & Nye 1974). The results obtained in such calculations depend markedly upon the flow rate of water assumed, and a serious change in water content may easily be predicted in a dry soil with a high flow rate. If this drying out is considerable it has important consequences for water supply and for the transport of other materials to or from the root. It is therefore important to consider whether it occurs in practice (Tinker 1976).

The mean uptake rate per unit length of root, $\bar{I}_w$, for a crop depends upon evapotranspiration rate per unit surface area of soil and the total length of root in the column of soil under this area, L_A. Table 8.2 contains typical values of root length per unit surface area for a variety of crops, and even a high transpiration rate of 1 cm d⁻¹ would indicate $\bar{I}_w$ values of much less than the value of 0·1 cm³ cm⁻¹ d⁻¹ which is sometimes assumed in model calculations (for discussion see Newman 1969).

Lawlor (1972) has given recent data obtained in large pot experiments with rye-grass, following a model proposed by Gardner (1964) for dealing with uptake of water by whole crop root systems in soils where θ and L_v varied with depth. He found that the root length per unit soil surface L_A was up to 1400 cm cm⁻², and that I_w was therefore small ($I_w = 0.006$ cm³ cm⁻¹ d⁻¹ or less). ϕ_{leaf} and ϕ_s were measured at different stages of drying

Table 2.1. Calculated values for R_p and R_s for rye-grass grown in progressively drier soil (after Lawlor 1972).

Day	Water loss (cm³/ day)	ϕ soil (bar)	ϕ root (bar)	R_s (bar day/cm³)	R_p	R_p/R_s
1– 5	25	−0·20	−0·21	0·0004	0·10	250
5– 8	16	−0·25	−0·26	0·0006	0·17	283
8–11	26	−1·0	−1·02	0·0007	0·10	143
11–13	9	−5·0	−5·10	0·011	1·03	94
13–18	12	−8·0	−8·30	0·025	0·60	24
18–22	6	−11·0	−12·00	0·160	0·83	5·2

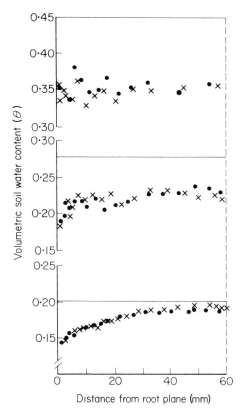

Fig. 2.8. Water content gradients at 6 days, in soil blocks in contact with absorbing onion roots. ● and x indicates replicate blocks on opposite side of root curtain. Note different original moisture contents in the three experiments (after Dunham & Nye 1973).

out of the soil (ϕ_s as an average weighted for K and L_v), and hence ($R_s + R_p$) was found. This increased rapidly with decrease in θ. R_s was calculated from the formulae of Gardner (1960) and Passioura & Cowan (1968), and was in most cases only a small fraction of R_p, though it increased relatively more rapidly than R_p (Table 2.1).

The evidence thus indicates that, at least for densely-rooted crops, there is no significant water content gradient near roots. This conclusion, based on calculations using $\bar{I}_w$, may not, however, be fully reliable. Local values of I_w may vary greatly from the mean and it is the local value which is important. It is essential that water contents at root surfaces should be directly measured, though it is technically very difficult. Wray (1971) was unable to observe any change in water content of a wet soil surrounding a root absorbing at a rate of 0.2×10^{-6} cm^3 cm^{-2} s^{-1}. Campbell (1968) did discover some drying out, using a γ ray absorption system, but the effect was small and the results rather irregular. Dunham & Nye (1973) have shown clear evidence of drying out near to roots in a linear system, when the bulk soil became dry. They measured the water content of soil close to the root by adapting a technique in which soil blocks are pressed against a layer or curtain of onion roots, and after a length of time the soil is frozen and sliced on a freezing microtome. They were able to determine the gradients in water content (Fig. 2.8), and in water potential. The gradients were large when the bulk soil was dry. with root surface potentials of less than -20 bars. No clear gradient of water content developed when the bulk soil was moist, though the water flow velocity into the roots was 1.7×10^{-6} cm s^{-1}, which is a high value. D_w and K were not independently measured, but all changes were in the direction expected, and were of the right order of magnitude. On the available evidence, it seems reasonable to accept the theoretical predictions as approximately correct, if the water uptake rate is known. No moderate figure for the latter will imply seriously dried-out zones around roots in soil which is moist. Further, root hairs may absorb water (Cailloux 1972) and thereby reduce gradients of θ particularly when I_w is large.

Larger values for I_w may arise because plants have short lengths of active root, or because their transpiration rates are exceptionally high. In practice, the water use of a complete crop rarely exceeds 15 cm per month, with a short term transpiration rate of up to 1 cm d^{-1}. Isolated plants which receive an exceptional amount of advective energy may well transpire more; Penman (1963) quotes up to 9 cm d^{-1} and Hudson (1965) records a loss of 2.5 cm d^{-1} on the upwind edge of an irrigated crop in the Sudan. It is commonly accepted that plants in arid areas have well developed root systems (Russell 1961), but it is not clear whether the value of this is to exploit a large soil volume, to reduce the required inflow of water into each length of root, or both.

Plants cannot decrease their water potential without limit—the majority wilt at -15 to -20 bars, (though specialized xerophytes may have much lower values) and the stomata will close at some higher potential. It is therefore certain that the evapotranspiration rate, and $\bar{I}_w$, will decrease as the soil dries out unless the potential evapotranspiration is very low. It is interesting to compare the actual and the potential evapotranspiration rate from a crop, such as is presented by Pereira (1957). However, in the field situation, it cannot be assumed that such a lessened rate of water loss represents an equally lowered uptake rate over the whole root system. Thus, Ogata, Richard & Gardner (1960) showed (Fig. 2.9) that drying out of the soil proceeds at varying rates in different soil layers, the top layers drying first, and it is obvious that a low evapotranspiration rate may still represent a high value of I_w at some point in the profile. For example, Reicosky *et al.*

(1972) found that roots near a water table absorbed at 1000 times the rate of roots in drier soil higher up.

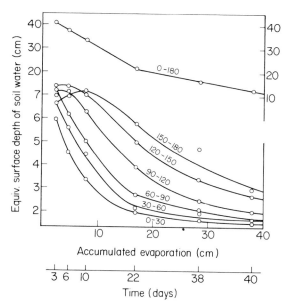

Fig. 2.9. Water depletion from various soil depths under alfalfa in relation to total evaporation from a free water surface and time. Figures on graph indicate soil depth intervals in cm (after Ogata *et al.* 1960).

Nearly all work on the detailed flow of water to roots has assumed that the root is in full contact with homogeneous soil of similar composition to that in bulk. There are however doubts about this, since roots often inhabit cracks or other voids in the soil (see Chapter 7), in which contact with soil will be at best partial. On drying beyond −5 bar, Cole & Alston (1974) found that wheat roots shrank markedly, losing 60 % of their original diameter by −10 bars. Gaps between root and soil could be a serious barrier, though root hairs or possibly mucigel, could bridge them, and Cowan & Milthorpe (1968) (see p. 18) calculated that vapour transport could carry a small but useful flux with moderate water potential gradients. Tinker (1976) considered that contact around part of the periphery was more probable than a complete gap, and used an electrical analogue to show that such partial contact was unlikely to alter potential gradients greatly unless the uptake rate was improbably large. Mass flow can accumulate salts near to roots (see p. 129) which could place an extra osmotic stress on roots, but such effects are also unlikely to be serious unless I_w is unusually large.

It is clear that almost all conclusions are subject to the absolute magnitude of I_w: if this is small, then the rhizosphere is unlikely to be a significant barrier to water transfer. Insufficient measurements have yet been made in the field to allow us to decide what local I_w values are, though $\bar{I}_w$ may easily be estimated.

2.4   Movement of water in roots

The movement of water through the root tissues (see Kramer 1969; Slatyer 1967) is not directly relevant to the main subject of this chapter, but flow in the roots leads to differences in water potential at different points in the root system. If this leads to different rates of uptake of water in different soil zones, it could be highly relevant.

It is generally assumed that water moves easily through the free space of the cortex, up to the endodermis. This is based on the belief that the hydraulic conductivity of the cell walls is relatively high (Weatherley 1963; Slatyer 1967; but see Newman 1976). The intercellular spaces are of the order of 10–μm diameter which should empty at a suction of 0·3 bars, and they are normally air-filled. The pores in cell walls, in contrast, are of radius 1–10 nm, which would only empty at 1500–150 bars suction, and they are consequently always water saturated (Jarvis 1975).

It is evident from the effects of temperature and metabolic inhibitors on flow (Brouwer 1965) that uptake into the root is not simply governed by viscous flow through pores, and it is highly probable that the pathway goes through the cytoplasm of the endodermis which constitutes the main resistance to flow. There is some evidence (see Dainty 1969b) that a layer of cytoplasm lines much of the xylem vessel wall, and that this could be an alternative, or additional, semipermeable barrier, and with low water conductivity. A factor of 1·3 between the effectiveness of equal hydrostatic and osmotic pressures in causing water uptake has suggested that there is also a minor direct 'mass flow' route into the xylem which bypasses the endodermal cytoplasm (Weatherley 1963), but Cowan & Milthorpe (1968) suggest that this could simply result from changes in the size of air-filled pores, or from a convection effect near semipermeable membranes. The factor of 1·3 is in any event not large in comparison with other uncertainties in the rest of the plant–soil water transfer system.

The local hydraulic conductivity of a short piece of root therefore depends largely upon the condition of the endodermis, and it is to be expected that it will diminish with progressive age and suberization of the latter. In general this is so, and Clarkson *et al.* (1974) found that the uptake rate of a segment of marrow root decreased by a factor of 8 during its ageing. In addition, the conductivity varies with flow rate, especially that of older portions of root (Fig. 2.10). The local conductivity always rises with increase in pressure gradient, possibly due to an alteration of membrane structure or shape.

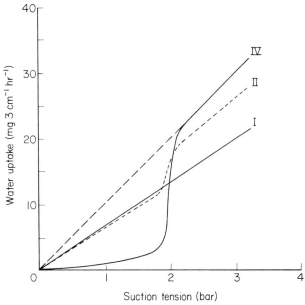

Fig. 2.10. The relation between suction tension and water uptake for different 3 cm long zones of a single root of broad-bean plant. I is nearest the tip (after Brouwer 1965).

Further, as Cowan & Milthorpe (1968) consider, the overall resistance to water uptake depends partly upon that across the root and partly upon that along the xylem. The flow rate up the root xylem is expressed by $f = L d\phi_x/dx$ where f is in cm³ s⁻¹, x is distance from the base of the root, L is the xylem conductivity and ϕ_x the xylem water potential. The flow across the cortex is given by $df/dx = S(\phi_a - \phi_x)$ where ϕ_a, ϕ_x are water potentials at the root surface and in the xylem, and S is the conductivity across the root. Assuming that xylem solute concentration is small and that ϕ_a is uniform, they give the overall conductivity of a single root, from the external medium to the stem of the plant, as $K_r = (SL)^{\frac{1}{2}} \tanh[(S/L)^{\frac{1}{2}}l]$ where l is root length. If $Sl^2/L \ll 1$, this becomes $K_r \sim lS$, since xylem resistance is then negligible. They give S as $0.4 - 2 \times 10^{-6}$ cm² s⁻¹ atm⁻¹ and L as $2 - 12 \times 10^{-4}$ cm⁴ s⁻¹ atm⁻¹ respectively, hence xylem resistance cannot be neglected if $l > 10$ cm. This effect will tend to increase the relative uptake of water by older parts of a root, i.e. those nearest the stem. A further complication is that the uptake rate of water into the root depends upon whether the suction in the root is increasing or decreasing at the time (see Cowan & Milthorpe 1968).

In brief, the youngest parts of a root are likely to absorb most rapidly as long as they are not too far from the stem base, but this cannot be assumed to be true in all cases. Where a root system contains a large fraction of suberized root, the latter may, in fact, absorb most of the water required by the plant (Kramer & Bullock 1966).

2.5   Uptake of water by field crops

This discussion should logically conclude with a consideration of uptake of water by crops and vegetation in the field. However, this is intimately linked with the distribution of roots in the soil, and also with various crop nutrition models which have been proposed. The latter are discussed in Chapter 8, and the discussion of water uptake by crops has therefore been deferred until then.

3

SOLUTE INTERCHANGE BETWEEN SOLID, LIQUID AND GAS PHASES IN THE SOIL

We noted in Chapter 1 that the concentration of ions in the soil solution is buffered by ions adsorbed on the soil surfaces. We also show in Chapter 4 that the overall mobility of labile ions* is related to their amounts and mobilities in the solid and solution. It is therefore fitting that in this chapter we should examine the general principles governing the interchange of solutes between all phases in the soil, dealing first with inorganic ions, specially the plant nutrients; and later with other solutes, such as pesticides, which may also occur in the vapour phase.

INORGANIC IONS

3.1 Exchangeable cations—composition of the soil solution

The total strength of the soil solution depends upon the concentration of unadsorbed anions that it contains. Among these, chloride, nitrate and bicarbonate are not adsorbed by the negatively-charged soil colloid surface—unless it contains positively charged sites, which are often associated with iron or aluminium oxides at pH below about 6. Sulphate also is not strongly adsorbed. These ions control the overall strength of the soil solution. The proportions in the soil solution of the different cations that balance these anions is determined by the ionic charge of the adsorbed cations, their proportions on the exchange complex, their ionic size and the properties of the exchanger. These factors are considered in turn in the next section. Thus to predict the effect of a treatment, e.g. drying the soil, on the solution concentration of a cation, e.g. potassium, the first question is, what will be the effect on the anion concentration? and then, what proportion of the anions in solution will be balanced by potassium?

Some typical soil solution compositions are given in Table 3.1. A useful summary is in Russell (1973, pp. 543–5).

---

* We use the term labile to include all those ions that will exchange rapidly with their own isotope in the soil solution. They will include the ions in solution and the conventional 'exchangeable' ions; but the nature of the exchanging ion is automatically specified.

33

Table 3.1. Representative soil solution compositions. m equiv l^{-1} of soil solution.

	pH	Ca^{++}	Mg^{++}	K^+	Na^+	NH_4^+	Total cation	NO_3^-	Cl^-	HCO_3^-	$SO_4^=$	Total anion	Ref.
Acid soil	4·2	1·0	1·4	0·4	0·4		3·2	3·8	0·2	—	0·8	4·8	Vlamis (1953)
Sandy loam	7·2	21·0	1·2	0·7	1·8	0·4	25·1	15·6	2·2	1·1	7·0	25·9	Eaton et al. (1960)
Average of 8 cropped soils	7·3	10·1	7·1	0·68	1·8		19·7	3·7	—	1·8	12·5	18·0	Burd & Martin (1923)
Average of 8 fallowed soils	7·0	27·9	10·9	1·6	2·8		43·2	29·6	1·4	1·0	9·7	41·7	Burd & Martin (1923)
Saline soils	8·3	43·5	48·0	9·6	21·7		114·8	(31·2) by difference	20·1	7·2	56·3	114·8	Reitemeier & Richards (1944)

3.1.1 Exchangeable cations—effect of ionic charge

Any change in the soil solution, such as increased salt concentration, that may change the electrical potential difference between the solid surface and solution, will alter the relative proportions in solution between ions carrying different charges, e.g. K^+ and Ca^{++}. Hence removing half the water from the soil solution does not double the potassium concentration. However, as shown in the next paragraph, the 'reduced activity ratio' $(A)^{1/z_A}/(B)^{1/z_B}$ (where $(\ )$ = activity, z = ionic charge) of a pair of cations is independent of the electrical potential difference between solid and solution, and remains constant as long as the proportions of the ions on the exchanger are not altered. Since there is usually a large reserve of exchangeable ions in the absorbed phase this tends to buffer the reduced activity ratios between the cations in solution.

The electrochemical potential of an ion A in the solution phase may be written (Guggenheim 1933 Chapter 10),

$$\mu_{el_A} = \mu_{1_A} + z_A F E_1 \qquad (3.1)$$

where μ_{1_A} is the 'purely chemical' potential,
E_1 is the electrical potential of the solution,
F is the Faraday,
z_A is the charge on the ion; positive for cations, negative for anions.
A similar equation may be written for any solid phase on which ion A is adsorbed:

$$\mu_{es_A} = \mu_{s_A} + z_A F E_s. \qquad (3.2)$$

Since the electrochemical potential of A must be the same in all phases in equilibrium, $\mu_{el_A} = \mu_{es_A}$, and therefore

$$\mu_{s_A} = \mu_{1_A} + z_A F(E_1 - E_s). \qquad (3.3)$$

Likewise for an ion B in the same two phases

$$\mu_{s_B} = \mu_{1_B} + z_B F(E_1 - E_s). \qquad (3.4)$$

The unmeasurable electrical potential difference between the phases may be eliminated between equations 3.3 and 3.4 giving

$$\frac{\mu_{s_A}}{z_A} - \frac{\mu_{s_B}}{z_B} = \frac{\mu_{1_A}}{z_A} - \frac{\mu_{1_B}}{z_B}. \qquad (3.5)$$

Hence the difference between the free energy per gram-ion equivalent of the two ions in any defined surface phase equals the corresponding difference in the equilibrium solution phase.

The chemical potential (free energy per mol) of a species in solution is conventionally written

$$\mu_{1_A} = \mu_{1_A}^o + RT\ln(A)_1 \tag{3.6}$$

where $\mu_{1_A}^o$ is the chemical potential in an arbitrarily defined standard state. Hence

$$\frac{\mu_{sA}}{z_A} - \frac{\mu_{sB}}{z_B} = \left(\frac{\mu_{1_A}^o}{z_A} - \frac{\mu_{1_B}^o}{z_B}\right) + RT\ln(A)_1^{1/z_A}/(B)_1^{1/z_B}. \tag{3.7}$$

Accordingly the reduced activity ratio $(A)_1^{1/z_A}/(B)_1^{1/z_B}$ is a measure of the difference between the free energies per gram ion equivalent of the two ions in the adsorbed phase.

We may write

$$(A)_1^{1/z_A}/(B)_1^{1/z_B} = ([A]_1 f_A)^{1/z_A}/([B]_1 f_B)^{1/z_B} \tag{3.8}$$

where [ ] = concentration

f_A, f_B = practical activity coefficients.

Since, at soil solution concentrations less than c. 0·003 M, the ratio $f_A^{1/z_A}/f_B^{1/z_B} \simeq 1$, for many practical purposes the reduced activity ratios may be replaced by reduced concentration ratios.*

---

* If the simplest expression for the activity coefficient derived from Debye & Huckel's (1923) expression is used:

$$\ln f_i = -az_i^2 I^{\frac{1}{2}} \tag{3.9}$$

where I = ionic strength of the solution = $\frac{1}{2}\sum m_i z_i^2$

m_i = molality of ion i
a = 0·509 at 25°C

Equation 3.9 is applicable to ionic strength less than about 0·01, i.e. 0·003 M in $CaCl_2$ solution.

At higher ionic strengths Guggenheim's (1935) expression is

$$\ln f_i = \frac{-az_i^2 \cdot I^{\frac{1}{2}}}{1+I^{\frac{1}{2}}} + 2 \sum B_{ij}(j) \tag{3.10}$$

where j represents ions of opposite charge to i.

B_{ij} is a constant depending on the i-j interactions. For a solution containing potassium, calcium and magnesium chlorides, Beckett (1965), using values of the B_{ij} terms derived from Guggenheim & Stokes (1958), has derived the expression:

$$\ln(f_K/f^{\frac{1}{2}}_{Ca+Mg}) = \frac{aI^{\frac{1}{2}}}{1+I^{\frac{1}{2}}} - 0·086\,[Cl] \tag{3.11}$$

When the potassium concentration is less than 50 mequiv l^{-1} and the calcium plus magnesium is less than 0·01 M, a useful range in practice, the reduced activity coefficient ratio lies between 1·0 and 1·3.

As a rule calcium and magnesium are the predominant cations in the soil solution. Thus if half the water in the soil solution is removed their concentrations will roughly double; and in order that the reduced activity ratio $(K)/(Ca)^{\frac{1}{2}}$ should remain steady, we may predict that the activity of potassium will increase by a factor of $\sqrt{2}$. Moss (1963) has verified on three soils that the reduced activity ratio does remain constant over the field moisture range.

Schofield & Taylor (1955) suggested that since calcium is usually the dominant cation in the soil solution it would be convenient to adopt it as a reference ion, and measure the concentration of other ions in a suspension of soil in M/100 $CaCl_2$. This would cause only slight disturbance to the adsorbed ions. Hence it has become conventional for practical purposes to measure soil pH in M/100 $CaCl_2$, rather than in water which gives rise to a variable calcium concentration depending upon the soil : solution ratio and the initial salt concentration in the soil solution. Methods which enable the cation activity ratios of a sample of soil to be determined accurately have been reviewed by Beckett (1971).

3.1.2   Relative proportions of exchangeable cations

The relation between the proportions of two cations on the exchange complex and their reduced concentration ratios in the soil solution has to be determined experimentally for each soil, since there are no general equations that account accurately for the complexity of real soils. The more theoretical aspects of exchange reactions are full of subtle questions of definition and interpretation and we recommend Bolt's review (1967) as being one of the few that treat the subject fundamentally. Some exchange equations that have proved useful both in interpreting experimental behaviour, and in providing the means of making approximate predictions are given below.

Cation exchange equations

The essential problem is to express the activities of the ions in the adsorbed phase as a function of their proportions in it. Of the many exchange equations that have been suggested only two are currently much used: the Gaines & Thomas, and the Gapon.

The Gaines & Thomas (1953) equation is derived as follows:
A is a mono and B a divalent cation. Writing formally:

$$(A)_s^2/(B)_s \times (B)_l/(A)_l^2 = K \tag{3.12}$$

where K is the thermodynamic equilibrium *constant*. They *choose* as standard states the homoionic exchanger at infinite dilution of solution. Writing formally:

$$(A)_s = f_A N_A, (B)_s = f_B N_B$$

where N_A, N_B are *equivalent* fractions of the total exchange capacity

$$K\frac{f_B}{f_A{}^2} = \frac{N_A{}^2}{N_B} \times \frac{(B)_1}{(A)_1{}^2}. \tag{3.13}$$

This is a purely formal equation, since the activity coefficients in the adsorbed phase are not functionally related to the concentrations units N_A, N_B chosen. Thus the equation does not carry the implication that the right choice of concentration unit is the equivalent fraction. For example the Vanselow equation which uses mol fractions instead of equivalent fractions is formally as correct. In both instances divergences between measured activities and fractions are taken care of by the activity coefficients. The main advantage of the equation is that it allows experimental data to be presented in a systematic and thermodynamically exact manner. The value of K may be calculated from a complete isotherm, such as that shown in Fig. 3.1, ranging from 0 to

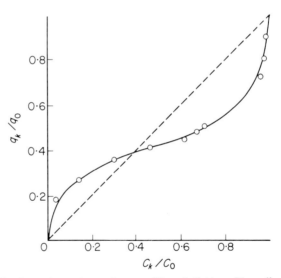

FIG. 3.1. Isotherm for exchange between K^+ and Ca^{++} on Harwell soil. q_k/q_0 is the fraction of the total negative charge satisfied by K^+ on the exchanger, and C_k/C_0 is the fraction of K^+ in the solution (after Deist & Talibudeen 1967a).

100 per cent of each ion on the exchanger. The calculation assumes that the cation exchange capacity remains constant over the range; this is often untrue for some soils and clays (e.g. Deist & Talibudeen 1967a,b). The standard free energy of exchange $\triangle G^0$, standard enthalpy of exchange $\triangle H^0$, and standard entropy of exchange $\triangle S^0$, may be determined from the relation

$$RT\ln K = \triangle G^0 = \triangle H^0 - T\triangle S^0 \tag{3.14}$$

Equation 3.14 is predictive in the sense that statements about the standard free energy, enthalpy and entropy of exchange between one pair of ions, e.g. sodium and ammonium, can be predicted from a knowledge of the corresponding values when sodium and ammonium separately exchange a third ion, e.g. barium (Laudelout *et al.* 1968). However, it is not possible to predict the form of only a part of the complete isotherm for a given ion pair in a similar way. For this reason the value of the Gaines & Thomas equation is limited at present.

The Gapon (1933) equation is

$$Kg = \frac{N_A}{N_B} \times \frac{[B]_1^{\ddagger}}{[A]_1}$$ (3.15)

where *Kg* is the Gapon constant. Though this was put forward as an empirical equation, it has been derived theoretically by Eriksson (1952) and Bolt (1955) from the Gouy–Chapman theory of the electrical double layer.

As will be seen the Gapon equation *predicts* the activity ratio in solution from the proportion of ions on the exchanger in terms of a single constant that may be determined experimentally. Bolt (1955) showed that the value of *Kg* for the exchange between sodium and calcium on illite remained constant from 1 to 70% sodium saturation. The exchange of potassium for calcium, however, behaved irregularly. N_{Na} has been found to be linearly related to the activity ratio $(Na)/(Ca + Mg)^{\ddagger}$ for many saline soils (US Sal. Lab., 1954).

In the theoretical derivation of the Gapon equation *Kg* is derived from the surface charge density of the exchanger, but this can only be determined reliably on pure clays. Beckett & Nafady (1967a,b) have shown that on the planar adsorption sites of several 2:1 type clay minerals N_K is linearly related to $(K)/(Ca + Mg)^{\ddagger}$. This is not true of the edge sites which show a strong preference for potassium. Since the theoretical derivation of the Gapon equation assumes a uniformity charged planar surface, its failure to explain behaviour in soils containing a significant proportion of their negative charge on edge sites is not surprising.

An example of exchange isotherms for K–Ca and Mg–Ca covering the whole range of saturation from 0 to 100% is shown in Figs 3.1 (Deist & Talibudeen 1967a), and 3.2 (Salmon 1964). It should be noticed that in cases when the ions have different charges the position of the curve depends upon the total concentration of the soil solution.

3.1.3 Effect of ionic size

Martin & Laudelout (1963) studied the complete exchange isotherm between the alkali cations and ammonium on montmorillonite. Their affinity relative to the ammonium ion was directly proportional to their polarizability.

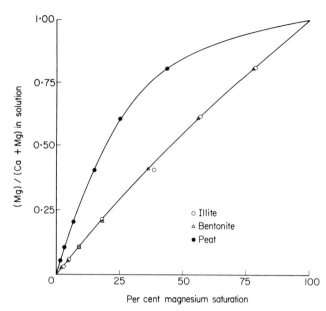

FIG. 3.2. Relation between magnesium saturation and the activity ratio (Mg)/(Ca + Mg) in solution for illite, montmorillonite and peat (after Salmon 1964).

Since polarizability increases with the volume of the ion, the affinity was in the order Cs > Rb > K > Na > Li. Laudelout *et al.* (1968) similarly found the affinity of the alkaline earth cations to be in the order Ba > Sr > Ca > Mg.

3.1.4   Effect of the exchanger

For a given activity ratio $(K)/(Ca + Mg)^{\frac{1}{2}}$ it is usually found that the proportion of potassium on the exchanger, N_K, is in the order: illite > montmorillonite > humus as in Fig. 3.3 (Salmon 1964).

3.2   Buffer power

In considering plant uptake we are rarely concerned with the complete isotherm, but only with a small section of it, usually in the range from zero to a few per cent of saturation. This is true even for a nutrient such as potassium, required in large amounts by plants. For elements required in trace amounts, e.g. zinc, or for heavy metals where problems of toxicity arise, the range is lower still.

A typical relation between the reduced activity ratio $(K)/(Ca + Mg)^{\frac{1}{2}}$ in solution and the potassium adsorbed or desorbed from the exchange complex of a field soil is shown in Fig. 3.4 (Beckett 1964). In many soils it is

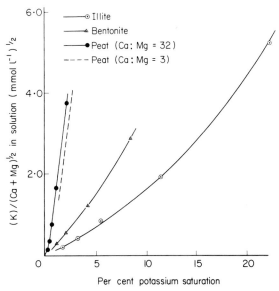

FIG. 3.3. Relation between percentage potassium saturation and the activity ratio $(K)/(Ca + Mg)^{\frac{1}{2}}$ for illite, montmorillonite and peat (after Salmon 1964).

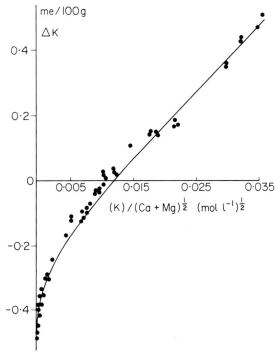

FIG. 3.4. Relation between the reduced activity ratio $(K)/(Ca + Mg)^{\frac{1}{2}}$ and potassium adsorbed or desorbed from Lower Greensand soil (after Beckett 1964).

very difficult to replace all the exchangeable ion by calcium, owing to release from slowly exchangeable sites. The amount of potassium replaced also varies with the replacing ion. Thus the zero point of an isotherm such as that illustrated in Fig. 3.1 is often indefinite.

From the slope at any point of the curve illustrated in Fig. 3.4 the buffer power dC/dC_1, may be derived, where C is the concentration of labile ions in the soil, and C_1 their concentration in the soil solution.

The nomenclature of the variables used in describing the sorption of ions on soils is confused, often inexact, and unstandardized; and it is necessary to draw attention to some definitions. Since the concepts of thermodynamics are often applied to these isotherms the definitions of terms should be those of thermodynamics, and not culled at will from the scientific literature. To quote from Guggenheim (1933):

'Extensive Properties'—'Any property, such as mass, whose value for the whole system is equal to the sum of its values for the separate phases is called an "extensive property" or a "*capacity factor*".' Examples are energy, entropy, volume.

'Intensive Properties'—'The density of a phase is clearly constant throughout the phase, because the phase is by definition, homogeneous. Further the density of a phase of a given kind and state is independent of the quantity of the phase. Any property of a phase with these characteristics is called an "intensive" property or an "*intensity factor*".' Examples are temperature, pressure, concentration, chemical potential.

'Buffer capacity'—This term derives from pH changes in electrolyte solutions. To quote from Glasstone (1940) 'The buffer capacity of a solution is defined by $db/d(pH)$, where $d(pH)$ (an intensity factor) is the increase of pH resulting from the addition of db of base (a capacity factor)'. Clearly if the amount of solution is doubled, so is its buffer capacity.

In the cation exchanges described, the amount of ion on the exchanger is expressed as a proportion of the exchange capacity, e.g. Fig. 3.1, or per unit weight of soil, e.g. Fig. 3.4. These are both measures of an *intensity factor* of the solid phase. The other axis may be plotted as a reduced activity or concentration ratio, or, related to these, as the difference between the free energies per equivalent of ion pairs, as described by equation 3.7. These are also *intensity factors*. Accordingly, it seems incorrect to describe the slope of such a curve as a buffer *capacity*; though if it is referred to unit mass of soil it might be called a specific buffer capacity. Since we always need the ratio dC/dC_1 (Chapter 4) under specified conditions, we use for it the simple term 'buffer power'.

The increase in buffer power at low concentrations of potassium shows that relative to calcium and magnesium ions the potassium ion is adsorbed with greater affinity, probably by sites having a special affinity for the potassium ion at the edges of the clay mineral lattices. Such curved isotherms for exchangeable ions present in low proportion have also been described for sodium (Tinker & Bolton 1966; Bolton 1971), aluminium (Nye *et al.* 1961), rubidium (Deist & Talibudeen 1967a). Though not usually presented in this context it is also true of the hydrogen ion, as shown in Fig. 3.5 (Farr *et al.* 1970). Here the hydrogen ion is covalently bonded to oxygen atoms which ionise progressively as the pH is lowered.

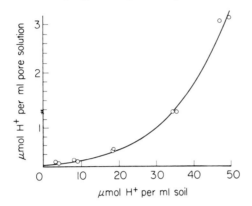

FIG. 3.5. The relation between the exchangeable hydrogen ion on a soil and the concentration of hydrogen ion in an equilibrium solution of calcium chloride (from Farr *et al.* 1970).

Given the diversity of clay minerals and other exchange materials in soils, and the corresponding variety of types of exchange site they offer, it seems very likely that most soils will have a small proportion of sites with a strong affinity for a particular cation. Thus trace amounts of micronutrients, e.g. copper, or toxic or unusual cations arising from industrial processes, e.g. mercury, or radionuclides of elements not already present in quantity, e.g. strontium, are likely to be relatively strongly adsorbed (Poelstra *et al.* 1974). Their effect on plant uptake and growth will be less than if they had been adsorbed with lower affinity, but on the other hand the rate at which they are leached out of the soil will also be reduced.

3.3 Insoluble compounds

The concentration in the soil solution of many cations absorbed in trace quantities by the plant, e.g. iron, aluminium, copper, manganese and zinc, may be controlled by insoluble salts, e.g. $ZnSiO_3$, oxides or hydroxides, e.g. $Al(OH)_3$, or ill-defined complexes, that may exist as separate crystals or adsorbed on crystal surfaces (Hodgson 1963). Ideally, the concentrations of these elements in the soil solution can be calculated from the solubility products of the solid forms if these can be identified; but in practice the formation of mixed crystals, and the presence of poorly ordered adsorbed and colloidal material makes accurate calculation impossible. Understanding of these ions is further complicated because they form soluble complexes with organic matter, and the proportion between free and complexed ion in the soil solution varies widely between different soils (Cottenie & Kiekens 1972).

We do not deal with the chemistry of these insoluble compounds here, because the transport of micronutrients has rarely been studied in detail. However, it is worth making the point that if an insoluble single component compound does control the concentration of an ion, then in principle the solution is infinitely buffered until all the solid phase has dissolved.

3.4   Cations with variable charge: oxidation–reduction reactions

Prediction of solid-solution equilibria is particularly difficult in poorly-drained soils where cations such as Fe (III) and Mn (IV), which form very insoluble oxides, may be reduced to Fe (II) and Mn (II). The reduced forms may occur in much higher concentration in the soil solution, where they are usually subject to the same controls as normal divalent exchangeable cations. Following submergence a wealth of redox and microbiological reactions occur that are very important both for understanding plant nutrition in badly drained and rice paddy soils, and in studying soil development. These have recently been reviewed by Ponnamperuma (1972) and a further abstract of his admirable digest would be superfluous here. At present, reliable data about these reactions is scarce largely because it has only recently been realized that redox potentials can be measured systematically and repro-ducibly only when the electrodes are in soil solution extracts rather than in soil. Rate processes are even more difficult to study than equilibria, because they are controlled by diffusion of oxygen, microbiological activity, and rates of chemical reaction in which pH changes are usual and have dramatic effects on solubilities. At present the simultaneous processes that occur are usually too complex to be analysed in detail. Some attempts that have been made are mentioned in succeeding chapters, but in general this very important branch of the subject is over-ready for a fundamental attack.

3.5   Adsorption of anions

As we have already emphasized at the beginning of this chapter, since the soil colloid carries a net negative charge there is a general tendency for anions to be repelled from solid surfaces (negative adsorption) and retained in the soil solution, where their amounts together with the soil moisture content control the overall salt concentration of the soil solution. Anions may, however, be adsorbed on the soil solid both by non-specific and specific bonding.

3.5.1 Non-specific anion bonding

Soils contain free hydrous oxides of iron (III) and aluminium that may become positively charged when one of the oxygen atoms accepts a hydrogen ion. Similarly at the edges of aluminosilicate clay mineral lattices, there are oxygen atoms whose electron orbitals are not fully co-ordinated with aluminium or silicon atoms. These oxygen atoms readily accept protons. The amount of positive charge developed depends upon the concentration of hydrogen ion, which is thus a surface potential determining ion. These localized positive charges attract anions by electrostatic forces. Such anions are freely exchangeable with anions in the soil solution; and since this is dominated by chloride, nitrate and sulphate it is these anions that predominate on the non-specific sites. Since there are few positive sites at pH greater than 6, the amount of bicarbonate ion held in this way is small.

3.5.2 Specific anion bonding

Many anions form strong bonds with the aluminium, iron (III) and other cations in hydrous oxides and alumino-silicate clay lattices by replacing the surface ligands attached to them; thus:

$$> Al - OH + H_2PO_4^- = > Al - H_2PO_4 + OH^-.$$

Examples of these anions are F^-, $H_2PO_4^-$, $HMoO_4^-$, $HSeO_3^-$, $H_2BO_3^-$, $H_3SiO_4^-$, HCO_3^-. Specific anion adsorption has been systematically studied on pure mineral hydrous oxides such as goethite, $FeO(OH)$, in the past decade by Posner, Quirk and co-workers in Western Australia, and a brief account of their findings and those of others in relation to soil has been given by Mott (1970). Many of these anions are conjugate bases of weak acids. At a given pH the relation between amount adsorbed and solution concentration follows a Langmuir adsorption isotherm. The maximum adsorbed is very sensitive to pH (see Fig. 3.6) (Hingston *et al.* 1968), and the curve relating the maximum adsorbed to pH shows a sharp break in slope at about the pK value of the acid.

Unlike cation adsorption, the maximum amount of anion adsorbed specifically differs markedly from one anion to another, as shown in Fig. 3.6.

A specifically adsorbed anion may be displaced by another specifically adsorbed anion, but not, to any appreciable extent, by a non-specifically adsorbed anion. It has been found that specific adsorbtion increases the negative charge on the mineral surface. When two anions are in competition, the ion that increases the negative charge to the greatest extent is adsorbed preferentially. Obihara & Russell (1972) have found the exchange between silicate and phosphate on four soils agrees with these generalizations.

The method (p. 35) by which we showed that any pair of exchangeable cations, A and B, tend to buffer the reduced activity ratio $(A)^{1/z_A}/(B)^{1/z_B}$ in

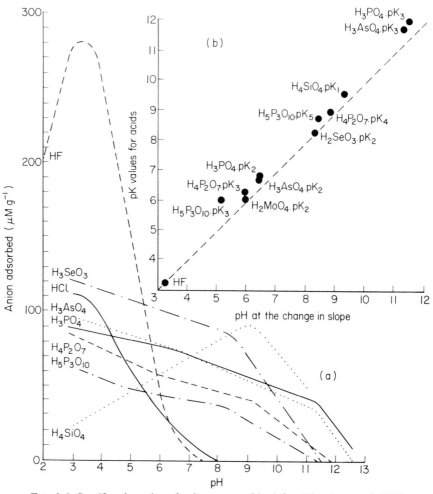

FIG. 3.6. Specific adsorption of anions on goethite (after Hingston *et al.* 1968).
(a) Adsorption 'envelopes' for anions with 0·1 M NaCl as supporting electrolyte.
The curve for Cl⁻ illustrates non-specific adsorption. (b) The plot of pKs for weak
acids against pH for breaks in slope of adsorption 'envelopes'. The broken line
indicates the ideal relationship.

the soil solution, can be applied to anions adsorbed by the soil in a definite
phase, simply by giving z_A and z_B negative signs. Likewise, an adsorbed
cation A with charge z_A and an adsorbed anion B with charge z_B will buffer
in the equilibrium soil solution the ratio $(A)^{1/z_A}/(B)^{1/z_B}$. This is the reduced
activity product $(A)^{1/z_A} \times (B)^{1/|z_B|}$ where $|z_B|$ is the valency of the anion
regardless of sign. Such a constant activity product does not require that an
insoluble salt $A_{z_B} B_{z_A}$ should exist in the soil. As an example, Aslyng (1954)

showed that the activity product $(Ca)^{\frac{1}{2}}.(H_2PO_4)$ remained constant in the soil solution when the calcium concentration was changed 10-fold.

Among the specifically bonded anions phosphate has received by far the most detailed study in soil, from the point of view both of solid-solution equilibrium, and transport characteristics.

3.5.3 Phosphate equilibria in soil

A great deal of effort has been expended on trying to relate the concentration of phosphate in the soil pore solution to specific insoluble phosphate compounds or surface complexes. This complicated story has been clearly reviewed by Larsen (1967). The difficulties arise because the rates of equilibration of phosphate in solution with the crystalline forms are extremely slow; and because the pure crystalline forms, particularly of hydroxyapatite, may have variable surface compositions. After prolonged equilibration of samples, or in field soils to which no phosphate has been added for many years, the concentration of phosphate in the soil solution is probably controlled by a surface complex of hydroxyapatite, not only in alkaline and neutral soils, but in slightly acid soils as well. In very acid soils, below pH 3, variscite, $Al(OH)_2H_2PO_4$, may control the solution concentration, but this dissolves incongruently above pH 3.1 to form a more basic solid phase of aluminium hydroxyphosphate, possibly as a surface complex on variscite. Bache (1963) concluded that strengite, $Fe(OH)_2H_2PO_4$ is never likely to be in equilibrium with any soil solution. Thus it seems that in long equilibrated acid soils surface complexes on both variscite and hydroxyapatite may coexist. There is also considerable evidence that phosphate is adsorbed on the surface of free iron oxides in acid soils.

Most soils have not reached this long-term equilibrium state, and in practice it is more important to know what insoluble complex controls the immediate phosphate concentration. Murrmann & Peach (1969) studied soils from old lime and phosphate fertilizer trials that had not been treated for at least 5 years. They shook samples for 24 h in solutions containing graded amounts of hydrochloric acid or lime water. Figure 3.7 relates the phosphate concentration in solution in one series of experiments to the pH. It will be noted that the concentration is at a minimum at about pH 6; in other experiments the minimum lay between pH 5 and 6.5. They concluded that the same form of phosphate complex, probably an alumino-phosphate, controlled the phosphate concentration in both the acid and limed soils. The concentration varied with pH in a way contrary to that expected of crystalline phosphates: the calcium forms being expected to maintain a low concentration under alkaline conditions, and iron and aluminium forms under acid conditions. The discontinuity in the curves shown in Fig. 3.7 at high pH is thought to be caused by precipitation of calcium octaphosphate. Thus it

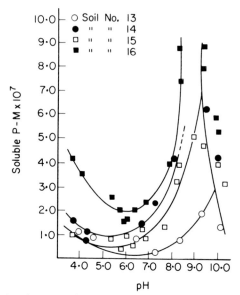

Fig. 3.7. Relation between phosphate concentration and pH of solution for 4 soils shaken with graded amounts of hydrochloric acid or lime water (after Murrman & Peech 1969).

would appear that in the present state of knowledge we cannot easily predict the relation between labile phosphate and the concentration of phosphate in solution.

In practice this relationship may be determined by equilibrating the soil with a calcium salt of an 'unspecifically' adsorbed anion, e.g. chloride, to which graded amounts of phosphate have been added. An example is shown in Fig. 3.8 (Bagshaw *et al.* 1972). As we have shown, the soil tends to buffer the activity product $(Ca)^{\frac{1}{2}} \times (H_2PO_4)$ in the equilibrium solution. The values of the phosphate concentration obtained will therefore depend on the strength of the calcium salt solution used. A strength equal to that of the soil pore solution will cause least disturbance to the adsorbed phosphate at the null point where the added solution neither gains nor looses phosphate. The equilibrium concentration of the soil at the null point ($\triangle P = 0$), and the slope of the curves is very sensitive to pH, as shown in Fig. 3.8.

A period of about half an hour is required for equilibration. All the difficulties that were associated with the determination of potassium ad-sorbtion isotherms are accentuated with phosphorus. In particular: the equilibrium is sensitive to CO_2 pressure (Larsen & Widdowson 1964), micro-bial effects (White 1964) and inequalities of calcium phosphate potential in the samples used (Beckett & White 1964). Thus it is only by taking great care that reproducible results can be achieved.

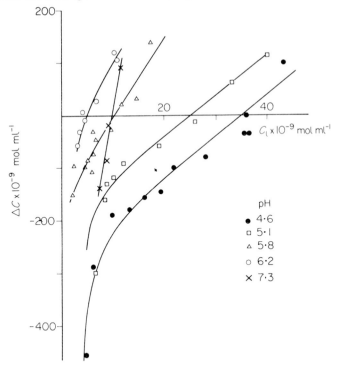

Fig. 3.8. The effect of pH of the equilibrating calcium chloride solution on the phosphate sorption isotherm of a soil (after Bagshaw *et al.* 1972).

It is useful to be able to characterize the adsorption relations of a soil by a minimum number of parameters; e.g. when writing computer programs. Although some workers have found that phosphate adsorption curves may be fitted to a Langmuir type isotherm (Olsen & Watanabe 1957), in general no maximum of adsorption is attained as the phosphate concentration is increased (Gunary 1970). Bache & Williams (1971) find a plot of the sorbed P against the logarithm of the concentration of P in solution is linear for their adsorbtion data on Scottish soils. We have found a similar equation to describe many desorption isotherms satisfactorily, in the form:

$\triangle C_s = $ kln C_l/C_{li}, where $\triangle C_s$ is the change in concentration of sorbed P and C_{li}, is the initial concentration of P in the soil solution.

This equation corresponds to the middle range of surface coverage of the Temkin isotherm, which may be derived theoretically if it is assumed that the energy with which each phosphate ion is bonded decreases linearly with the number already adsorbed. The Langmuir isotherm assumes that the bonding energy is unaffected by the number of ions already adsorbed. The characteristics and assumptions of various types of isotherm are discussed by Hayward & Trapnell (1964).

3.6   Rates of ionic interchange between solid and solution

3.6.1   Cations

Although we have seen a fairly stable equilibrium is usually established within minutes between exchangeable cations and cations in solution, we have to consider slower processes that may occur over longer periods comparable to the life of a root or a crop or a stand of natural vegetation.

It is convenient to divide these reactions into rapid, with half-times of the order of an hour or less; intermediate, with half-times of the order of a day; and slow, with half-times of the order of a week or more.

Rapid reactions

The exchangeable cations are held on the external surfaces of clay or humus particles, or in the interlayers of expanded clay minerals in which the aluminosilicate sheets are separated by at least two and usually three molecular diameters of water molecules. The half-time of a reaction whose rate is controlled by diffusion is *c.* L^2/D where L is the average distance a particle travels. If the diameter of a clay particle is 1 μm, and the apparent diffusion coefficient of an interlayer cation is 10^{-8} cm^2 s^{-1}, e.g. in vermiculite, then the half-time for exchange with the pore solution would be of the order of 1 s. In the pores themselves the diffusion coefficient of cations is about 10^{-5} cm^2 s^{-1}, and the diameter of a very large pore might be 1 mm, so the half-time for equilibration across such a pore could be of the order of 1000 s. Thus it would seem that attainment of exchange equilibrium in a structurally intact soil with no water movement is often limited by diffusion across pores rather than release from the solid phase. Exchange when a soil is shaken with solution will be more rapid. Malcolm & Kennedy (1969) found that 75% of the exchange between potassium and barium on kaolinite, illite and montmorillonite occurred in less than 3 s.

The rates of rapid reactions are needed to decide what should be described as a 'diffusible' ion—a matter to be discussed in the next chapter.

Intermediate reactions

For most cations the fraction showing an intermediate rate of exchange is small or negligible. However, for potassium and other ions of similar size and charge, like ammonium, rubidium and caesium the amount exchanging at an intermediate rate is significant.

Beckett (1969) has summarized knowledge of the release of potassium from soils under field conditions as in Fig. 3.9.

The fraction showing an intermediate rate is from $\frac{1}{4}$ to 5 times the potassium exchangeable to ammonium acetate. The rate of release appears as a rule to obey first-order kinetics. This fraction, described as intermediate

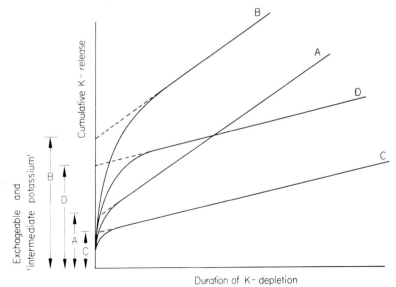

FIG. 3.9. Representation of release of potassium from soils under continuous cropping Unfertilized A and fertilized B plots on a virgin soil. Unfertilized C and newly-fertilized D plots on the same soil, following prolonged exhaustion (after Beckett 1969).

potassium, is probably held within the interlayers of weathered illitic clays (Arnold 1970). In unweathered illites the alumino-silicate sheets do not readily expand to allow interchange of potassium with ions in solution. However, weathering of illites is accompanied by a lowering of the charge density on the lattice, which consequently tends to expand more easily.

Stanford & Pierre (1947) have found that ammonium was fixed in a similar way to potassium on eight Iowa soils. Such reactions are described in greater detail by Black (1968).

The rates quoted above have been obtained in laboratory experiments under controlled conditions. The rates of release or fixation are accelerated by the degree of drying of the soil that occurs in the field. The effect of drying is only likely to be large if the matric suction exceeds 10 bars, so that as a rule only the top soil will be affected. Under these conditions it seems unlikely that it will be possible to characterize the reaction by any simple rate equation.

Slow reactions

We do not propose to review the subject of mineral weathering, but would refer to the book by Loughnan (1969) and the symposium edited by Hallsworth & Crawford (1965).

In Fig. 3.9 above the amount of potassium released at the slow rate would not satisfy crop requirements. The rate approximates to zero order, i.e. it is nearly constant with time, though there are some reports that it decreases with time.

Though such slow reactions may not be significant during one growing season, under natural vegetation they are the main means by which exchangeable potassium lost to drainage is replaced in the profile.

3.6.2 Anions

The only anion whose rate of exchange in the soil has been investigated in any detail is phosphate.

Although a great deal of work has been done on the rate of exchange of soil with ^{32}P, we must emphasize that for practical application we are concerned with the rate of adjustment in the solution pores of the phosphate concentration as it is disturbed by plant uptake or percolating solutions. ^{32}P exchanges more rapidly, and finally to a much greater extent, with solid phase phosphorus than will the non-specifically adsorbed ions that predominate in solution. In the simplest natural situations the change in the phosphate concentration in solution will be accompanied by small changes only in the other ions. The effects on the kinetics of changes in pH, bicarbonate ion, calcium ion, etc., must be investigated separately.

A typical rate of exchange with ^{32}P is shown in Fig. 3.10 (McAuliffe *et al.* 1947). To explain the exchange rate, the exchange is usually decomposed into two or more reactions: for example Arambarri and Talibudeen (1959) distinguish a fast reaction with half-times between 0·3–1·6 h, which involves exchange with phosphate in pore solutions and on readily accessible surfaces;

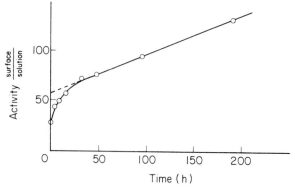

Fig. 3.10. Exchange of ^{32}P in solution with 'surface' phosphate (^{31}P) of Caribou soil (after McAuliffe *et al.* 1947).

an intermediate reaction with half-times of 1·8–8·6 h, and a slower reaction with half-times of 25·8–46·1 h. These slower reactions represent exchange with phosphate in micropores of diameter < 1 nm, and in the interior of crystals and complexes which are accessible to exchange through micro-cracks and crystal defects. In fact there is no definite limit to the amount exchanged and given sufficient time most of the soil phosphate would exchange. Larsen (1967) offers a different interpretation. He suggests the isotopically exchangeable phosphate is present as very small crystals of hydroxyapatite with a varying content of phosphate, attached to the surface of soil particles. He could distinguish no breaks in the rate of exchange up to 1000 h when adequate precautions were taken to ensure initial equilibrium between soil and solution.

Study of the kinetics of the release of soil phosphate in an otherwise undisturbed ionic environment has been most nearly achieved by measure-ment of the release of soil phosphate to a strong-base anion exchange resin saturated with chloride ion, a technique due to Amer *et al.* (1955). One of their curves is shown in Fig. 3.11. Vaidyanathan & Talibudeen (1968) have

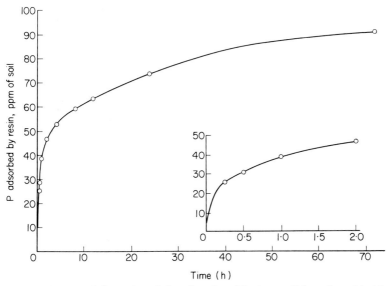

Fig. 3.11. Rate of absorption of phosphate from Waukegan silt loam by a chloride saturated anion exchange resin (after Amer *et al.* 1955).

interpreted similar curves in terms of an initial rapid first-order reaction together with a slower reaction varying as the square root of time, which was considered to be due to slow diffusion of phosphate from within soil particles. However, in their latest paper Vaidyanathan & Talibudeen (1970) reinterpret their data as being controlled by the rate of phosphate diffusion within the resin beads. Hence evidence for the slow release in soil is questionable.

Attempts to determine the change in buffer power with time by shaking samples of soil with dilute calcium chloride solution have been made. The procedure of Beckett & White (1964) involving shaking for 2 h measures the rapid equilibration. Drew (1966) found no change between 6 and 24 h. Vaidyanathan & Nye (1970) found no change between 16 and 80 h. White (1964) found an increase of about 50% between 2 and 24 h, which he attributed to microbial effects.

The present balance of evidence seems to be that an equilibrium is attained within about 2 h and that this is stable over periods of days unless disturbed by pH changes or microbial activity.

Slow reaction of added phosphate

Larsen *et al.* (1965) find the half-time for the reversion of soluble phosphate added to mineral soils into nonexchangeable forms is between one and six years.

Fitter (1974) found that readily extractable phosphate added to colliery shale soil, declined rapidly for five days, steadily till two months, and then exponentially till the end of the sixth month. The important feature of this work lay in the finding that the exponential decay constant was proportional to the buffer power of the soil. We do not know how rapidly the reverse process—mobilization of residual phosphate—occurs (Larsen 1973).

3.6.3   Relaxation and hysteresis in sorption isotherms

Adsorption and desorption isotherms do not always follow the same curve. An example for potassium is shown in Fig. 3.12 (Arnold 1970). Similar observations have been made for phosphate by Muljadi *et al.* (1966) and sulphate by Sanders (1971), when only a short time—less than 24 h, has been allowed for equilibration. Laudelout *et al.* (1968) have noted that to determine accurately the exchange equilibrium between two cations it is essential to approach the equilibrium point from both directions.

If the difference between the two curves is due to insufficient time being allowed for equilibration the phenomenon is relaxation (Everett & Whitton 1952). But in many instances it seems likely that the difference would persist however long a time was left for equilibration, in which case it is known as hysteresis.

In most instances the distinction between relaxation and hysteresis is not made and may not in practice be relevant. Newman (1970) instances the sodium–potassium exchange in vermiculite as an example of true hysteresis in solid-solution reactions. He says 'Apparently, hysteresis in this system is a consequence of the different structures of the two homo-ionic forms: Na-

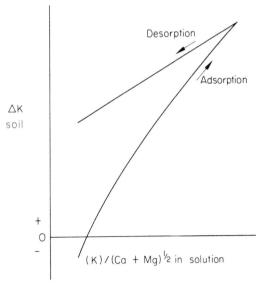

FIG. 3.12. Hysteresis in the sorption isotherm for potassium on a soil (after Arnold 1970).

vermiculite, in contact with aqueous solution, contains two layers of water molecules whereas K-vermiculite contains none. When Na- is introduced into K-vermiculite, the structure expands to accommodate interlayer water and the swelling pressure required to initiate expansion is greater than the pressure at which collapse begins when K enters Na-vermiculite'. It seems that a similar effect is likely to operate in the reactions between 'intermediate' soil potassium and the solution.

From a practical point of view, if the isotherm is being used to interpret a desorption process, e.g. removal of solute by plant uptake—then the desorption limb must be used, and likewise the adsorption limb to follow an adsorption process.

3.7 Mineralization and immobilization in organic forms

These processes are particularly important in determining the level of nitrate in the soil solution. This not merely controls the nitrogen supply to plants, but in fertile soils the overall strength of the soil solution. It therefore partly determines the concentration of the cations in solution, and hence their availability and ease of leaching. Supply of phosphate, sulphate and organically bound trace elements by mineralization is also of some importance.

3.7.1 Mineralization

Unfortunately it is only possible to make the most general statements about rates of mineralization since large fluctuations in average rates are caused by such factors as temperature, wetting and drying, freezing and thawing, and cultivation techniques—these are discussed at length in standard text-books on Soil Science such as Black (1968) and Russell (1973).

According to a review by Gasser (1969) the most widely used method of estimating the organic nitrogen that will be mineralized during the growing season of an annual crop is by rewetting air-dry soil samples and incubating aerobically from 7 to 28 days according to temperature. A variety of chemical extractants have given good correlations with crop performance, among the best being an estimate of reducing sugars extracted with 0.1 N barium hydroxide (Jenkinson 1968), and nitrogen extracted with boiling water and potassium sulphate (Keeney & Bremner 1966).

There is a 'flush' of mineralization in the spring in temperate arable soils, and after the early rains following a dry season in tropical soils. Thereafter the amount of nitrogen released by mineralization over the growing period may be small compared with the amount already present plus that added as fertilizer, and a rough indication of mineralization rate may be sufficient. For example the coarse sandy loam soil of the National Vegetable Research Station, Wellesbourne contains about 50 kg/ha of mineral N in the surface 20 cm before fertilizers are applied, and well over 100 kg/ha are usually applied as fertilizer to most vegetable crops.

McLaren's (1970) review of attempts to express rates of nitrogen mineralization in terms of enzyme activities and microbial growth kinetics reveals the difficulties of this approach even under controlled conditions.

Over sufficiently long periods the simplest equation for representing the organic C balance is

$$dC/dt = -k\mathbf{C} + A \tag{3.16}$$

where k is a measure of the rate of mineralization, and A the annual addition of humus C. Though there are wide variations in the ratios of C : N : S : P in soil humus they tend to be of the order of 100 : 10 : 1 : 1. Equation 3.16 may be used for each element.

Thus if a soil contains 4000 kg ha^{-1} nitrogen and k is 0.03, 120 kg ha^{-1} nitrogen are likely to be formed as nitrate during the year. Release of nitrogen of this order maintained yields of corn when virgin prairie was ploughed, but the amounts are inadequate for high yielding crops today.

The decomposition constant k is an average over many organic compounds and equation 3.16 could clearly be refined if experimental data were available. Jenkinson (1966), for example, has shown that the humus resulting from addition of ^{14}C labelled rye-grass, was still decomposing at four times the average rate for the soil humus present initially, four years after the addition.

Stevenson (1965) has reviewed the use of equations more complex than equation 3.16.

Values of k for nitrogen derived from long-term arable systems in North America lie in the range 0·02–0·10 (Bartholomew & Kirkham 1960). After clearing tropical forest and cropping continuously k for organic C range from 0·018 to 0·09 (Nye & Greenland 1960) and in savannas from 0 to 0·068. At Rothamsted a field under continuous barley for over 100 years gave $k = 0·025$ for nitrogen (Jenkinson 1966).

Under undisturbed vegetation decomposition rates are slower. In Southern Australia under pasture $k_N = 0·029$ and 0·013 (Greenland 1971). Under tropical forest $k_C = 0·02 - 0·05$ and under savanna 0·005 – 0·012 (Greenland & Nye 1959). In uncultivated Iowa top soils under grass the average age of the organic matter was determined by ^{14}C dating as 210–440 years ($k_C \simeq 0·005–0·002$). In a woodland the sodium hydroxide soluble fraction had an average age between 50 and 250 years (Broecker & Olson 1960), though residual material was about 2000 years old. In peat bogs of the Northern English uplands aerobic peat (0–12 cm) yielded $k_C = 0·07$, while wet anaeroboic peat yielded $k_C \simeq 0·0014$ (Gore & Olson 1967). Cooke (1967) has summarized much of the available data on long-term organic matter and nitrogen changes under both arable and perennial crops.

Levels of nitrate under natural vegetation are usually low, and in accord with this it is found that nitrate levels in drainage water from natural vegetation are also low. For example Cole *et al.* (1967) found only 4·8 kg ha^{-1} nitrogen were leached annually below the forest floor, and only 0·6 kg ha^{-1} nitrogen below the rooting zone (1 m deep) of a 36-year-old stand of Douglas Fir in the State of Washington under a rainfall of 1360 mm yr^{-1}.

In regions remote from airborne sources of chloride and sulphate, and in the absence of nitrate, the bicarbonate ion becomes the major anion in the soil solution. In the Douglas Fir stand already mentioned, the bicarbonate ion concentration nearly balanced the total cation concentration in leachates (McColl & Cole 1968). It was found that during a dry period mineralization increased the pressures of CO_2 and hence the concentration of bicarbonate, a process that rose with temperature as shown in Fig. 3.13.

3.7.2 Immobilization and denitrification

Nitrate (and phosphate) may be removed from the soil solution by microorganisms stimulated by additions of energy-rich material, when this contains less than about 2% nitrogen. Removal of nitrate may also occur under anaerobic conditions. Under normal free drainage conditions losses as nitrous oxide may still occur, presumably because of temporary anaerobic pockets. Long-term soil nitrogen balance sheets usually show a small unaccountable loss of nitrogen (Allison 1965). Greenland (1971) concludes

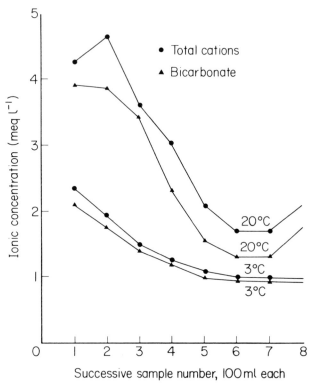

FɪG. 3.13. Concentration of cations and bicarbonate ion in leachate from soil under Douglas Fir following a dry period (after McColl & Cole 1968).

from experiments in South Australia in which nitrogen and nitrous oxide in the soil atmosphere were analysed, that under 3 years of wheat and pasture the annual losses would not have exceeded 23 kg ha^{-1}.

3.8   The composition of the soil solution

The concentration of solutes in the soil solution plays such an important part in the interpretation of experimental results mentioned in this book that further discussion of their measurement is in order.

3.8.1   Measurement

The method of displacing the pore solution from a column of soil with ethanol introduced by Ischtscherikow (1907) has been most recently examined by Moss (1963, 1969). He found that the activity ratios $(K)/(Ca + Mg)^{\frac{1}{2}}$ and $(K)/(Ca)^{\frac{1}{2}}$ determined on the displaced solutions remained constant over considerable changes in soil moisture level to the point of saturation.

He also found that the activity ratio $(K)/(Ca + Mg)^{\frac{1}{2}}$ in the extracts from a wide range of soils agreed well with the activity ratio determined by the null point method of Beckett & Craig (1964). In this method the soil is shaken with dilute $CaCl_2$ solution containing graded amounts of potassium, and the activity ratio at which the soil does not gain or loose potassium to the solution is determined.

Ethanol appears to displace solution from the fine as well as the coarse pores, and successive fractions, devoid of alcohol, have the same composition.

Suction methods are useful for following changes in composition of moist soils. These and other methods have been discussed by Nielsen (1972). He was able to extract soil solution through a layer of silica gel on a sintered glass filter, to tensions up to 0·8 bar.

Methods of expressing the soil solution under pressure should be used with discrimination since they change the pressure of carbon dioxide and hence the concentration of bicarbonate ion. In these and centrifuge methods it is also necessary to ensure that the support at the bottom of the soil column does not absorb the often very low concentrations of solution ions, especially phosphate. Any dispersed clay may tend to clog the pores of the porous support, and lead to a change in concentration of the displaced solution because of 'salt sieving' caused by exclusion of anions from the negatively charged surfaces of the clay particles. Some workers use 'saturation extracts', obtained by saturating the soil with water and extracting at a suction of 0·5 bar. It is not possible to estimate the concentration of ions in the soil pores from analyses of such extracts because the concentration in the portion removed is greater than in the portion remaining owing to negative adsorption of anions at the negatively-charged surfaces (Bower & Goertzen 1955). Preparation of samples usually leads to enhanced microbial activity, and the production of nitrate. If samples are initially dry they must be incubated for about a month at the required moisture level before a steady level is attained (Larsen & Widdowson 1968).

3.8.2   Seasonal variation

When cropped soils at Woburn, Saxmundham and Rothamsted were sampled systematically at monthly intervals over the year, changes in the concentration of phosphorus and potassium on shaking with 0·01 M $CaCl_2$ commonly amounted to 2–3 fold and sometimes 5 fold (Garbouchev 1966; Blakemore 1966). Levels were highest in the spring, or following application of fertilizer, and lowest after harvest. Soluble phosphorus and potassium was greater in air-dried than in fresh samples. Typical examples of fluctuation of ammonium and nitrate N through the year are shown in Fig. 3.14 for forest, grassland, and cultivated soils (Greenland 1958).

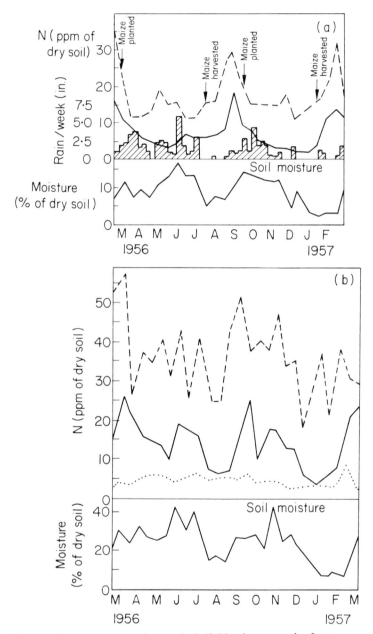

FIG. 3.14. Changes in nitrate, ammonium and nitrifiable nitrogen under forest, grassland and cultivated conditions in Ghana (after Greenland 1958). (a) Cultivated plot, forest region; (b) natural forest.

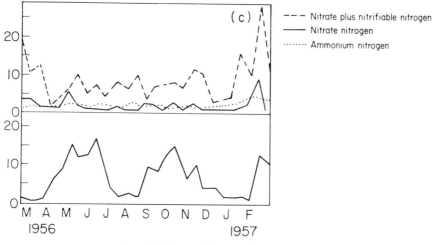

FIG. 3.14 (cont.) (c) natural savannah.

ORGANIC SOLUTES

3.9   Sorption reactions of organic materials

An already great and ever-increasing variety of herbicides, insecticides, fungicides, stimulants and repressants, pollutants and wastes are added to the soil; and their uptake by plants, and their movement within the soil and out of it in drainage urgently need to be rationalized. We have much less detailed knowledge about them than we do about the inorganic ions, but in principle the same ideas govern their behaviour and the same approaches are appropriate. Many of them are indeed ions. The sorption reactions of the remainder, which are uncharged, are thermodynamically more straightforward than the ions. The uncharged forms often have the complication that they are volatile, whereas only a few inorganic ions in solution are in acid–base equilibrium with uncharged volatile forms: e.g. $NH_4^+ - NH_3$, $H_2CO_3 - (H_2O + CO_2) - HCO^-_3$, $H_2S - HS^-$.

3.9.1   Solution-gas equilibria

Solution of gases in liquids is controlled by Henry's Law: 'the mass of gas dissolved by a given volume of solvent, at constant temperature, is proportional to the pressure of gas with which it is in equilibrium'. Since the pressure of gas is, ideally, proportional to its concentration in the gas phase the solubility is most conveniently expressed by the ratio, $C_l : C_g$, which is thus the distribution coefficient between liquid and gas phases at a given

Table 3.2. Distribution of organic solutes between liquid and gas phases (from Hamaker & Thompson 1972).

Compound	Temp.	Water Solubility (ppm)	$\beta =$ conc. in liquid/ conc. in gas
Methyl bromide	20°C	16,000	4·1
Chloropicrin	20°C	0·19	10·8
DDT	20°C	$1·2 \times 10^{-3}$	$3·3 \times 10^2$
Trifluoralin	25°C	0·58	$3·2 \times 10^2$
Aldrin	25°C	0·2	$2·26 \times 10^3$
Disulfoton	20°C	15	$5·56 \times 10^3$
Lindane	20°C	10	$1·96 \times 10^4$
Chlorpropham	25°C	80	7×10^5
Dimethoate	20°C	30,000	$2·81 \times 10^8$

temperature. This distribution coefficient was proposed by Ostwald and is sometimes listed in tables as Ostwald's solubility coefficient, β.

Table 3.2 illustrates the wide range of volatility of organic solutes now added to the soil. We may here anticipate Chapter 4 to point out that if the solubility coefficient is much less that 10^4, diffusion of the solute is predominantly in the gas phase, but if it exceeds 10^4 diffusion through the liquid phase predominates. Thus DDT, aldrin, trifluoralin, and the common fumigants such as methyl bromide and chloropicrin move as vapour; insecticides like lindane and disulfoton diffuse in both phases; while an insecticide like dimethoate and the many herbicides of low volatility such as the s-triazines and substituted ureas diffuse largely through solution.

The vapour pressure of a volatile compound may be much reduced in moist soil. Thus, although DDT has a low value of β, it is very insoluble in water and its vapour pressure over aqueous solution is only 1.9×10^{-7} mm Hg at 20°C.

3.9.2 Solid–solution equilibria

Organic materials exhibit a tremendous range of sorption behaviour. The main properties controlling their sorption are their charge, their molecular weight and the possibility of hydrogen bonding. We discuss them in order of decreasing affinity with the soil. An admirable and more extensive account, containing summary tables of available data has been given by Hamaker & Thompson (1972).

Cations

The bipyridylium (quaternary ammonium) herbicides (e.g. paraquat) have received most attention. The flat pyridyl ring, containing a positively charged

N atom, is very strongly held on the flat surface of clay minerals by normal ionic and also Van der Waals forces. Figure 3.15 shows the sorption isotherm of paraquat, and the extremely low concentration maintained in solution at normal applications is evident (Knight & Tomlinson 1967). In contrast to non-cationic compounds, the bipyridyls are more strongly adsorbed by clays, which offer a flat surface, than by soil organic matter. The effect of Van der Waals forces is apparent in the adsorption of methyl substituted ammonium cations on calcium montmorillonite. The strength of adsorbtion increased regularly from $MeNH_3^+$ to Me_4N^+ (Theng *et al.* 1967), see Fig. 3.16.

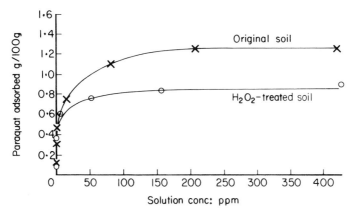

FIG. 3.15. Paraquat adsorption isotherm for Jealott's Hill soil (after Knight & Tomlinson 1967).

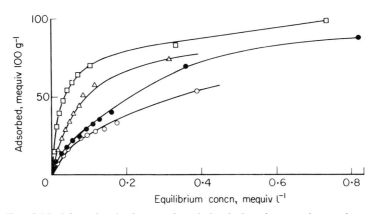

FIG. 3.16. Adsorption isotherms of methyl substituted ammonium cations on Ca–montmorillonite (after Theng *et al.* 1967). $\bigcirc$, $MeNH_3^+$; $\bullet$, $Me_2NH_2^+$; $\triangle$, Me_3NH^+; $\square$, Me_4N^+.

Uncharged bases

Such compounds can accept a proton to become cations $(B + H^+ = BH^+)$. The extent to which this occurs depends on the base strength of the organic compound and the concentration of hydrogen ions, which is considerably greater near the negatively-charged clay surface than in the free solution. As for any other cation pair, at equilibrium the activity ratio $(H^+)/(BH^+)$ will be constant throughout the system. Consequently BH^+ is concentrated on and near the clay surfaces (Bailey *et al.* 1968). The triazine herbicides and amitrole are important examples in this group, and also the systemic pyrimidine fungicides, e.g. ethyrimol. As would be expected titratable soil acidity is well correlated with the adsorption of simazine and atrazine, and amitrole, over a series of soils (Nearpass 1965).

Uncharged compounds

Greenland (1970) says about sorption of organic compounds 'Considering soil profiles as a whole, the inorganic surfaces are usually more extensive than those associated with organic materials, but in surface soils and peats the organic sorption sites are the more important.'

Sorption isotherms are typically curved and usually best represented by the empirical Freundlich equation: $x/m = KC_1^n$ where C_1 is concentration in solution, x is amount of adsorbate, m is amount of soil, and n and K are constants. It is usually necessary to allow from 1 to 4 h for equilibrium to be established. A good example is provided by the adsorption of a series of substituted urea herbicides studied by Hance (1965). The adsorption increased with the molecular size of the herbicide (Fig. 3.17). It was very closely correlated with the organic matter content, but not clay content (Fig. 3.18). Values of n can be as low as 0·6. However, at the very low concentrations at which many pesticides are used the isotherms are often linear $(n = 1)$. Hamaker & Thompson (1972) have tabulated the constants in the Freundlich equation for a wide range of compounds (Table 3.3).

In surface soils it is not clear to what extent the preferential absorption on organic matter compared with clay surfaces is due to direct preference or to the fact that the soil organic matter is shielding the clay surfaces from the herbicide.

Adsorption on the siloxane surfaces of alumino-silicate clay minerals is better understood. The following interactions leading to adsorption have been identified (Greenland 1970):

(a)   H-bonding between water in clays and added organic molecules, and between the soil organic matter and added organic molecules.

(b) Non-specific Van der Waals bonding. Though the molecules are uncharged they are often highly polarizable in the aromatic ring.

(c)   When a single organic molecule displaces several water molecules from the surface of a clay there is a net increase in entropy. The effect increases with the size of the organic molecules.

In general the much lower degree of adsorption for the physically bound than the ionically bound molecules will be noted (*cf.* Figs 3.17 and 3.15).

Table 3.3. Adsorption of uncharged organic compounds by soil and soil organic matter (from Hamaker & Thompson 1972).

Compound	K_{soil}	$K_{organic carbon}$
s- Triazines		
Simazine	1–4	54–275
Prometryne	6–8	310–580
Ureas		
Fenuron	0·8	12–37
Linuron	6–18	154–1618
Carbamates		
Chlorpropham	4–14	245–590
Amino nitrophenyl sulphone		
Nitralin	26–50	500
Halo-hydrocarbons		
Ethylene di-bromide	0·4–0·5	17–32
Lindane	9–20	961–1193
DDT	$1·3 \times 10^4$	$1·3 \times 10^5$– $3·5 \times 10^5$
Carboxylic acids		
Picloram	0·29–0·67	6–23
2, 4–D	0·8–6	16–32
2, 4, 5–T	0·8–1·1	23–42

$$K_{soil} = \frac{x/m}{C_1} \text{ (values for soils } < 8·7\% \text{ organic C)}$$

$$K_{\substack{organic \\ carbon}} = \frac{x/m \text{ organic carbon}}{C_1}$$

These figures are a guide only, because they assume a linear relation between x/m and C_1; and because the solution concentration is often higher than would occur in the soil solution in practice. (x/m in $\mu g/g$; C_1 in ppm).

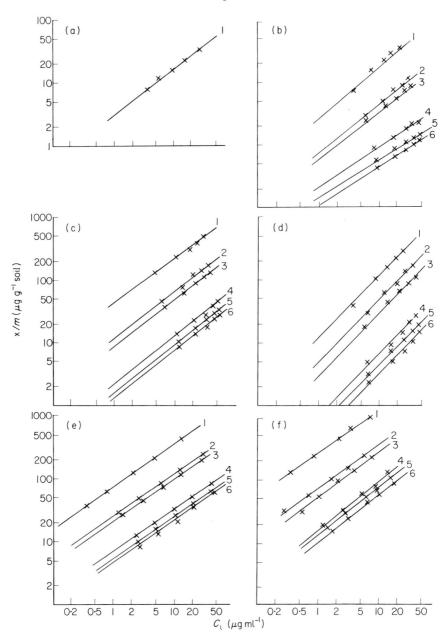

Fig. 3.17. Adsorption isotherms of substituted urea herbicides showing effect of molecular weight of herbicide and organic matter content of soil (after Hance 1965). (a) urea; (b) methyl urea; (c) phenyl urea; (d) fenuron; (e) monuron; (f) diuron. The molecular weight increases from (a) to (f). The soil organic matter decreases from 1 to 6.

Organic anions

Species with a net negative charge are unlikely to be adsorbed to an appreciable extent, though there is some evidence (Greenland 1965a) that they may be attached as ligands to polyvalent exchangeable cations. They may also be attracted electrostatically to positive sites on iron and aluminium hydrous oxides, which occur in acid soils by protonation.

Many anions are derived from weak acids, and the unionized form may be adsorbed by non-specific forces already discussed. An example is Picloram ($pK_A = 3.7$), which was adsorbed more strongly as the pH of the soil was changed from $pH = 10$ to $pH = 2$ (Hamaker *et al.* 1966).

3.9.3 Rates of equilibration

In experiments in which pesticides are added to stirred soil slurries most of the sorption occurs within minutes and equilibrium is attained within a few hours (Graham-Bryce & Briggs 1970). Hamaker & Thompson (1972), however, record several examples of a further slow reaction over weeks or

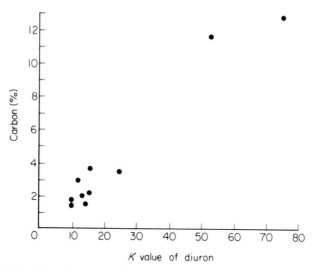

FIG. 3.18. Relation between soil organic matter and K value in Freundlich Equation ($x/m = KC_1^n$) for diuron (after Hance 1965).

months. The process is largely reversible, though desorption may be considerably slower than adsorption, particularly on soils with much organic matter; and a fraction of the adsorbed material may be very slowly removed, particularly if the soil has been dried and rewetted.

3.9.4 Decomposition rates of organic materials sorbed by soil

Adsorption by the soil may be expected to decrease the accessibility of a compound to microbes; on the other hand there is a greater concentration of microorganisms on surfaces than in the pore solution. Hance (1970) concludes that it is not possible to predict the net effect of adsorption on decomposition rates.

It may seem that the situation is so complex that quantitative approaches are premature. This is not so. Often, detailed data on degradation rates is available. Frequently we are concerned with transport phenomena that are rapid compared with the half-life of the biocide, such as movement to a root or redistribution in the top soil for the first few weeks after application. And it must be remembered that some biocides are extremely persistent, e.g. simazine or BHC. The great range in persistence of biocides is illustrated in Table 3.4 (Graham-Bryce & Briggs 1970).

The overall kinetics for disappearance are often approximately first-order at the usual rates of application in the field. If the decomposition is largely due to microorganisms there is usually a lag phase while the microorganisms multiply, followed by rapid decay. When decay is slow it is difficult to distinguish between direct chemical and microbial decomposition. Losses of material are not only due to decomposition. Volatilization, leaching and uptake and removal in crops are also involved. These matters are taken up in Chapter 8. Detailed information on the persistence of herbicides has been assembled by Kearney & Kaufmann (1969); and a good recent summary review has been given by Edwards (1974).

Table 3.4. Persistence of materials in soils: approximate times for at least 75% of added dose to disappear (from Graham-Bryce & Briggs 1970).

Compound	Use	Persistence	Normal rate of use per application (kg ha^{-1})
Methyl bromide	Fumigant for soil sterilization	< 1 week	1000
Ammonia	Fertilizer	3 weeks	up to 100
2, 4-D	Foliar herbicide	1 month	1
Disulfoton	Systemic organo-phosphorus insecticide	6 weeks	1–2
Chlorfenvinphos	Organophosphorus soil insecticide	6 months	1–4
Diuron	Soil applied substituted urea herbicide	8 months	0·5–2
Simazine	Soil applied triazine herbicide	1 year	0·5–1·5
Dieldrin	Soil and crop organo-chlorine insecticide	> 3 years	1–5
DDT	Soil and crop organo-chlorine insecticide	> 5 years	1–5

4

LOCAL MOVEMENT OF SOLUTES IN SOILS

In the last chapter we dealt with the distribution of solutes between gas, liquid and solid phases in the soil at equilibrium; and with rates of re-distribution between these phases within soil pores. In this chapter we consider movement on a larger but still local scale—from one volume of soil to another. Such movements occur largely by diffusion and mass flow in the liquid and vapour phases, and by mechanical movements of the body of the soil. Major movements involving the balance and amount of solutes in the whole soil profile, including plant uptake and drainage losses, are treated in Chapter 8.

4.1 Diffusion

The process of diffusion results from the random thermal motion of ions, atoms or molecules. Consider a long column of unit cross-section orientated along the x axis, and containing a mixture of components in a single phase at constant temperature and external pressure. If the concentration of an uncharged component is greater at section A than at section B then on average more of its molecules will move from A to B than from B to A. The net amount crossing a unit section in unit time, which is the flux, is given by the empirical relation known as Fick's first law:

$$F = - D \, dC/dx \qquad (4.1)$$

where F is the flux, and dC/dx is the concentration gradient across the section. The minus sign arises because movement is from high to low concentration in the direction of increasing x. The diffusion coefficient, D, is thus *defined* by this equation as a coefficient relating two quantities, F and dC/dx, which can be measured experimentally.

The diffusion coefficient of the molecules in a phase is directly propor-tional to their absolute mobility, u, which is the limiting velocity they attain under unit force. D and u are related by the Nernst–Einstein equation:

$$D = u k T \qquad (4.2)$$

where k is the Boltzman constant, and T is the temperature on the Kelvin scale. The Nernst-Einstein equation is derived as follows. For a system with concentration gradients in the x direction only, the force on a molecule is the gradient of its chemical potential, $d\mu/dx$. Therefore the flux across a plane normal to the x axis is:

$$F = -C\,(d\mu/dx)u. \qquad (4.3)$$

Now for a single molecule

$$\mu = \mu_0 + kT \ln Cf \qquad (4.4)$$

where f is the activity coefficient. Therefore, differentiating equation 4.4 with respect to x and substituting in equation 4.3

$$F = -ukT\,(dC/dx + d\ln f/dx). \qquad (4.5)$$

The term $d\ln f/dx$ is usually very small for uncharged particles, and can be neglected. So comparing the definition of D in equation 4.1 with equation 4.5 we find $D = ukT$.

Because ions are charged their movements depend not only on the chemical potential gradients they would have as uncharged particles, but on the electrical potential gradients set up in the solution by movements of other ions or by external electric fields; i.e. their movements depend on their electro-chemical potential gradients (see p. 166). Thus the *interpretation* of D for ions may not be simple, but the *definition* of it is applicable to them in any given situation. The interactions between ions in the diffusion of salt solutions are described by Robinson & Stokes (1959, Chapter 11) and their application in soils by Nye (1966a).

Table 4.1 gives an idea of the enormous range of diffusion coefficients that are of interest to us. Molecules in the vapour phase are approximately 10,000 times as mobile as they are in the aqueous phase; e.g. compare O_2 in air and water. Adsorbed ions show a wide variation in mobility. In pure clays, for example, the mobility of the sodium ion in montmorillonite ($D = 4 \times 10^{-6}$ cm^2 s^{-1}) is one third of its mobility in free solution. In montmorillonite the hydrated sodium ion lies within a layer of water a few molecules thick between the alumino-silicate sheets of the clay mineral lattice. On the other hand, when there is no water between the sheets, as in the unexpanded mineral illite, the self-diffusion coefficient of the potassium ion is only 10^{-23} cm^2 s^{-1}.

The physical significance of these diffusion coefficients may be appreciated more clearly if we fix our attention on one of the particles moving randomly. Then, on average, in a linear system it would move in time, t, a distance of $(2\,Dt)^{\frac{1}{2}}$ from its starting point (Jost 1960, p. 25). Thus a molecule of gas will move about 100 cm in 10^5 s (1.2 d). In solution it will move about 1 cm in 10^5 s. The potassium ion in the example discussed ($D = 10^{-23}$ cm^2 s^{-1}) will move only 1 nm in 5×10^8 s (16 y).

Table 4.1. Some typical diffusion coefficients.

		T (°C)	D (cm² s⁻¹)
	Gas phase		
	O_2 into air	25	0·209
	CO_2 into air	25	0·163
	Chloropicrin into air	25	0·088
	Liquid phase		
	O_2 in water	25	$2·26 \times 10^{-5}$
	CO_2 in water	25	$1·66 \times 10^{-5}$
	NaCl in water	25	$1·61 \times 10^{-5}$
	Glucose in water	25	$0·67 \times 10^{-5}$
	Solid phase		
Lai & Mortland (1962)	Na in montmorillonite gel	25	4×10^{-6}*
Lai & Mortland (1968)	Na in vermiculite	20	6×10^{-9}*
de Haan *et al.* (1965)	K in illite	lab	10^{-23}*
	Soil		
Rowell *et al.* (1967)	Cl in sandy clay loam (40% H_2O by vol)	lab	9×10^{-6}*
Rowell *et al.* (1967)	Cl in sandy clay loam (20% H_2O by vol)	lab	$2·4 \times 10^{-6}$*
Rowell *et al.* (1967)	Na in sandy clay loam (40% H_2O by vol)	lab	$2·2 \times 10^{-6}$*
Rowell *et al.* (1967)	Na in sandy clay loam (20% H_2O by vol)	lab	$0·5 \times 10^{-6}$*
Rowell *et al.* (1967)	PO_4 in sandy clay loam (40% H_2O by vol)	lab	$3·3 \times 10^{-9}$*
Rowell *et al.* (1967)	PO_4 in sandy clay loam (20% H_2O by vol)	lab	$0·3 \times 10^{-9}$*

* indicates self-diffusion coefficient

Statisticians may be interested to note that since movement of such a molecule results from a succession of collisions that may move it at random in a positive or negative direction, the probable distribution of such molecules, i.e. their concentration, follows the normal (Gaussian) distribution curve, with root-mean-square displacement of $(2\,Dt)^{\frac{1}{2}}$:

$$C = \frac{M}{2(\pi Dt)^{1/2}} \exp - x^2/4Dt \qquad (4.6)$$

where M is the amount of substance deposited at the plane $x = 0$ at $t = 0$.

The factors affecting diffusion of particles in gas, liquid and solid phases will be discussed in turn.

4.1.1 Gases

When, at a constant overall external pressure, molecules of gas A diffuse into those of gas B, there is a counter movement of molecules of gas B, and the diffusion rate is expressed in terms of an interdiffusion coefficient D_{AB}. The interdiffusion coefficient can be predicted approximately from the kinetic theory of gases (Bird *et al.* 1960, p. 508). The simplest case arises in self-diffusion when gas molecules diffuse into others of the same kind. This is nearly true of the interdiffusion of isotopes, with a diffusion coefficient $D_{AA^{\cdot}}$.

From the kinetic theory for pure rigid spheres at low density:

$$D_{AA^{\cdot}} = \tfrac{1}{3}\,\bar{u}\,\lambda \tag{4.7}$$

where $\bar{u}$ is the average molecular speed, and λ is the mean free path; also

$$\bar{u} = \left(\frac{8\,k\,T}{\pi\,m}\right)^{\tfrac{1}{2}} \tag{4.8}$$

where m is the mass of molecule A, and

$$\lambda = \frac{1}{\sqrt{2}\,\pi\,d^2 n} \tag{4.9}$$

where d is the molecular diameter, and n is the molecular concentration.

Hence

$$D_{AA^{\cdot}} = \tfrac{1}{3}\left(\frac{8\,k\,T}{\pi\,m}\right)^{\tfrac{1}{2}}\left(\frac{1}{\sqrt{2}\,\pi\,d^2 n}\right) \tag{4.10}$$

Using the ideal gas equation, $P = CRT = nKT$, to substitute for n, we obtain,

$$D_{AA^{\cdot}} = \frac{2}{3\,\pi^{3/2}}\left(\frac{(kT)^{3/2}}{Pd^2\,m^{\tfrac{1}{2}}}\right). \tag{4.11}$$

It will be noted that the self-diffusion coefficient varies inversely not only as the square root of the molecular weight, but as the square of the diameter. Graham's so-called Law of Diffusion, stating that 'at a given temperature and pressure the rate of diffusion of a gas is inversely proportional to the square root of its density', and hence to the square root of its molecular weight, applies to the rate of *effusion* of gases through a small hole in their container into a vacuum, and not to the systems we are considering.

A similar but more complex expression may be derived for D_{AB}:

$$D_{AB} = \frac{2}{3\,\pi^{3/2}}\,\frac{(kT)^{3/2}}{P\left(\dfrac{d_A + d_B}{2}\right)^2}\left(\frac{1}{2m_A} + \frac{1}{2m_B}\right)^{\tfrac{1}{2}}. \tag{4.12}$$

A more accurate prediction of D_{AB}, within 5%, may be made from the Chapman–Enskog kinetic theory of gases (Chapman & Cowling 1951), which takes account of intermolecular forces between the molecules.

For multi-component systems, such as occur when oxygen is consumed and carbon dioxide liberated in air containing nitrogen, prediction of inter-diffusion coefficients is complex.

Curtiss & Hirschfelder (1949) derive the formula

$$D'_{12} = D_{12} \left\{ 1 + \frac{x_3[(M_3/M_2) D_{13} - D_{12}]}{x_1 D_{23} + x_2 D_{13} + x_3 D_{12}} \right\} \tag{4.13}$$

where D'_{12} is the diffusivity of gases 1 and 2 in the 3 component system,
D_{12} etc are the diffusivities of gases 1 and 2 etc in a binary mixture,
x_1 etc are the mol fractions,
M_2 etc are the molecular weights.

This formula has been applied to gaseous diffusion in soils by Wood & Greenwood (1971).

4.1.2 Solutions

In solution, the force opposing the motion of a rigid sphere is, by Stokes Law, $6 \pi \eta r v$, where η is the coefficient of viscosity, r is the radius of the sphere, and v is its velocity. The force opposing motion is balanced by the force impelling the sphere. If this force is unity, by definition of mobility $v = u$, the mobility. Hence substituting for u in equation 4.2

$$D = kT/6 \pi \eta r \tag{4.14}$$

and D is inversely proportional to the radius.

The simple nutrient cations and anions have self-diffusion coefficients in solution in the range $(0.5–2.0) \times 10^{-5}$ cm^2 s^{-1} at 25°C, as shown in Table 4.2. It will be noted that the mobility increases in the order Li < Na < K < Rb < Cs, and Mg < Ca < Sr < Ba, because the cations of higher atomic weight have the smaller hydrated radius. The hydrogen ion, with $D = 9.4 \times 10^{-5}$ cm^2 s^{-1}, has an exceptionally high diffusion coefficient, because it is carried along by a chain mechanism from one water molecule to the next (Glasstone *et al.* 1941, p. 559). A similar explanation accounts for the abnormal mobility of the hydroxyl ion, $D = 5.3 \times 10^{-5}$ cm^2 s^{-1}.

The effective radius of organic molecules in solution depends on their molecular weight and shape. For spherical molecules the molecular weight

$$M = \frac{4\pi r^3 \rho}{3},$$

where ρ is the density. Hence the diffusion coefficient is approximately inversely proportional to the cube root of the molecular weight. For a range of lignosulphonates from spruce, Goring (1968) finds

$$D = 4 \times 10^{-5}/M^{1/3}$$

cm^2 s^{-1}. The globular proteins also fall into this category. Spherical soil bacteria are about 0·5 μm in diameter (Russell 1973 p. 165). This is about 10^3 times the effective diameter for diffusion of simple cations, so that the diffusion coefficient of bacteria free in the soil solution should be of the order 10^{-8} cm^2 s^{-1}.

Table 4.2. Self-diffusion coefficients of ions in aqueous solution at 25°C.

Ion	Li	Na	K	Rb	Cs
D (cm^2 s^{-1} × 10^5)	1·0	1·3	2·0	2·1	2·1
Ion	Mg	Ca	Sr	Ba	
D (cm^2 s^{-1} × 10^5)	0·7	0·8	0·8	0·8	
Ion	F	Cl	Br	I	
D (cm^2 s^{-1} × 10^5)	1·5	2·1	2·2	2·2	
Ion	H	OH			
D (cm^2 s^{-1} × 10^5)	9·4	5·3			

From Robinson & Stokes 1959, Appendix 6.1 'Limiting equivalent conductivities of ions at 25°C in water', and the relation

$$D° = \frac{RT}{F^2}\frac{\lambda°}{z} = 2.66 \times 10^{-7}\frac{\lambda°}{z} \text{ (equation 11.49 p.317).}$$

Large long chain molecules commonly assume in solution a randomly-coiled shape that is roughly spherical. Solvent molecules are trapped between the components of the coil and move with it. The radius of the equivalent sphere is $\frac{2}{3} R_G$, where R_G is the radius of gyration of the random coil. Since $R_G \propto M^{\frac{1}{2}}$, for such molecules $D \propto 1/M^{\frac{1}{2}}$ approximately. This relation was found for methylmethacrylate polymers in acetone for molecular weights ranging from 60,000 to 5,000,000 (Meyerhoff & Schulz 1952).

It may be calculated that for thin rod-shaped molecules of constant diameter, $D \propto 1/M^{0.81}$. Collagen is an example. A selection of diffusion coefficients of substances with high molecular weight including enzymes is given in Table 4.3, extracted from Tanford (1961, Chapter 6) who deals with this topic in detail.

Table 4.3. Diffusion coefficients of macromolecules in water (after Tanford 1961).

Shape		MW	D (cm^2 s^{-1} × 10^7)
Random coil			
	Ribonuclease	13,683	11·9
	Haemoglobin	68,000	6·9
	Urease	480,000	3·5
Rod			
	Collagen	345,000	0·7

4.1.3 Solid surfaces

The mobility of exchangeable ions in pure clays depends mainly on the expansion of the clay mineral lattice, and the consequent thickness of the water layer between the alumino-silicate sheets. Some typical values are given in Table 4.4.

Table 4.4. Self-diffusion coefficients of interlayer ions in alumino-silicate clays.

Ref	Ion	Temp	System	Interlayer water molecules thick	D (cm^2 s^{-1})
de Haan *et al.* (1965)	K	lab	Illite	nil	1 × 10^{-23}
Graf *et al.* (1968)	Mg	25°C	Vermiculite	2	7 × 10^{-13}
van Olphen (1957)	Ca	25°C	Montmorillonite	3–4	1 × 10^{-6}
Mott (1967)	Na	25°C	Montmorillonite	3	4 × 10^{-6}

It seems likely that anions specifically adsorbed on clays or oxides have a negligible surface mobility since they are covalently bonded. Non-specifically adsorbed anions held at positively charged sites (see Chapter 3) may be mobile, but there is no experimental information available.

It seems likely that organic molecules have a very low surface mobility when adsorbed on clay or other organic particles, but again experimental information is lacking.

4.2 Diffusion in soils

Following this brief review of diffusion in gas, liquid and adsorbed phases, we may now turn to soils, where a substance may be present in one, two or all three phases.

4.2.1 Diffusion of non-volatile solutes

Here we are concerned only with diffusion in solid and solution phases; conditions that obtain for most plant nutrients. For porous systems with well-defined simple geometry—e.g. close-packed spheres—it is possible to express the total flux of solutes in terms of their concentration gradients and diffusion coefficients in the solution and solid phases, together with shape factors. The flux is not simply the sum of the solution and solid pathways because the diffusate, as we have seen, passes rapidly between the surface and solution. The resulting equations are complex and may be consulted in the reviews of Meredith & Tobias (1962), Goring & Churchill (1961) and Barrer (1969); Nye (1968) has discussed their application to soils.

In a medium so heterogeneous as soil a more fruitful approach is to treat the soil as a quasi-homogeneous body to which Fick's first law, $F = -DdC/dx$, may be applied. This is legitimate as long as we are concerned with diffusion between volumes large enough to average over microscale variations of particle and pore size; i.e. to include a representative sample of gas and liquid filled pores and the adjacent adsorbed phases. C is then the concentration of diffusate in the whole soil system, i.e. those ions, atoms or molecules that are in, or pass through, a mobile phase during a time that is short in comparison with the time of the overall process to which the diffusion coefficient is to be applied. Thus we have seen in Chapter 3 that equilibration between exchangeable cations and the adjacent pore solution may be limited by release from the exchange site and diffusion through hydrated interlayers of a clay mineral. In *stirred* suspensions cation exchange is usually rapid, so that release from the solid is not normally rate limiting. In *undisturbed* soil, equilibration by diffusion across the adjacent pore liquid may be the slowest process. A large pore 1 mm wide may require 1000 s to achieve 95 % equilibration. Hence it seems unwise to apply the diffusion coefficient to processes taking much less than an hour. Solutes that do not exchange within the chosen period are not defined as diffusible, but as having a finite rate of reaction. They would then appear in the source term $f(C)_{x,t}$ of equation 1.7 in the solution of practical problems.

In practice the solute may move in the direction of diffusion both through the pore solution, and in the adsorbed form by surface diffusion. Although mobility in the adsorbed form may be much lower than in solution, a much greater proportion of solute may be in this form, so that it cannot be neglected; and to take account of it we may write

$$F = -D_1\,\theta f_1\,dC_1/dx + F_E \qquad (4.15)$$

where D_1 is the diffusion coefficient of the solute in free soluttion;
     θ is the fraction of the soil volume occupied by solution; and gives the cross-section for diffusion through solution;

f_1 is an impedance factor (see below);

C_1 is the concentration of solute in the soil solution;

F_E is the excess flux created by the possibility of surface diffusion.

Addition of the term F_E does not imply that surface diffusion and solution diffusion are independent and additive. In reality, solute molecules or ions will constantly move from one phase to the other, and F_E is simply the extra flux that arises if the solute can move in the direction of diffusion while it is adsorbed on the solid phase.

Since, by definition, $F = -D \, dC/dx$, substituting for F in equation 4.15 gives

$$D = D_1 \, \theta f_1 \, dC_1/dC + D_E \qquad (4.16)$$

where D_E allows for any effect that mobility in the solid phase has on D. If the mobility of the adsorbed solute in the direction of diffusion can be neglected, the flux will be due to diffusion through the solution alone and we may write

$$D = D_1 \, \theta f_1 \, dC_1/dC \qquad (4.17)$$

The terms in equation 4.16 may now be discussed.

θ—The concentration of the soil solution is that of an equilibrium dialysate; and θ refers to the water associated with this solute; thus it excludes water associated with adsorbed exchangeable cations, or in the anion exclusion layer, (negative adsorption—p. 44). The product $\theta \, C_1$ is thus the amount of solute in solution per unit volume of soil.

f_1—The impedance factor takes account primarily of the tortuous pathway followed by the solute through the pores. This has the effect both of increasing the path length to be traversed, and of reducing the concentration gradient along this path length. It may also include the effect of the increase in the viscosity of water near charged surfaces, which will affect the mobility of all solutes, though it is unlikely to be significant except in dry soils. Kemper *et al.* (1964) found the viscosity of the first three molecular layers of water on charged surfaces to be 10, 1·5 and 1·1 times that of free water. Another effect that may be included is the negative adsorption of anions. These may be excluded from very narrow pores and thin water films, which may thus cut-off connections between larger pores. In a soil all these effects are difficult to separate experimentally. The special case of macromolecules is discussed at the end of this section.

The relation between the impedance factor and soil moisture for chloride ion in a sandy loam is shown in Fig. 4.1 (Rowell *et al.* 1967). It will be seen that in very dry soil f_1 is very low: it was 2×10^{-4} at a water potential of -100 bar, and 10^{-2} at -15 bar. At these low potentials much of the water in the pores will be linked by very thin films. At potentials between -1 bar and $-0·1$ bar, f_1 increases approximately linearly with the moisture content. Thus over the field moisture range, $-0·1$ to -10 bar, the product θf_1 may decrease by two orders of magnitude.

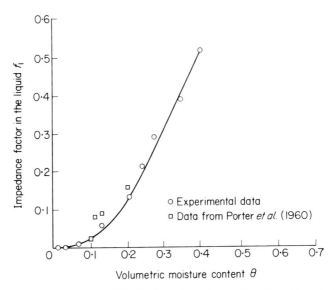

FIG. 4.1. Relation between diffusion impedance factor, f_1, and moisture content, θ, for chloride ion in a sandy loam soil (after Rowell *et al.* 1967).

At a given water potential clay soils usually have a higher value of f_1 than sandy soils, because they hold more water. But at a given moisture content sandy soils have a higher value of f_1 than clay soils, probably because a greater proportion of the water in the clay soils is held near surfaces. Porter *et al.* (1960) found that f_1 approached zero at a moisture content that was close to that required to cover the surface area in the soil with a monolayer of water.

Aggregation may affect the relation between f_1 and θ. Graham-Bryce (1965) found the self-diffusion coefficient of rubidium increased rapidly with increasing moisture at intermediate moisture levels, and slowly at high moisture levels (see Fig. 4.2). At intermediate moisture levels the water is held within the aggregates, but at high moisture levels the extra water is held between the aggregates.

In saturated soils f_1 values between 0·4 and 0·7 have been obtained. For a medium containing a mixture of different sized spherical particles Bruggeman (1935) derived theoretically the relation, $f_1 = \theta^{0.5}$, which accords with experimental values obtained by Dakshinamurti (1959) and Mott & Nye (1968).

It is not clear what effect compacting the soil has on the impedance factor when θ is kept constant. Graham-Bryce (1965) found the self-diffusion coefficient of rubidium was increased about three-fold when a sandy loam was compacted from 1·36 to 1·95 g dry soil per ml (Fig. 4.2). On the other hand Warncke & Barber (1972) determined f_1 from the self-diffusion of chloride ion in five silt loams, and found it decreased two- to three-fold when the density increased from 1·3 to 1·6 g dry soil per ml. Phillips & Brown (1965) found the self-diffusion coefficient of strontium increased by about 50%, and of rubidium decreased by about 20%, when Dundee silt loam was compacted from a density of 1·3 to 1·8. It is difficult to assess this experiment since the soils were prepared 'salt-free'. In practice an unknown amount of bicarbonate is likely to be formed in the soil solution by hydrolysis of clay and microbial activity.

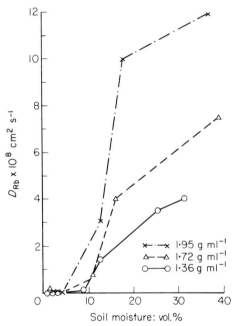

Fig. 4.2. Effect of moisture content and compaction on the self-diffusion co-efficient of rubidium in a sandy loam (after Graham-Bryce 1965).

The diffusion of a molecule through a channel of pores will be significantly reduced if its diameter is within an order of magnitude of the pore diameter—for two reasons. (a) The effective cross-section of the molecular pathway is reduced from πr_p^2 to $\pi(r_p - r_m)^2$, where r_p and r_m are the radii of the pore and the molecule. (b) Stokes Law (Section 4.1.2) applies only to particles moving in a medium that can be considered infinite. The viscous resistance to motion increases near the pore wall, and reduces the mobility by a 'drag factor' calculated by Faxen (1922) to be $(1 - 2{\cdot}09\, r_m/r_p + 2{\cdot}14$ $(r_m/r_p)^3 - 0{\cdot}95\,(r_m/r_p)^5)$. The combined effect of these two factors (Fig. 4.3(a)) has been shown by Renkin (1954) to account for the reduced diffusion of molecules of various sizes through cellulose membranes.

Barraclough (1976) has compared the *self*-diffusion coefficients of chloride ion, poly ethylene glycol of molecular weight 4000, and poly vinyl pyrollidone

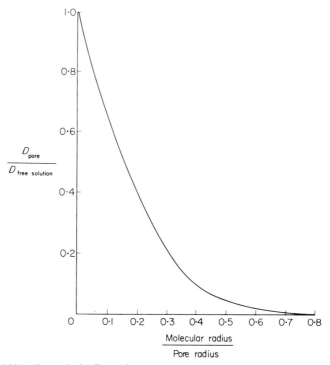

$$\frac{D_{pore}}{D_{free\ solution}}$$

Molecular radius
Pore radius

FIG. 4.3(a) Theoretical effect of pore radius on the effective cross-section for diffusion of macromolecules (after Renkin 1954).

of molecular weight 40,000 in an aggregated sandy loam at a range of moisture contents. Chloride ion has a radius of 0·25 nm; the poly ethylene glycol forms a random coil of radius 1·9 nm; the poly vinyl pyrollidone is rod-shaped with a long axis of 36·6 nm and diameter 1·6 nm; it is subject to Brownian rotation so that its effective radius in a narrow pore is 18·3 nm. The impedance factors at different water contents are shown in Fig. 4.3(b). In moist soil little poly ethylene glycol diffused in pores of radius less than 100 nm, or poly vinyl pyrollidone in pores less than 200 nm,—at least over 10 d. In dry soil, below − 7 bar water potential, diffusion of poly vinyl pyrollidone was restricted by its inability to move through intra-aggregate pores of radius less than about 200 nm; whereas the poly ethylene glycol and chloride ion diffused slowly

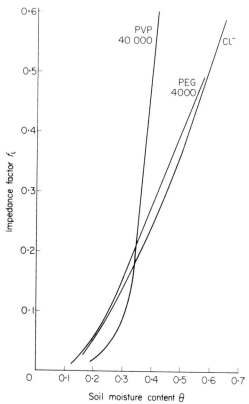

Fɪɢ. 4.3(b) The effect of soil moisture level on the impedance factor f_1 of PVP (MW 40,000), PEG (MW 4,000) and chloride ion (after Barraclough 1976).

until the soil dried to a water potential below -25 bar, when their movement was probably restricted by the thin films joining aggregates, which are only a few nm thick (Collis-George & Bozeman 1970; Kemper & Rollins 1966).

Further evidence for restricted movement of large molecules was given by Williams *et al.* (1966, 1967) who found poly vinyl alcohol was adsorbed by water saturated aggregates ($\frac{1}{2}$ − 1 mm diameter) of a clay subsoil more slowly as the molecular weight increased from 25,000 to 100,000. Since the disaggregated soil in suspension absorbed more of the polymer with higher molecular weight the reverse trend would have been expected if movement were unrestricted. The apparent diffusion coefficient of PVA (MW = 70,000) in penetrating initially polymer-free water-saturated aggregates was about 10^{-10} cm²s⁻¹. In free solution $D_1 \simeq 10^{-7}$ cm²s⁻¹. From Fig. 1 of Williams *et al.* (1967) $\theta \triangle C_1/\triangle C \simeq 0.1$; hence $f_1 \simeq 0.01$. Since nearly all the pores in the aggregates were less than 6 nm across it is understandable that movement of PVA should be slow. In such small pores adsorption of PVA on their walls constricts them further, and there is evidence that pores less than 3nm across are blocked.

These experiments suggest that the common herbicides and pesticides, whose molecular weights are usually less than 5000, should diffuse with impedance factors similar to simple ions. But humus compounds, synthetic polymers, plant and microbial exudates, and enzymes, of high molecular weight or linear shape—and living bacteria—will diffuse largely in pores more than about ten times their size; they will penetrate smaller pores very slowly, and may so reduce the pores' size by adsorbtion on their walls as to block them completely.

The buffer power dC/dC_1—The great range in values encountered has been discussed in Chapter 3. A correspondingly wide range of diffusion co-efficients has been measured. For a solute that is not usually adsorbed by the soil solids, such as the nitrate ion, $\theta dC_1/dC = 1$. Hence in equation 4.16, $D = D_1 f_1$, and in a moist soil the diffusion coefficient is about $2 \cdot 10^{-5} \times 0.5 \simeq 10^{-5}$ cm^2 s^{-1}. In contrast dC/dC_1 for the phosphate ion may be 1000, giving a diffusion coefficient, $D_1 \theta f_1 dC_1/dC$ of $10^{-5} \times 0.3 \times 0.5 \times 10^{-3} \simeq 10^{-9}$ cm^2 s^{-1} in a similarly moist soil. Thus in one day the root-mean square displacement of nitrate in a moist soil may be as much as 1 cm, while that of phosphate may be as low as 10^{-2} cm.

Since the relation between C and C_1 is usually non-linear the diffusion coefficient will vary with concentration. Figure 4.4 shows the slope of the adsorption isotherm of a clay soil at different levels of potassium and the corresponding diffusion coefficient measured.

The formula $D = D_1 \theta f_1 dC_1/dC$ has been found experimentally to express the diffusion coefficients of simple inorganic ions, such as calcium and magnesium (Tinker 1969a), potassium (Vaidyanathan *et al.* 1968; Nielsen, 1972), hydrogen (Farr *et al.* 1970) strontium (Mott & Nye 1968) and phosphate (Rowell *et al.* 1967; Olsen & Kemper 1968); and non-volatile organic solutes such as dimethoate (Graham-Bryce 1969).

D_E—The importance of this term has been examined in the work mentioned in the last paragraph. It has been found to be negligible for hydrogen, strontium, potassium, sodium and phosphate in moist sandy loam; and for magnesium and calcium in a moist clay. The relative importance of D_E would be expected to increase when the proportion of diffusible ions in the soil solution decreased. In accordance with this it has been found that the diffusion coefficient of sodium (Rowell *et al.* 1967) in a dry soil was significantly greater than would have been expected if the adsorbed ions were immobile.

The reason for the apparently low mobility of exchangeable ions in the few soils that have been critically studied is unknown, though it is

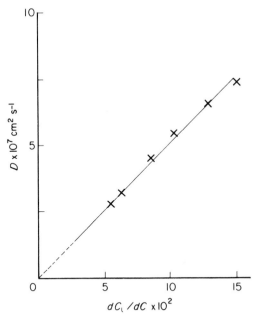

Fig. 4.4. The relation between the diffusion coefficient of potassium in Coral Rag clay soil and the reciprocal buffer power dC_1/dC (after Vaidyanathan *et al.* 1968).

suggested that hydrous oxides adsorbed on clay surfaces may block the diffusion pathways in soils, whereas such effects would be absent in purified clays.

For many practical purposes the diffusion coefficient estimated from equation 4.17, which assumes D_E can be neglected, is sufficiently accurate. Care must be taken that the sorption isotherm is so determined that it yields dC/dC_1 for the soil as it behaves in the diffusion process. The effects of the concentration of other ions, of disaggregation resulting from shaking, and of hysteresis phenomena all need careful attention.

4.2.2 Diffusion of volatile solutes

We have noted that the mobility of molecules in the gaseous phase is about 10,000 times greater than in aqueous solution. Hence, in a moist soil in which volumes of water and air are about equal, gaseous diffusion may be expected to dominate if the distribution coefficient, $C_1/C_g \ll 10^4$. (This distribution coefficient is Ostwald's 'coefficient of solubility', β). For example in water at $0°C$ β for $O_2 = 0.048$; $CO_2 = 1.71$; $SO_2 = 80$. When the distribution coefficient is about 10^4 the gaseous phase may replace the solution without greatly affecting the overall diffusion coefficient in the steady state. This is illustrated in Fig. 4.5 which shows the diffusion coefficient over a range of moisture levels of disulfoton—a volatile organophosphorus insecticide with a distribution coefficient of 5.5×10^3 (Graham-Bryce 1969).

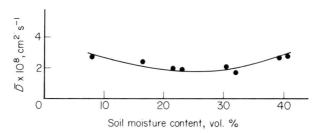

FIG. 4.5. Effect of soil moisture content on the diffusion coefficient of the volatile insecticide disulfoton (after Graham-Bryce 1969).

We may estimate the diffusive flux of a volatile solute, if the gaseous and liquid pathways are assumed to be independent, and any solute adsorbed by the solid to be immobile. We may then write

$$F = -(D_g v_g f_g \, dC_g/dx + D_1 \, \theta f_1 \, dC_1/dx). \tag{4.18}$$

Assuming Henry's law is obeyed, $C_1/C_g = \beta$. Hence the flux in terms of the concentration gradient in solution is:

$$F = -(D_g v_g f_g/\beta + D_1 \, \theta f_1) \, dC_1/dx \tag{4.19}$$

or in terms of the concentration gradient in the gaseous phase:

$$F = -(D_g v_g f_g + D_1 \, \theta f_1 \, \beta) \, dC_g/dx. \tag{4.20}$$

In reality the gaseous and solution pathways, while usually continuous, are not independent, and the true flux will be greater than these estimates because there will be an extra term for the interaction between them.

Gaseous diffusion—the impedance factor f_g

In dry materials the Bruggeman equation, $f_g = v_g{}^n$, satisfactorily accounts for the relation between f_g and porosity. The value of n which can be derived theoretically, depends on particle shape: it is 0.5 for spheres and larger for more complicated shapes. Figure 4.6 shows values calculated from Currie (1960) for diffusion of hydrogen and air through a range of materials, including dry soil crumbs for which $n = 1$.

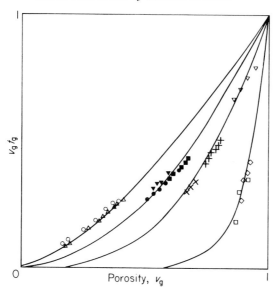

FIG. 4.6. Diffusion of gases through a range of porous materials—relation between $f_g v_g$ and porosity v_g. Curves are plotted according to the Bruggeman equation $f_g = v_g^n$ (after Currie 1960). $n = 0.5$: ○ glass beads, △ sand; $n = 1$: ●, ▼, ■ soil crumbs; $n = 2$: × pumice, ▽ diatomaceous earth, + kaolin; $n = 9$: ◇ vermiculite, □ mica.

When water is added to a dry material with a unimodal distribution of pore sizes, e.g. sand, it has a greater effect in lowering f_g than would consolidation of the dry material to the same v_g. Figure 4.7 taken from Currie (1970), shows this striking difference. The reason for it is that the water is held at the narrowest constrictions of the channels between the pores, just where it exerts the greatest blocking effect. We may expect that f_g will depend upon whether a given moisture level has been reached by wetting or drying the soil, since this will affect the distribution of water in the pores. Although such changes in f_g have not yet been demonstrated, Poulovassilis (1969) has measured differences in the hydraulic conductivity at a given moisture level depending upon the wetting and drying history of the soil.

For several types of material the behaviour on wetting is described by the equation derived from Currie (1961):

$$f_g = v_g^n [v_g/(v_g + \theta)]^{2.5}. \qquad (4.21)$$

In soils having well marked structural aggregates the pore size distribution is bimodal, corresponding to the micropores within the crumbs and the macropores between them. When water is added to these soils and the air-filled porosity decreases, the value of f_g may actually increase until the micropores are saturated (see Fig. 4.7). On further wetting f_g then follows the same

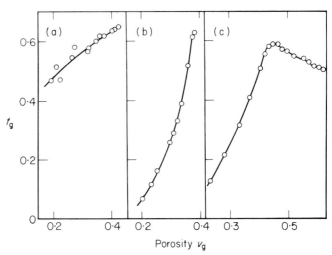

Fig. 4.7. Effect of (a) consolidation, (b) adding water on f_g for a sand. (c) Effect of adding water on f_g for an aggregated soil (after Currie 1961).

course as in sand. The reason is that the diffusion path is simpler when the air-filled pore space contains only macropores, than when it contains micropores as well. Considered in isolation dry crumbs have low values of f_g: Currie (1970) gives $f_g = 0.33$ for a crumb from permanent pasture and 0.17 for a crumb from over-cultivated arable soil, compared with 0.67 for dry unconsolidated packings.

Millington (1959), following a model of Childs & Collis-George (1950) originally used for hydraulic conductivity, has tried to calculate the impedance factor by considering the probability of two pores being opposite each other when adjacent 'layers' of soil are superposed. In effect, he calculates the continuous pore space across the two layers. Millington & Shearer (1971) have extended this approach to include an assembly of soil aggregates. They obtain very reasonable agreement with Currie's experimental data on packings of moist soil crumbs.

The predicted difference brought about by aggregation in the impedance factor for diffusion through the solution and gas phase is illustrated in Fig. 4.8, derived from the model of Millington & Shearer (1971). Aggregation reduces the value of f_l, but increases that of f_g.

4.3   Mass flow and dispersion in solution

In practice, solutes are moved both by diffusion and by convection (mass flow) of the soil solution caused by such processes as transpiration, evaporation or drainage; and the relative importance of diffusion and convection

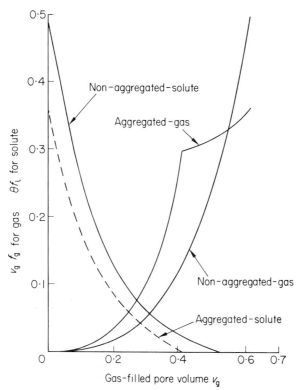

FIG. 4.8. The theoretical effect of aggregation on the relation $v_g - v_g f_g$ for gases and $v_g - \theta f_l$ for solutes (after Millington & Shearer 1971).

will be an important theme in subsequent chapters. Here we are concerned with the effect of water movement through the soil on the dispersion of the solute, since this augments the molecular diffusion discussed in Section 4.1.

When water flows steadily through a tube by laminar (in contrast to turbulent) flow its velocity profile across the tube is given by $v_r = K(a^2 - r^2)$ where K is a constant, a is the radius of the tube and r is the radial distance from the centre. The velocity is zero at the wall and greatest at the centre. If at time zero, t_0, a thin layer of dye were injected across the tube, it would after successive times t_1, t_2, t_3 have assumed the parabolic forms shown in Fig. 4.9. Thus even in a simple tube convection spreads a solute. In a porous medium the microscopic flow pattern of the water is far more complex. Philip (1969) has pointed out that in a spherical pore there are even zones where eddies create a flow in the opposite direction to the main stream.

A complete theory for dispersion is lacking, though the many approximate theories have been reviewed by Philip (1969) from a hydrodynamic viewpoint; and by Biggar & Nielsen (1963) and Frissel & Poelstra (1967a) from a soil

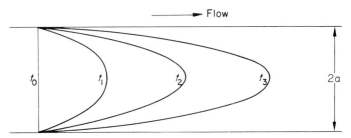

FIG. 4.9. Sketch of movement of a thin layer of dye by water flowing in a tube.

viewpoint based on theories of ion exchange chromatography. In these approaches the molecular diffusion coefficient, D_1, due to the random thermal motion of the solute molecules, is replaced in equation 4.15 by a longitudinal dispersion coefficient, D_1^*. The term 'longitudinal' is used because the dispersion in the direction of mass flow differs from that induced across it.

Experimental data on dispersion in porous media with uniform pore size has been summarized by Pfannkuch (1963). In Fig. 4.10 D_1^*/D_1 is plotted against the dimensionless quantity

$$\frac{v}{\theta}\left(\frac{d}{D_1 f_1}\right)$$

where d is the diameter of the particles in the media, and v/θ will be the average velocity through the pores in the direction of flow. It may be useful to note at this point that the water flux, v, induced by transpiration at a root surface rarely exceeds 5×10^{-6} cm s^{-1}. Drainage rates are often much greater, 10 cm d^{-1} being c. 10^{-4} cm s^{-1}.

Frissel and Poelstra (1967b) found the equation

$$D_1^* = D_1 + \frac{v\,d\lambda}{\theta f_1} \tag{4.22}$$

in which the dispersion coefficient increases linearly with the flow rate, described the dispersion of strontium in saturated columns of sand–resin and sand–clay mixtures, over a range of v from 2×10^{-6} to 20×10^{-6} cm s^{-1}. λ is a packing factor: it is one for spheres, but may increase to ten for irregularly-shaped particles.

In soil it is impossible to separate $d\lambda$, and the combined effective size and packing of the aggregates must be determined experimentally. In natural soils the dispersion coefficient may be expected to exceed that measured in columns of uniformly-packed solids or aggregates because of cracks due to shrinkage, and channels created by growth and decay of roots and the passage

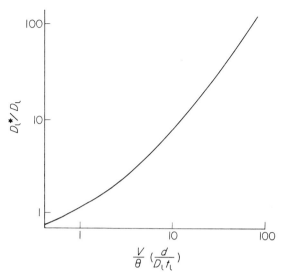

FIG. 4.10. Plot of D_1^*/D_1 against $\dfrac{v}{\theta}\left(\dfrac{d}{D_1 f_1}\right)$ (after Pfannkuch 1963).

of worms and other fauna. Frissel *et al.* (1970) found the product $d\lambda$ had values of 0·7, 0·8 and 6·0 cm in sandy, clay, and loess loam soils—over a range of v from $0·6 \times 10^{-6}$ to 230×10^{-6} cm s^{-1}. Their experiments were made by passing tritiated water through vertical *undisturbed* soil columns 100 cm long. The base of the columns was saturated and the matric suction increased to 0·1 bar at the top. Fried & Ungemach (1971) traced saline water seeping laterally from a well. They found $d\lambda$ to be as high as 1100 cm compared with an average laboratory value of 0·1 cm. Elrick & French (1966) measured $D_1^* = 39 \times 10^{-5}$ cm^2 s^{-1} for movement of chloride through a core of wet silt loam with $v = 10^{-4} - 3 \times 10^{-4}$ cm s^{-1}, compared with $D_1^* = 2·7 \times 10^{-5}$ cm^2 s^{-1} for the same soil after it had been dried, sieved (2 mm), packed into a column and brought to the same moisture level—$\theta = 0·52$. However, at a lower moisture content, $\theta = 0·42$, in which the larger pores were drained, the undisturbed and artificially aggregated soils behaved similarly.

Porous aggregates tend to increase dispersion, since there is not instantaneous equilibrium between the intra- and inter-aggregate pore solution. This last effect has been shown theoretically by Gluekauf (1955) and Passioura (1971) to increase the dispersion coefficient, D_1^* by an amount

$$\frac{1}{15}\frac{\varepsilon_A}{\varepsilon_T}\frac{(v^2)}{(\varepsilon_M)}\frac{a^2}{D_A f_{1M}},$$

where ε_A, ε_M, ε_T are the porosities of the aggregates, the interaggregate spaces, and total porosity; a is the aggregate radius; D_A is an intra-aggregate molecular diffusion coefficient $= (D_1 f_1)_A$. This expression has been tested by Passioura and Rose (1971). For values of va/ε_M of less than 2×10^{-5}cm^2 s^{-1} it is negligible, but it becomes comparable with the linear velocity term,

$$\frac{v\,d\lambda}{\theta f_1},$$

in equation 4.22 when va/ε_M is c. 10^{-4} cm^2 s^{-1}. This will only be attained with large aggregates and rapid drainage: e.g. for crumbs of 2 mm diameter and $\varepsilon_M = 0\cdot 2$ v would be 10^{-4} cm s^{-1} or c. 10 cm d^{-1}.

In natural soils, at flow rates that occur during infiltration and drainage, it seems clear that the dispersion coefficient will usually greatly exceed the molecular diffusion coefficient. It is not yet clear whether D_1^* is appreciably greater than D_1 for solute convected to root surfaces by transpiration. For example, if v at the root surface $= 10^{-6}$ cm s^{-1}, $\theta f_1 = 0\cdot 1$, and $d\lambda = 1\cdot 0$ cm, then in equation 4.22

$$\frac{v\,d\lambda}{\theta f_1} = 10^{-5} \text{ cm}^2 \text{ s}^{-1},$$

which is comparable with D_1. However, since the water flow is radial v decreases away from the root. Also the value of $d\lambda$ seems likely to be lower over short distances around roots, than in columns in which long channels and cracks are influential. Unfortunately, quantitative experiments with roots, described in Chapter 6, have been made with carefully-packed aggregates, so there is no direct evidence about the situation in the field.

4.4   Gaseous convection and diffusion

Transfer of gases within the soil is mainly by diffusion; and convection resulting from hydraulic pressure gradients is of minor importance. The hydraulic conductivity of the soil for gaseous movement is so large that the overall pressure within the soil adjusts rapidly to changes in the external barometric pressure. An increase in atmospheric pressure forces an amount of gas into the air filled pore space in the soil that can be predicted from Boyle's law. Buckingham (1904) estimated that air would enter and later be expelled from only 3 to 5 mm depth of a soil 3 m deep because of fluctuations in atmospheric pressure.

Pressure fluctuations caused by wind have a more important effect because they mix the air in surface layers, and hence aerate soils which have a surface crust that hinders the escape of carbon dioxide. Farrell *et al.* (1966), in a theoretical study, treated the turbulent air at the soil surface as a train of sinusoidal pressure waves. They predicted that a wind speed of 15 mph would cause mixing of air in a coarse soil or a surface mulch to a depth of several cm. The effective diffusion coefficient of air in coarse mulches could be as much as 100 times the molecular diffusion coefficient. Currie (1970) has verified that passage of turbulent air does decrease the accumulation of carbon dioxide beneath a surface crust.

Temperature gradients within the soil, and temperature differences between the soil surface and the atmosphere also cause mass movement of soil air, but in comparison with movement by gaseous diffusion their effects are small (Romell 1922). Infiltration, following rainfall or irrigation, also results in displacement of soil air. More detailed accounts of mass flow of gases in soil have been given by Keen (1931).

4.5 Mechanical movement

In most of the examples we treat, the soil matrix is regarded for simplicity as fixed; but in any applications of theory to field situations due weight must be given to mechanical disturbances arising from a wide range of sources that are difficult to quantify. These are fully discussed in standard textbooks such as Russell (1973) and Baver *et al.* (1972). Here, we note without elaborating that shrinkage and swelling caused by moisture changes create cracks and channels through which air and water pass readily. An extreme situation arises in the so-called self-mulching soils (Vertisols), in which the whole solum gradually turns over as the top soil at the edges of the cracks falls to their base.

During the growth of a crop solutes may be redistributed by cultivation treatments, e.g. earthing up of potatoes. Erosion by wind or water may redistribute the immediate surface layers. Fauna such as worms and termites carry soil from subsoil to surface, and create channels for the passage of roots, gases and water. In spite of these effects it is noteworthy that sharp horizontal boundaries are maintained between adjacent plots under different treatments in long-term trials such as Park-grass and Broadbalk, at Rothamsted Experimental Station, where no major surface transfer takes place.

5

THE UPTAKE PROPERTIES OF THE ROOT SYSTEM IN SOLUTION

This book deals essentially with soil processes, and how they relate to the growth of plants. Our treatment of plant and root properties is therefore highly selective. We are interested in the gross morphology and geometry of root systems, in methods of quantifying plant demand at the root surface, and how such demands vary with position and environment. Other topics, such as membrane structure or influx–efflux analysis, are not of direct interest at present, though we deal briefly with some, in an introductory review of ion uptake.

5.1 Root Morphology

Only a brief description of the structure of the individual root is given here, since this is available in a number of standard texts (e.g. Esau 1960; Cutter 1971; Fahn 1967; Troughton 1957). The general structure is shown in Fig. 5.1 (a), (b), (c) and (d) and our discussion is mainly concerned with points having a special bearing upon the process of ion uptake, or root behaviour

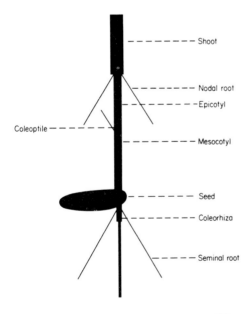

Shoot

Nodal root

Epicotyl

Coleoptile

Mesocotyl

Seed

Coleorhiza

Seminal root

Fig. 5.1(a). Schematic drawing of young grass seedling, showing root development (after Troughton 1957).

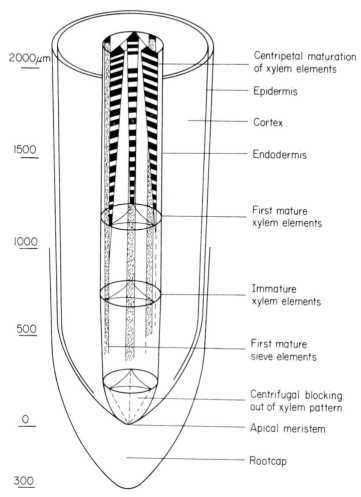

2000μm

1500

1000

500

0

300

Centripetal maturation
of xylem elements

Epidermis

Cortex

Endodermis

First mature
xylem elements

Immature
xylem elements

First mature
sieve elements

Centrifugal blocking
out of xylem pattern

Apical meristem

Rootcap

FIG. 5.1(b). Diagram of internal structure of root tip of pea (after Esau 1960).

in soil. Byrne (1974) has commented upon the rarity of anatomical studies of soil-grown roots, and the latter may differ from roots grown in solution culture in their development and gross morphology.

5.1.1 Root-tip structure

Root cells originate in an apical meristem a fraction of a mm behind the visible root tip. Cells which divide at the front of the meristem (Clowes 1969) form the root cap material, which is normally rubbed off in soil and

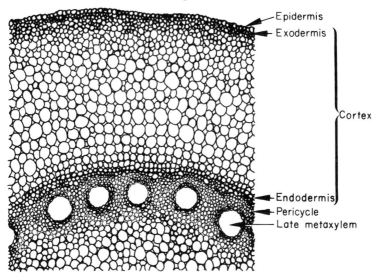

Fɪɢ. 5.1(c). Cross-section of root, showing cortex structure and endodermis (after Esau 1960).

forms part of the rhizosphere substrate. The root cap is also the site of the geotropic response of the root (Juniper *et al.* 1966; Audus 1969). The mechanism of this response to gravity is not yet fully understood. The behaviour of roots in this respect varies markedly with species, and there are cases in which the mechanism appears to reside in tissues behind the meristem, or depend upon a light stimulus in order to function (Wilkins 1975). Polygalacturonide mucilage, sometimes called 'mucigel', is formed in the root cap cells and is released around the cap base (Jenny & Grossenbacher 1963). The cells formed behind the quiescent centre rapidly differentiate into the various tissues of the root (Fig. 5.1(b)).

5.1.2 Epidermis and root hair structure
The outer surface of the cortex is formed by the epidermis, a single layer of thin-walled cells. The root hairs are extensions of the epidermal cells, which extend out at right angles for distances of up to 10 mm as an absolute maximum. Typically, root hairs are around 100–1000 μm long and 5–17 μm in diameter (Dittmer 1949). Their size and density on the root varies greatly with species (Cormack 1962); and the numbers reported vary from 2 per mm^2 of root surface on loblolly pine to 50–100 per mm of root length in various members of the Gramineae (Drew & Nye 1970; Newman 1974). In some species they appear only on specialized root surface cells, the trichoblasts. It is commonly stated that the root hair zone on roots is relatively short, implying rapid disappearance of root hairs. This is certainly not true, since quite old roots may carry hairs. Sparling (1976) found up to one third of

all roots carried root hairs in natural grassland in northern England, and the ratio varied little with seasons.

The frequency and size, and probably the duration of root hairs depend greatly upon environmental factors, including water, phosphate (Bhat & Nye 1974), calcium (Cormack *et al.* 1963), microorganisms (Bowen & Rovira 1961), pH (Ekdahl 1957) and soil impedance. Bergmann (1958) tested the effect of calcium carbonate and aeration and concluded that any factor which increased the bicarbonate concentration also increased root hair production. Some of these effects are shown in data of Sparling (1976) on the percentage of total root length carrying dense root hairs, in plants

Table 5.1. Root hair frequency (% root length) in grass and clover (from Sparling 1976).

	Cynosurus cristatus (soil pH 5·3)	*Trifolium repens* (soil limed to pH 7·0)
Natural soil	41·6	44·9
Natural soil plus P	62·5	75·1
γ-irradiated soil	84·2	57·1
γ-irradiated soil plus P	84·0	72·4

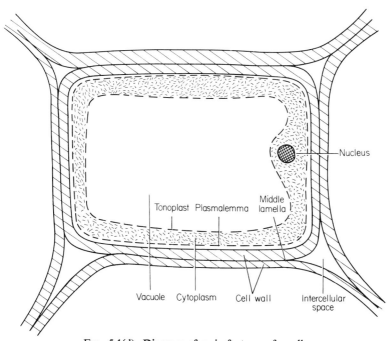

FIG. 5.1(d). **Diagram of main features of a cell.**

grown for 6 weeks in an acid brown earth with a very low phosphate level (Table 5.1).

It is normally taken for granted that the root hairs absorb water and ions, and the former point has been directly proved by very delicate potometer experiments (Rosene 1943). Their detailed uptake properties are not known quantitatively (see p. 110), though they are structurally part of an epidermal cell, filled with cytoplasm, and presumably have a plasmalemma with similar properties. Vigorous cytoplasmic streaming in root hairs has been noted (Nutman & Dart 1967) suggesting rapid translocation of absorbed water or solutes into the main body of the cell.

5.1.3   Cortex structure

The cortex is the part of the root which is most important for our topic. It is composed of elongated cells, which become highly vacuolated during extension growth. Intercellular spaces are conspicuous (Fig. 5.1(c) and (d)), and in some specialized roots these are further increased by the death and breakdown of some cells to give air-filled spaces called aerenchyma (Macpherson 1939; Armstrong 1972). The thickness of the cortex varies widely, being even as little as a single-cell layer.

5.1.4   Endodermis structure

The cortex is bounded on the inner surface by the endodermis, which usually comprises a single annular layer of cells. These are distinguished by a layer of suberin impregnating the radial walls of each cell (the Casparian strip), which is believed to prevent movement of water and solutes within the cell wall towards the stele (see Section 5.1.5 and Fig. 5.1(b)); in effect, these impregnated areas of wall contain no free space. The cell plasmalemma (Fig. 5.2.(a)) is joined to this strip, thus forming a cylindrical barrier to free

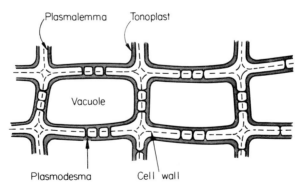

FIG. 5.2(a). Diagram showing main features of a vacuolated cell in the cortex, with symplasmic connections. (Not to scale).

water movement which encloses the stele. It is usually assumed that the endodermis is: (a) More or less semi-permeable, which prevents outward leakage
of solute; the restriction of solute within the stele can lead to water uptake by
osmosis and excess pressure within the root xylem (root pressure), and the
integrity of the endodermis is necessary for the latter to develop (O'Leary &
Kramer 1964). (b) The source of most of the impedance to water movement
into the root (see Slatyer 1967 p. 204; Newman 1974), since the permeability
of roots to water appears to be of the order of $1-10 \times 10^{-6}$ cm s^{-1} bar^{-1},
which is similar to that for single cells, and suggests that most of the resistance
lies within a single-cell layer. However, Newman (1976) has argued that the
permeability of the free space (cell wall) pathways must always be quite low,
and that most water moves from cell to cell in the cortex via the plasmades-
mata. If this is so, the resistance to flow may be distributed uniformly in the
cortex and endodermis. The amount of suberin deposited in the endodermal
cells increases with age of the root, and the fully developed cell layer is probably an excellent semi-permeable membrane. There are certainly points of
discontinuity; so-called passage cells exist in which little suberin is deposited (Clarkson 1974), and the endodermis may be ruptured during the
formation of lateral roots but in general transport through the endodermis is
via the symplasm (see Anderson 1975).

5.1.5   Stele structure

The root within the endodermis is named the stele. This contains relatively
undifferentiated tissue, the pericycle, and two systems of connecting vessels,
the xylem and the phloem. The phloem forms part of the internal translocation system of the plants, and carbohydrate with some mineral nutrients,
particularly potassium, is moved through it to the roots from the shoot.
The mechanism and quantity of this transport is still a subject of controversy
(Peel 1974; Kirkby 1974).

The xylem vessels can be regarded for most purposes as simple tubes, of
a few μm, sometimes rising to as much as 100 μm, in diameter. Water and
mineral nutrients pass up to the shoots, by a process of simple convective
flow. There may be either a positive pressure in the root xylem (root pressure)
or a suction, depending upon the relative balance of the transpirational
demand for water by the shoot and the osmotic process by which water is
drawn into the stele from the free space of the cortex.

One point, of major interest for the operation of the root, is at present the
subject of controversy. Xylem elements are formed from specialized elongated
cells which lie adjacent to each other in single files. Within a short distance
of the tip the walls between the cells break down, and a continuous vessel is
formed. It has often been assumed that the cytoplasm disappears when the
end-walls disintegrate and that the cell is then dead, but other work (House

& Findlay 1966; Dainty 1969a) indicates that the open xylem vessel still retains a lining of cytoplasm for a considerable period. If so, the interior of the xylem may be comparable to a vacuole, and bounded by a tonoplast (Fig. 5.2(a)).

5.1.6   Changes with age

With age, various changes take place in the root structure. The endodermis and sometimes the epidermis accumulates suberin, a hydrophobic material found as lamellae on the cell walls, and composed of long-chain fatty acids (Martin & Juniper 1970). The thickness of the Casparian strip of endodermal cells increases with age. In some grass species the whole root cortex may slough off (Troughton 1957). In dicotyledons the root diameter may increase with age, with the loss of the cortex and formation of a secondary cambium below the corky outer layers; in the trees the old xylem forms the woody centre to the root. It is commonly assumed that roots in this condition cannot absorb ions, though there is clear evidence (see p. 32) that water may be absorbed by old tree roots with cork deposits on the surface, possibly via lenticels or other breaks.

5.1.7   Root branching

Lateral initials arise in the pericycle (Fig. 5.1(c)) and these meristems steadily expand, gradually breaking out through the endodermis and cortex of the main axis. The endodermis of the main root and the lateral finally become continuous, and vascular connections are established in the xylem and phloem. The mechanism inducing meristem formation at a particular point on the root is unknown, but environmental factors affecting the frequency of branching are discussed in Chapter 7 (p. 196 *et seq*). As in the case of shoots, removing the apical meristem of the root causes increased lateral branching further back along the root. The growth and branching is under complex hormonal control, and a number of hormones are produced by the roots, of which the cytokinins are particularly interesting (Fox 1969), because they promote cell division. Increased concentrations of ethylene in the surrounding medium cause intense and repeated branching, and endogenous ethylene may be involved in control of branching.

5.1.8   Root classes

It is usual to describe an individual root in terms of its origin on the plant (e.g. adventitious or seminal), its position with respect to the root system as a whole (e.g. main axes, tap root, lateral) or the type of plant from which it springs (monocotyledon, dicotyledon or gymnosperm), but the main structure of roots varies rather little from that described here. The most important difference is the secondary thickening usually found in dicotyledons and gymnosperms which produces a main root framework quite different to that found in perennial monocotyledons. There are a variety of different arrangements of the vessels in the stele, which may be used to classify roots, but they are of no interest for our subject.

5.2 Ion uptake mechanism

5.2.1 General

This section discusses the sequence of mechanisms or processes which cause solutes to be transferred from the exterior of the root into the xylem. Purely passive uptake of organic solutes is mentioned on p. 128. Work on inorganic ion uptake has frequently been reviewed, recent authors being Epstein (1972), Gauch (1972), Moore (1972), Anderson (1972), Higinbotham (1973), Dainty (1969a), Hodges (1973), Clarkson (1974), Baker & Hall (1975), Torrey & Clarkson (1975), Luttge & Pitman (1976). There is as yet no agreement on the fundamental processes whereby ions are transferred into the xylem of plant roots, usually against a concentration gradient, and there is still uncertainty on the exact sites of the various processes and on the path followed by ions.

The point of most pressing importance is the site of the 'active' stage or stages in the uptake of ions, where they move against an electrochemical potential gradient, before further transport within the root and final release into the xylem sap. The reason is that, from the physicochemical point of view, the first active uptake stage can be considered to define the 'boundary' conditions for 'passive' movement of the nutrients through the soil and all or part of the cortex 'free space', i.e. that part which can be penetrated without passing a living cell membrane. This first active uptake step has been variously thought in the past to occur at the root epidermis, the endodermis, or the plasmalemma or tonoplast of cortical cells (see Gauch 1972). It is clear that most ions are or may be actively transported across cell plasmalemmas, and that this may also occur at the tonoplast; and most writers now regard the plasmalemma of the cortical cells as being the site of this first uptake step, though this does not exclude other active transport steps elsewhere. The precise point in the root cortex at which this occurs may however vary. In symplasm transport ions are actively absorbed into the cells at or near the root surface, and then transported across the cortex, endodermis and pericycle within cells, being passed on from one to the next via the cytoplasmic connections between cells, the plasmodesmata (see Clarkson 1974 p. 216). In the alternative mechanism of 'free space' or 'apoplasm' transport ions move passively in cortical cell wall spaces by diffusion and/or mass flow, though they have to be actively absorbed before transport through the endodermis. In practice it seems certain that both mechanisms operate, and the uncertainty

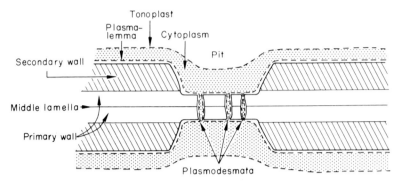

FIG. 5.2(b). Diagram showing main features of the plasmodesmatal connections between cells, through which symplasmic transport occurs.

concerns their relative contributions to transport in different situations and with different ions. We assume here that ions move from the surface of the roots to the endodermis by the parallel symplasm and free space pathways (Fig. 5.2(a), (b) and (c)), with progressive transfer of ions from the free space into the cell cytoplasm; though there is evidence that occasionally one mechanism may predominate—thus calcium is moved very largely by the free space pathway up to the endodermis (Clarkson 1974). Weatherley (1969) considered the accumulation of salts at a semipermeable endodermis if water were absorbed much more rapidly than solute, and gave a simple 'steady-state' mathematical treatment predicting a solute accumulation at the endodermis to be exponentially related to water uptake rate.

It is therefore convenient to consider uptake as a sequence of four processes: transport in the free spaces in the cortex of the root; uptake into the living parts of cells; transport in the living parts of cells, i.e. movement in the cytoplasm and release into the xylem; transport in the xylem (Fig. 5.2(c)).

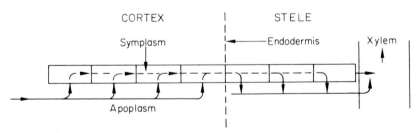

FIG. 5.2(c). Schematic diagram of the interaction of symplasm and free space (apoplasm) pathways in cortex and stele.

5.2.2 Transport in the free space

We are here concerned with the space in and between the cell walls, and the properties of the surface surrounding these spaces. There is ample evidence that ions move through this 'apparent free space' (Briggs *et al.* 1961). The cell wall is composed of cellulose microfibrils, of the order of 10–20 nm in diameter (Albersheim 1965; Preston 1974). These microfibrils only comprise a fraction of the total cell wall weight, the rest being mainly pectic substances (polygalacturonic acids with varying amounts of neutral calcium pectate). The secondary cell wall, deposited later than and within the primary wall, also contains pectin. The non-cellulosic fraction is not fully crystalline, and is best regarded as a gel (Roelofsen & Kreger 1951). The free carboxyl groups of the pectic material are the cation-binding sites of the cell wall.

It is obvious that solutes can diffuse freely into the intercellular spaces (Fig. 5.1(d)) which can be seen in section under the microscope, and which frequently appear to lie between three adjoining cells, if these are water filled; in fact, they are usually air-filled in soil-grown roots. It is not as clear just how water and solutes move in the cell wall itself. The wall contains 10–20% of polysaccharide to wet weight, and it may be that the amount of water in the gel is sufficient to allow solutes to diffuse through all of it inless the molecular weight is very large. In a range of poly ethylene glycol polymers, it seems that cell walls can be penetrated by Carbowax 400, but not by Carbowax 6000 (Michel 1971). Small ions can therefore probably penetrate all of it readily, and if so, all the carboxyl groups will potentially be ionizable, and will contribute to the root cation exchange capacity at appropriate pH values.

Much attention was at one time given to this root cation exchange capacity, which was regarded as an important factor in ion uptake (Drake 1964). There is, however, no evidence that ions must be adsorbed on ion exchange sites before they enter the active uptake process, and we regard the ion exchange properties of the cell wall as irrelevant to ion uptake, except in so far as they modify the transport parameters, e.g. the diffusion coefficient, of the cortex for ions.

Some confusion arose in early works between the 'apparent free space', 'water free space' and 'Donnan free space' (Briggs *et al.* 1961). A clear discussion is given by Dainty (1969a) of these terms. We take 'free space' to mean the fraction of the volume of a root accessible to non-charged small molecular weight solutes by normal diffusion. The 'free space' comprises up to 10–20% of most root volumes (Gauch 1972; Fried & Broeshart 1967), but recent measures have tended to lie in the lower end of this range. This is in the cortex (Crowdy 1972) and the surface film of water on the root. Inspection of root sections shows that the visible intercellular spaces are often less than this in volume, and a large fraction of the cell wall volume must be in rapid equilibrium with the external solution. Shone (1966) measured anion exclusion volumes in roots, and found that only 0·03ml g^{-1} fresh weight was unaffected by anion exclusion due to diffuse double-layer formation. This must be in intercellular spaces, or external water films. The rest of the free space was considered to be within the cell walls, and Shone deduced that the cation exchange surfaces there were on average separated by 20 nm. We therefore visualize the free space as a combination of macropores between cells, and micropores within cell walls.

Two interesting points concerning diffusion and mass flow transport in the root cortex are raised by this work, though it is a subject which has received rather little attention. One would expect a quite marked change of the cortex diffusion coefficients for anions as the electrolyte concentration becomes less, and the electrical double layers around the charged sites expand. Shone's data (see his Fig. 1) indicate appreciable exclusion in 0·01 M potassium iodide, and it is surprising that no clear effects on ion uptake have been found resulting from this. This could be due to the rather large total ionic concentrations often used in plant nutrition experiments; or calcium or other polyvalent cations present may form an ion pair with the ionized carboxyl groups, and thus prevent diffuse double-layer formation. This could possibly be one origin of the effect of calcium on potassium uptake (Viets 1944), and the finding by Franklin (1969) that phosphate was absorbed more rapidly in the presence of multivalent cations. The cortex spaces which are not affected by ion exclusion in the water-saturated root, will be air-filled in soil at field moisture tensions, and not available for transport. Consequently, any ion exclusion effects should be much more marked in soil than in partially or wholly water-saturated excised roots. Roots of freely-transpiring plants in solution culture are not usually water-saturated.

Pitman (1965) made direct measurements of sodium and potassium self-diffusion coefficients in the barley root cortex, in 0·01 M iodide solutions, and found both were about 3–4 $\times$ 10^{-7} cm^2 s^{-1}. It may be deduced from his data that dC/dC_1 was about 0·44, and the free space volume was 0·24. If the latter is equivalent to θ (see p. 78), then f_1 must be about 0·065, which seems a very low value. Collins & House (1969) reported diffusion coefficients of the order 10^{-7} cm^2 s^{-1}, but the fact that their material was dead, and problems in the interpretation of their results, makes it difficult to put much weight on them. Anderson (1975) states that diffusion can never be rate-limiting in the cortex, but this will depend upon ions and conditions, and further work seems called for. The values reported by Pitman could certainly imply important concentration gradients in the cortex (see pp. 119, 132).

5.2.3   Active uptake mechanisms

Flows of ions across membranes may be driven by, or be against an electro-chemical potential gradient. Many flows of the first (passive) type are known, and the measured electrical resistance of the plasmalemma ($c.$ 3000 ohm cm^2) implies that ions can cross it under an electrical gradient (Higinbotham 1973). Other flows are against the gradient (active), but the decision for any particular ion flow is technically difficult, since it demands measurement of ion fluxes, and the electrical potentials and ion concentrations on both sides of the membrane, to allow application of the Ussing–Teorell flux ratio calculation. Isotopic labelling experiments show that simultaneous influx and efflux of an ion can occur.

The potential difference between the external solution and either the interior of a cell, or the xylem sap, is composed of a diffusion potential, which is shown in presence of metabolic inhibitors, and a potential due to transfer of electric charge by electrogenic active transport processes. Some transport processes may be coupled, so that there is no net charge transfer, e.g. $K^+ - Na^+$ and $NO_3^- - HCO_3^-$ exchanges across the plasmalemma. Charge and pH equilibria are finally maintained by proton transport and by organic acid anion formation. The interaction of all these simultaneous ion transport processes is extremely complicated, and the situation is not fully understood even for a single cell, still less for whole roots and plants. For technical reasons, most information about anion transport refers to the halides, and not to the important nitrate and phosphate ions. It seems that the external solution is usually positive with respect to the root, and all anion entry must be active. Potassium may, particularly at high concentrations, be in near equilibrium across the plasmalemma, but an active uptake mechanism certainly exists. An active efflux seems required for sodium and possibly hydrogen.

The actual mechanism by which active transport occurs is becoming slightly clearer. 'Carriers' have been postulated for a long time (see Epstein 1972), and are now visualized as compounds that complex the ion and render it more 'soluble' in the membrane, such as the specific alkali-metal binding cyclic compounds (Clarkson 1974; Baker & Hall 1975 p. 71, Truter 1975). Specific ion-binding proteins have also been found. These could aid both active and passive transport. There is now evidence—supported by many analogies with animal cells—that ion-activated ATPases are involved as the energy source (Hodges 1973), but precisely how is still unknown. There are still other theories in the field, such as endocytosis—the pinching off of small pieces of membrane, with bound ions, into the interior of a cell, where it rearranges and releases the ions. How these processes are controlled by the condition of the whole plant is unknown. There is as yet too much uncertainty in this fascinating subject for it to give much aid to whole-plant nutrition studies. It has been reviewed recently by Hodges (1973), Baker & Hall (1975) and Luttge & Pitman (1976).

5.2.4 Symplasm transport

The idea of symplasm transport depends upon there being direct contact between the cytoplasm in adjacent cells. This occurs through fine pores (plasmodesmata), of the order of 5 nm in diameter, containing a plug of cytoplasm reaching from one protoplast to the next. These are grouped in 'pit fields' where the cell wall is very thin (Fig.5.2(b)). The ultrastructure of plasmodesmata is still a subject of controversy, and the transfer across the symplasm has not been studied in detail. However, Tyree (1970) has calculated, on the basis of a model of the plasmodesmata structure, that diffusion alone could easily transport the measured potassium fluxes, with a concentration difference of roughly 10^{-4} M between adjacent cells, if complete mixing by cyclosis (cytoplasmic streaming) is assumed to occur in the cytoplasm of each cell. This concentration difference is small compared with the concentration of potassium in cells (though all of this may not be free to diffuse), and suggests that transport is not rate-limiting in the symplasm. Cyclosis can occur at speeds of several centimetres per hour, hence mixing in the cells of dimensions of the order of 10 μm may be rapid. The simplest theory is that ions leak passively into the xylem, but Lauchli (1972) has shown, with the electron beam microprobe, that there is an increased concentration of potassium within the cells of the pericycle, which he regards as evidence against the hypothesis of diffusion down a concentration gradient within the symplasm and passive transfer into the xylem. If there is living cytoplasm lining the xylem, direct excretion into the xylem vessels is possible (Anderson 1972), since this would be equivalent to movement of ions into the vacuole of a normal cell (but see Anderson 1975).

5.2.5 Transport in the xylem

There is no reason to doubt that transport in the xylem is by simple mass flow of water up to the shoot. In the absence of transpiration from the leaves there is probably still a slow circulation of water in the plant, since it has been estimated that about 5% of water normally ascending in the xylem descends again in the phloem (Zimmermann 1969). The concentration of ions in the xylem stream in inversely related to transpiration rate (Russell & Barber 1960), as would be expected for separate water and ion-absorbing mechanisms.

5.3 Ion Uptake Kinetics

In this section we aim to consider the ion uptake relations of a single short piece of uniform root which is still part of an intact plant, but surprisingly few of the results in the literature refer directly to this situation. Most information is from excised root studies (see Epstein 1972; Wyn Jones 1975), but it may not be directly relevant because (a) roots are not attached to the final sink, the shoot; (b) the uptake period is often so short (e.g. 20 min) that a steady state could not be set up, even if the plant were whole; (c) the roots have often been pretreated in a dilute solution of calcium sulphate; (d) in some experiments there may be leakage from the cut ends of roots into the solution from which uptake is occurring; (e) the entire root system is usually used, so that the kinetic parameters determined are the averages of those at each position along the roots. Some work (e.g. Torii & Laties 1966) used short pieces of tip or specified older root, but there is then the increased possibility of leakage from the cut ends. Several workers have stressed the differences in short-term uptake and retention in excised roots at different points on the root (Table 5.2). It is obvious that low-salt roots used in such experiments will act as very strong sinks for ions for short periods (Pitman 1965), but the uptake rate, or sink strength, rapidly diminishes as the cell vacuoles accumulate the ion. The initial behaviour of such excised roots may however be fairly similar to that of low-salt entire plants in dilute nutrient solution, in which the shoot acts as a powerful sink, and prevents accumulation of the nutrient ion in the root.

Table 5.2. Sodium uptake by barley roots in 10 min. at different distances from the apex, at different sodium concentrations (from Eshel & Waisel 1973).

Distance from apex (cm)	p moles cm^{-1}							
	1	2	3	4	5	6	7	8
Low salt roots, 0·5 M NaCl	190	800	610	350	500	260	70	60
Low salt roots, 20 M NaCl	2400	3200	2500	2400	2500	2000	1800	1700
High salts roots, 0·5 M NaCl	57	25	22	21	28	29	33	35
High salt roots, 20 M NaCl	750	240	220	240	300	420	440	300

It is noted above that 'carrier' molecules—possibly ATPase enzyme molecules embedded in membranes—are believed to carry out the transfer of ions against electrochemical potential gradients. In consequence, it might be expected that uptake would show the characteristics of an enzyme-catalysed reaction, in that the relation between concentration on the input side of the membrane and the flux was governed by Michaelis–Menten-type kinetics. If so, the flux F of ions into a root cell should be given by

$$F = F_{\max} \frac{C_1}{K_m + C_1}$$

where $F_{\max}$ is the maximum flux into the cell, and K_m is the Michaelis constant. From this, it is likely that the uptake rate V into whole roots will be governed by a similar relationship:

$$V = V_{\max} \frac{C_1}{K_m + C_1}. \qquad (5.1)$$

The results of many experiments have been treated in this way, and Fried & Broeshart (1967), Clarkson (1974) and Wyn Jones (1975) list K_m values for uptake of several ion species into excised roots (Table 5.3). The treatment of uptake data, and derivation of $V_{\max}$ and K_m is usually by the Lineweaver–Burke plot of $1/V_{\max}$ against $1/C_1$, or the Hofstee plot of V against V/C_1 (Fig. 5.3) (see discussion in Wyn Jones 1975). There is however, much evidence that this view is oversimplified. It has been reported frequently that there are two separate uptake mechanisms (see Epstein 1972) which are effective at low and high concentrations respectively (Fig. 5.4), and two straight lines on the Hofstee plot, with a break of slope where they join, is often taken as evidence for two uptake mechamisms. There has been much controversy over these ideas. It has been suggested (Barber 1972) that mechanism 2 (at high concentration) is a passive diffusion of ions. Cartwright (1972) confirmed that both mechanisms appear in sterile culture, though Barber (1972) stressed possible errors associated with the presence of microorganisms at low concentrations. It has become clear that, whereas mechanism 1 (up to about 1 mM, or less for K^+ and $H_2PO_4^-$) is a single process with Michaelis–Menten kinetics, mechanism 2 is poorly defined, and may have no clear $V_{\max}$ (Epstein & Rains 1965). Indeed, a whole sequence of mechanisms would be needed to account for the irregular curves sometimes obtained (Nissen 1971, 1974) (Fig. 5.4).

It is fortunate that soil solution concentrations of some of the most important nutrient ions (potassium and phosphate) are such that normally mechanism 1 alone is operative, and the debate over mechanism 2 may often be academic for uptake of these ions by plants in the field unless they are near recently applied fertilizer or in saline soils. To a first approximation, we assume Michaelis–Menten kinetics apply within this range for a given piece of root. The uptake rate Q for a whole root system, or assembly of roots at various ages, would then be

$$Q = \Sigma \frac{V_{\max} C_1}{C_1 + K_m} \qquad (5.2)$$

Table 5.3. Ion uptake characteristics for excised roots and for long-term whole-plant experiments, derived from Hofstee-type plots. Note V_{max} for excised roots corresponds to S_{max} for whole plants, both in mol g^{-1} s^{-1} (fresh wt), and also that

$$\frac{S_{max}}{K_m} = \frac{2\alpha_{max}}{a}.$$

Where uptake parameters for mechanism 1 and 2 are given in the original reference, the values quoted are for mechanism 1. Where no root radius is given, it is assumed to be 0·015 cm.

Species	Ion	a cm	K_m $\times 10^{-6}$M	S_{max} mol g^{-1} s^{-1} $\times 10^{-10}$	α_{max} cm s^{-1} $\times 10^{-4}$	Reference
a) *Excised roots*						
Barley	$H_2PO_4^-$	—	8	1	0·94	Noggle & Fried, 1960
Barley	K^+	—	21	0·34	0·12	Epstein *et al.* 1963
Barley	Rb^+	—	16	0·25	0·12	Epstein *et al.* 1963
Rice	NH_4^+	—	20	2	0·75	Fried *et al.* 1965
Rice	NO_3^-	—	600	1·5	0·02	Fried *et al.* 1965
(b) *Whole plants*						
Onions (8–16 days)	$H_2PO_4^-$	0·03	6·7	2	4·5	Brewster *et al.* 1975b
Onions (16–32 days)	$H_2PO_4^-$	0·03	10	1·5	2·2	Brewster *et al.* 1975b
Rape (5–10 days)	$H_2PO_4^-$	0·0125	30	7·5	1·6	Brewster *et al.* 1976
Rape (10–15 days)	$H_2PO_4^-$	0·0135	44	4·8	0·75	Brewster *et al.* 1976
Clover	$H_2PO_4^-$	—	1·1	1·2	5·5	Loneragan & Asher 1967
Lupin	$H_2PO_4^-$	—	6	2·6	2·0	Loneragan & Asher 1967
Capeweed	$H_2PO_4^-$	—	1	1·2	6·5	Loneragan & Asher 1967
Cenchrus ciliaris	$H_2PO_4^-$	—	13	4	2·1	Christie & Moorby 1975
Dactylis glomerata	K^+	0·008	2	6	12	Wild *et al.* 1974
Anthoxanthum odoratum	K^+	0·008	0·7	4	24	Wild *et al.* 1974
Trifolium pratense	K^+	0·009	2·2	5·5	11	Wild *et al.* 1974
Medicago lupulina	K^+	0·018	1·5	5	40	Wild *et al.* 1974

Table 5.3. (*Continued*)

Species	Ion	a cm	K_m $\times 10^{-6}M$	S_{max} mol g^{-1} s^{-1} $\times 10^{-10}$	α_{max} cm s^{-1} $\times 10^{-4}$	References
Maize	NH$_4^+$	0·011	170	30	2	Warncke & Barber 1973
Maize	NO$_3^-$	0·011	110	25	2·2	Warncke & Barber 1973
Onion (6 days old)	NO$_3^-$	0·025	25	20	10	Bhat *et al.* (to be published)
Onion (21 days old)	NO$_3^-$	0·023	53	15	3·2	Bhat *et al.* (to be published)
Rape (6 days old)	NO$_3^-$	0·015	30	94	24	Bhat *et al.* (to be published)
Rape (21 days old)	NO$_3^-$	0·010	200	40	1·0	Bhat *et al.* (to be published)

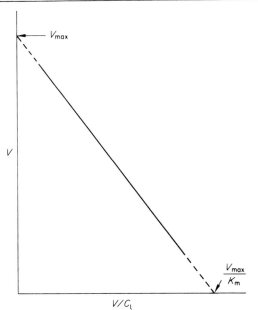

Fig. 5.3. Hofstee plot of the Michaelis–Menten equation.

when V_{max} and K_m vary at different positions on the root. This equation does not fit Michaelis–Menten kinetics if K_m varies widely, and it is perhaps surprising that excised roots should obey the equation as closely as they do at low concentrations. It suggests that K_m does not vary very greatly with distance from the root apex (see p. 105, 119).

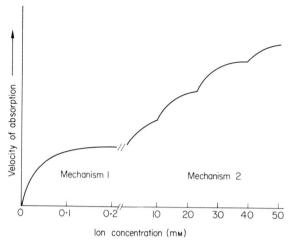

F IG. 5.4 Generalized uptake isotherm for ions into roots, showing mechanism 1 and mechanism 2 (after Hodges 1973).

It would be most relevant to our topic to discuss uptake rates of ions into short, defined lengths of root on intact, growing plants. There are several reports of uptake measurements with these conditions (Brouwer 1965; Clarkson *et al.* 1968) but in no case has a complete uptake isotherm been measured, though this would be of great interest. On the other hand, a number of investigators have measured the uptake rate of nutrient ions into entire plants in solution culture. The aims and interpretation of this type of work is rather different, and it is therefore discussed in the next section. We are there mainly concerned with the practical effect of solution concentration on the uptake rate for a whole root system, which allows us to specify the 'boundary condition' (see Chapter 6) for the transfer of ions in the soil up to the root surface.

5.4   Root boundary condition—the effect of concentration on uptake rate by whole root systems in solution culture

5.4.1   Uptake rates

There are three primary factors in considering nutrient supply to entire plants: the external concentration at the root surface C_{la}, some measure of the nutrient uptake rate (such as the unit absorption rate S, based on root fresh weight; the flux F, based on root surface area or the inflow I, based on root length), and the growth rate of the plant. The relationships with plant growth are more relevant when considering the plant as a whole, and are dealt with in Chapter 7. We consider here the relation of C_{la} and uptake rate, which define the boundary conditions for the root system as a whole, whilst bearing in mind that the condition of the whole plant has very strong effects on this relationship. It is essential to define the latter in mathematical terms, but the present state of plant physiological knowledge makes it impossible to do this accurately. On the preceding argument, it seems reasonable to assume that there is a relation between F (the flux into the root surface) and C_{la} for any short root segment approximating to

$$F = \left(\frac{F_{max}}{K_m + C_{la}} \right) C_{la} \tag{5.3}$$

at fairly low concentrations. Where $K_m \gg C_{la}$, this reduces to $F \simeq (F_{max}/K_m) C_{la}$.

The 'root absorbing power' α, defined by $F = \alpha C_{la}$ (Nye 1966b), is widely employed in this book. It is accepted as the best compromise between the need to represent root properties with accuracy, and the need for a boundary condition which allows relatively simple analytical solutions of diffusion equations. α will rarely be a constant over a range of concentrations as is evident from equation (5.3), and its numerical value must therefore be chosen carefully after considering the probable mean concentration at the root surface in any particular situation. A constant value of α can often be used, even in a situation where it in fact varies with concentration, to yield approximate information of value, particularly when it is as high as α_{max} in Table 5.3. Further problems in the use of α in relation to nutrient uptake by whole plants are discussed in Chapter 7.

Alternative methods of expressing this relationship, which will be used frequently in later chapters, are

$$I = 2\pi \alpha a C_{la} \text{ and } S = \frac{2\alpha C_{la}}{a}$$

when I, the inflow, is the uptake rate per unit length of root, and S, the unit absorption rate, is the uptake rate per unit fresh weight of root (Brewster & Tinker (1972). The last equation assumes a specific gravity of one in the roots, and S is really uptake per unit volume. Where these parameters are to be averaged over a whole root system, this is indicated by placing a bar over the symbol, e.g. $\bar{S}$. The term $\overline{\alpha a}$ is of particular importance, since it occurs frequently in the application of diffusion theory to root uptake (Chapter 6), and it is therefore given the name 'root demand coefficient' (Nye & Tinker 1969).

It is clear from the discussion above that the flux is that across the surface of the root cylinder itself. If the roots carry root hairs, and ions enter via the surface of the latter, this will increase the flux over the surface of the root proper. Hence root hairs will simply increase the apparent value of α in solution culture, though roots and root hairs in soil require a more elaborate treatment (Chapter 6 and 7).

Equation 5.3 implies that there is no lower concentration limit to uptake of nutrients, but this point is not clearly settled. It appears unlikely that there is a finite concentration below which the carrier mechanism ceases to function, but it is possible that a low level of ion efflux could balance uptake at some given concentration so that the net flux was zero (Clarkson 1974 p. 103). Bieleski (1973) noted that algae in non-renewed culture solution establish an equilibrium concentration at some level of total inorganic phosphate, but little equivalent work has been done with intact plants. Scheffer *et al.* (1961) tested the rate of loss of phosphorus from 21-day-old oat seedlings cultured at different levels of phosphate in solution. They found a loss of phosphate when plants were transferred to much lower concentrations, with an 'equilibrium concentration' where there was no net loss or gain, and which depended on the phosphorus percentage in the plant. These values, of 0.7×10^{-6} M or above, seem large, but the very high phosphorus concentrations (0.4–1.3% to DM) in the plants may explain this. Jungk & Barber (1975) measured the minimum phosphorus concentration to which maize plants could reduce a nutrient solution over periods of a few hours, and found that it varied with age of plant between 0.1 and 0.3×10^{-6} M. Their plants were cultured in 10^{-4} M P solution. These results contrast with the growth of plants at concentrations of the order of 10^{-7} M in Asher & Loneragan's (1967) experiments, and suggest that either the history of the roots, or the immediate phosphorus concentration in the plant, greatly modifies the lowest concentration allowing net uptake. Barley (1970) and Barber & Loughman (1967) state that plants can show a net absorption of phosphate from solutions of respectively 10^{-9} M and below 10^{-8} M. Further study of this point could be interesting. Rovira & Bowen (1970) showed that ^{32}P was lost from along the whole length of wheat roots after absorption of the isotope, but it is uncertain in which form the phosphorus was combined, how much was lost, and whether the root treatment enhanced the loss.

5.4.2 Comparisons between different experimental systems

The possible uncertainties in the use of excised roots to measure uptake rates, in comparison with intact plants, have been noted above. A further source of difficulty, in both excised root and intact plant experiments, lies in the unstirred layers of liquid which surround individual roots in solution culture. Their thickness depends on the rate of flow or stirring of the solution, and it appears likely that in experiments where dense masses of fine root are formed there will be little movement of liquid within the latter. The mean concentration at the liquid surface may therefore not be that measured in solution. The technique of growing plants in large volumes of rapidly circulated solution (Asher *et al.* 1965; Clement *et al.* 1974) is a considerable advance, though it is aimed at rapid replacement of absorbed solute in the bulk of very dilute solutions, rather than the concentration at the root surface as such.

In the limit, the unstirred layer thickness δ around a single root in rapidly moving liquid will be around 10–100 μm (Helfferich 1962 p. 253). This is thin enough for the root surface to be regarded as a plane without serious error. If the root surface flux of a major ion, such as NO_3^-, which is required by the plant is 5×10^{-12} mole cm^{-2} s^{-1}, $D_1 \sim 10^{-5}$ cm^{-2} s^{-1} and δ is 50 μm, then the concentration difference needed to drive the flux is $2\cdot5 \times 10^{-6}$ M. It therefore appears that the problem is not serious in well-stirred cultures at concentrations above, say, $2\cdot5 \times 10^{-5}$ M. This perhaps overstates the case, since such fluxes may not occur at concentrations of the order of 10^{-6} M. However, the *measured* value of $\bar{\alpha}$ will always relate to the bulk solution concentration C_1. In the limit, the concentration at the root surface itself is zero, and then

$$F = \frac{D_1 (C_1 - 0)}{\delta} = \alpha\, C_1. \quad \text{In this case,} \quad \alpha \to \frac{D_1}{\delta},$$

or about 2×10^{-3} cm s^{-1}. It is therefore unlikely that α values greater than this will ever be found, and any results in the region of 10^{-3} cm s^{-1} are probably in error due to boundary layer effects. If stirring is not adequate, the errors will of course be much greater, and will occur at higher values of C_1 and lower values of $\bar{\alpha}$.

In work on uptake of nutrients by whole plants, a further distinction must be made between short-term and long-term experiments. In the former case, the results can be regarded as analogous to excised root work. However, uptake from solutions with different C_{1a} values over a long period by a growing plant is obviously not comparable to short-term measurements made on either whole plants or excised roots. In the latter case, one normally starts with uniformly pretreated plants or roots, and the uptake curves (Fig. 5.4) represent the response of the uniformly conditioned uptake mechanism to a varying solute concentration. In practice it is more usual to grow plants continuously in different solutions or in soil with different levels of nutrient. The roots themselves then assume different properties due to their different past history, ion concentrations, growth rates and general metabolism (see p. 195 *et seq*). We may define $\bar{\alpha}$ for any whole plant as the ratio of mean flux to solution concentration; but in this situation it is not reasonable to consider the variation of $\bar{\alpha}$ with C_{1a} solely in terms of ion carrier saturation, as for single roots or similar whole root systems (p. 109), since other factors such as nutrient concentration in the shoot may be involved. We then regard $\overline{\alpha a}$ or $\bar{\alpha}$ simply as a physiological parameter.

The comparison of different types of experiment is therefore complicated. Loneragan & Asher (1967) noted that results for phosphate uptake by barley roots, measured by different techniques, agreed well at high and moderate solution concentrations, but that the uptake rate was much greater at low concentrations in flowing culture experiments (Fig. 5.5). They attributed this entirely to the impaired maintenance of solution concentration, since results for excised roots and for normal solution culture work with intact plants (by different authors) agreed well, but it seems possible that the physiological differences discussed above may also have been involved. We are not aware of any published experiments which could settle this point.

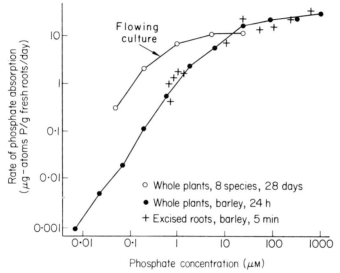

FIG. 5.5. Effect of phosphate concentration in solution on uptake rate by roots. (From Loneragan & Asher 1967, with additions).

In practice, we are interested in plants which grow in roughly uniform concentrations for considerable periods of time, and we take long-term flowing culture experiments as the best indication of how plants in soil will behave. Results of this type are discussed below.

5.4.3 Experimental results for uptake parameters

ā and K_m'

In practice, uptake rates and C_{la} are related by curves of the same general shapes as for excised roots. Figure 5.6(a) shows a good example of uptake data for potassium obtained in experiments with high volume flowing

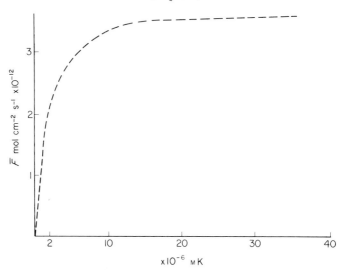

FIG. 5.6(a). Uptake isotherm for *K* into *Dactylis glomerata* in flowing culture solution. (after Wild *et al*. 1974).

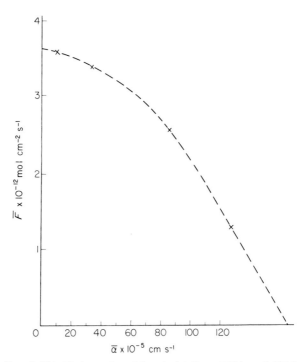

FIG. 5.6(b). Hofstee plot of data in (a) (from Wild *et al*. 1974).

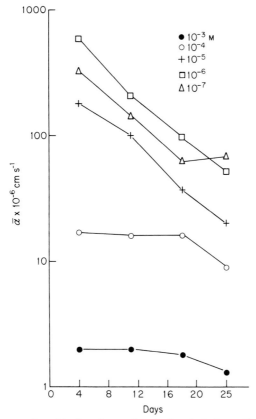

FIG. 5.7. Change of $\bar{\alpha}$ with time, in uptake of phosphate from *Cenchrus ciliaris* (Christie & Moorby 1975).

solution culture. These can be interpreted as Michaelis–Menten type uptake curves, and it is convenient to use the Hofstee-type plot again (Fig. 5.6(b)), though it must be stressed that the theoretical justification for using Michaelis–Menten kinetics is slight, as pointed out above.

This plot is particularly useful, because the horizontal axis variable is (uptake rate)/C_{la}, i.e. it is

$$\bar{\alpha}, \ \overline{a\alpha}, \ \text{or} \ \frac{2\bar{\alpha}}{a}$$

depending upon whether F, I or S is used as the measure of uptake rate. If the Michaelis–Menten type of relationship exists, then the line is straight, the intercept on the vertical axis gives the maximum uptake rate, and the intercept on the horizontal axis gives $\bar{\alpha}_{max}$, or the equivalent parameter. A value corresponding to K_m can also be derived (Fig. 5.3)—to indicate the rather different interpretation in this case, we write it as K_m'.

This $\bar{\alpha}_{max}$ is the theoretical $\bar{\alpha}$ at zero solution concentration, though no plant could in fact be grown at this level. The minimum concentration to allow considerable (e.g. 50% of maximum) growth rate is discussed by Brewster *et al.* (1975a). Asher & Loneragan found it to be below 10^{-6} M for phosphorus, and Wild *et al.* (1974) and Williams (1961) both found over half of maximum growth at 10^{-6} M potassium.

Good straight lines are rarely found in these Hofstee plots (Fig. 5.6(b)); often they are curved, or completely irregular, though the latter may be the result of technique which is not sufficiently good for the demands of this type of experiment. A sharp break in the Hofstee plot is, in excised root work, often taken as proof of a dual uptake mechanism; but any curvature or break in similar graphs for whole plant uptake often seems to be in the opposite direction to that expected, and not much weight should be given to such an interpretation in this case. This means, in effect, that $\bar{\alpha}$ is more constant than would be predicted from the Michaelis–Menten equation at very low concentrations, and less so at higher concentrations. As discussed above, the boundary layer effect will preclude measured $\bar{\alpha}$ values of much over 2×10^{-3} cm s^{-1}, and the tendency for $\bar{\alpha}$ to be constant at low concentrations may in part be an artefact due to this effect. Table 5.3 (p. 107) gives a selection of data from whole plant experiments in which the Hofstee plot was approximately linear at low concentrations, and from excised root experiments which have been added for comparison. The species differ, but in general the values of $\bar{\alpha}_{max}$ indicate a more rapid uptake rate in the flowing culture than excised root experiments at low concentrations.

The data of Bhat *et al.* (priv. comm.), Brewster *et al.* (1975), Warncke & Barber (1974), and Christie & Moorby (1975) showed clearly that the $\bar{\alpha}$ parameters tend to decrease with time particularly at low solution concentrations (Fig. 5.7). It is an interesting but unsolved question whether such changes are to be ascribed to changes in the age distribution of individual root segments, ontogenetic changes in the hormonal distribution and growth pattern of the whole plant, or changes in the internal concentration of the nutrient ion in the plant acting via a feedback mechanism. These changes occur in plants which are in uniform growth, and much sharper variations might occur where there is a major change in plant growth pattern, such as during the change from vegetative to reproductive growth.

5.4.4 Uptake rates

It is most usual to consider root uptake in relation to solution concentration at the surface as the boundary condition. However, in certain cases the uptake rate can be specified, and in that case the rate itself forms a boundary condition. Brewster & Tinker (1972) summarized $\bar{S}$ values for plants which were growing freely under reasonably normal conditions, and they therefore indicate the orders of magnitude of fluxes, inflows and unit absorption rates which must occur in practice in crop plants (Table 5.4). The magnitude of the fluxes of ions into roots in this type of experiment is usually of the same order as those found in excised root experiments (see Brewster & Tinker 1972; Tyree 1970; Noggle & Fried 1960), e.g. about 5×10^{-12} mol cm^{-2} s^{-1} for potassium and nitrate ions. Higinbotham (1973) gives the total sum of fluxes across the plant cell plasmalemma as lying in the range $0\cdot2 - 10 \times 10^{-12}$ equivalents cm^{-2} s^{-1} in experiments on tissues. The mean fluxes across the plasmalemma of cortical cells in a normal absorbing root are consequently rather low by this standard, since unit root surface area corresponds to a much larger area of plasmalemma.

5.5   Factors affecting the root absorbing power α

Apart from ion concentration, it is clear that a number of other factors will affect the local rate of ion uptake, and these are discussed below.

5.5.1   Root radius

It is necessary to relate the uptake rate of a nutrient ion into a root to its size, and we have utilized the external surface area as a measure of the latter quantity by the definition $F = \alpha C_1$, where F is the flux per unit surface area. This is convenient mathematically, but it may carry the implication that the active uptake process is localized in the external surface, and is limited by its extent. As stated above, the active process is probably located throughout the cortex, and it is therefore possible that the volume of the root, rather than the surface area, could be the best measure of its activity, i.e. if its uptake rate per unit volume is really given by some constant (k) times solution concentration, then

$$I = \pi a^2 k \, C_1 = 2\pi a \alpha \, C_1; \text{ hence } \alpha = \frac{a\,k}{2}.$$

There is thus a simple relation between the constants, once the radius is known.

The discussion of diffusion in the cortex (p. 101 *et eq.*) makes it likely that in many cases C_1 will diminish with distance from the root surface. In extreme cases this reduction may be so large that only the surface layer of cells is in contact with significant concentrations of the ion, and in such cases the root will behave as though the uptake process is located solely in the surface. Nye (1973) has dealt with this problem in detail. Large values of k and a, and small values of D tend to reduce penetration of ions into the inner parts of the cortex. When $a(k/D)^{\frac{1}{2}} > 3$, α becomes practically independent of root

Table 5.4. Uptake rates of ions into roots in solution culture, with value of α, for plants growing in concentrations which give maximum growth rate. For whole root systems, S, F, I and α have a bar over.

Species	Sol. conc. M	Element	S $\times 10^{-10}$ mole g⁻¹ s⁻¹	F $\times 10^{-12}$ mole cm⁻² s⁻¹	I $\times 10^{-13}$ mole cm⁻¹ s⁻¹	α $\times 10^{-5}$ cm s⁻¹	Root radius (cm)	Author
Barley	4×10^{-4}	K	2·1	1·5	1·5	0·37	0·015	Hackett (1969)
	1×10^{-4}	P	1·1	0·8	0·78	0·82	0·015	
	4×10^{-4}	K	5·6	7·0	10	1·8	0·025	Hackett (1969)
	1×10^{-4}	P	0·86	1·0	1·68	1	0·025	
	$2·1 \times 10^{-3}$	N	6·1	7·1	11·9	0·3	0·025	
Leek	4×10^{-4}	K	3·8	5·7	10·7	1·4	0·03	Brewster (1971)
	1×10^{-5}	P	0·37	0·56	1·05	5·6	0·03	
	22×10^{-3}	N	4·5	6·8	12·8	0·03	0·03	
Barley Seminal root segment	3×10^{-6}	P	0·12	0·12	0·15	0·4	0·02	Russell & Sanderson (1967)
Lateral root segment	3×10^{-6}	P	0·4	0·2	0·12	0·7	0·01	
Maize	1×10^{-3}	N (18 days)	33	20	13	2·0	0·011	Warncke & Barber (1974)
	$0·5 \times 10^{-3}$	P (18 days)	15	9·1	6	1·8	0·011	
	1×10^{-3}	K (18 days)	18	11	7	1·1	0·011	
	1×10^{-3}	N (74 days)	3·0	1·8	1·2	0·2	0·011	
	$0·5 \times 10^{-3}$	P (74 days)	0·9	0·53	0·35	0·1	0·011	
	$1·0 \times 10^{-3}$	K (74 days)	3·0	1·8	1·2	0·2	0·011	

Table **5.5.** Root radius above which α becomes effectively radius-independent (cm).

α (cm s^{-1})	$D \times$ (*cortex free space*) (cm^2 s^{-1})	
	10^{-7}	10^{-6}
10^{-4}	0·001	0·01
10^{-5}	0·01	0·1
10^{-6}	0·1	1·0

radius, and Nye suggested that the root radius at which the change from radius—dependence takes place should be taken as roughly $a(k/D)^{\frac{1}{2}} = 1·6$. With this criterion, the critical value of a for different values of α and D are in Table 5.5.

Further development of these ideas requires more precise data on D and k in the cortex, and probably demands consideration of whether a flux of water moving through the cortex can aid transport significantly.

5.5.2 Position on root

One of the most difficult problems in defining root uptake properties is the variability between different parts of the root, but this may be less than is sometimes thought. We do not distinguish here between uptake and retention, which is largest near the apical meristem, or uptake and transport, which is greatest further back where the vessels are fully differentiated; but

Table **5.6.** Uptake rates of P and Sr by portions of barley roots at different distances from the apex (from Clarkson, Sanderson & Russell 1968).

Content of phosphorus, mole per plant per day $\times$ 10^{-12}				
Concentration of solution, M	3×10^{-6}		10^{-4}	
Distance of treated part from apex, cm	1	44	1	44
In treated part	395	335	2070	770
In rest of plant	53	241	330	560
Total	448	576	2400	1330
Content of strontium, mole per plant per day $\times$ 10^{-12}				
Concentration of solution, M	$1·2 \times 10^{-6}$		10^{-4}	
Distance of treated part from apex, cm	1	44	1	44
In treated part	89	58	2910	2240
In rest of plant	51	2	1100	200
Total	140	60	4010	2440

regard all absorption by the root as uptake into the whole plant. The work of Clarkson *et al.* (1968) and Harrison-Murray & Clarkson (1973) shows clearly that the rate of absorption of phosphate is biggest in the youngest tissue, but that uptake continues for a considerable distance from the tip, up to at least 44 cm, at a comparable rate (Table 5.6). The variation was larger for Sr, but still not extreme. However, Troughton (1960) found that basal sections of nodal roots of *Lolium perenne* translocated about five times more phosphate to the shoot than did mid or apical sections.

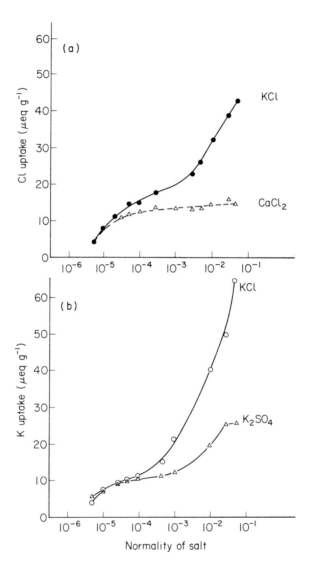

Fig. 5.8. Effect of counter ion on uptake of (a) chloride, and (b) potassium, by excised barley roots (after Hiatt 1968).

The great variability of uptake found by Rovira & Bowen (1970) along the root may be a consequence of their technique rather than a true variation with position on root. They measured ions contained in the root, which will include bacteria on the root surface, and also the large retention in lateral root initials due to incorporation in the meristem tissue, but the measurements give no indication of total uptake and translocation.

5.5.3 Temperature

The effect of temperature on uptake is, as expected, an increase of rate up to about 30°C, with Q_{10} of about 2, followed by a reduction at still higher temperatures (Epstein 1972). Low temperatures may greatly alter root membrane permeabilities (Zsoldos 1972). This point is discussed in more detail in Chapter 7 in relation to whole plant uptake.

5.5.4 Competing ions

The influence of one ion species on the uptake of another varies widely. It may range from the increase caused by calcium on potassium uptake (Viets (1944) effect), to the marked antagonistic effect of potassium ions on magnesium uptake. Pairs of very similar ions ($SeO_4 - SO_4$; $K - Rb$) often compete strongly. Other pairs show no competition at all; for example uptake of chloride by *Agropyron* was not affected by 500 times the concentration of sulphate (Epstein 1972). The counter-ion in a salt may be extremely important; e.g. potassium will be more rapidly absorbed from a salt solution with a rapidly absorbed anion, such as nitrate, than with a slowly absorbed one such as sulphate (Kirkby & Mengel 1967). This is usually attributed to the tendency to maintain electrical neutrality in the cell. The situation may be rather complicated, since Hiatt (1968) has found little effect of counter ion in the system I concentration range ($< 10^{-4}$ M), but strong dependence above that, with excised roots (Fig. 5.8).

5.5.5 Oxygen availability

The uptake process has been shown to be metabolic and oxygen-consuming by the application of respiratory inhibitors. Lack of oxygen stops it very rapidly, and continued anaerobic conditions lead to loss of ions from the root (Hopkins 1956) (Fig. 5.9). This has considerable implications for behaviour of roots in waterlogged soils, where the availability of O_2 may depend upon transport down the air-filled spaces in the root itself (Luxmoore *et al.* 1970).

5.5.6 Effects of other plants

Studies of allelopathy (the detrimental effect of one plant on another by the production of harmful chemicals) has a long history, but few conclusive

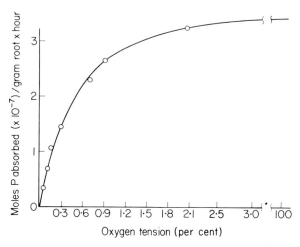

Fig. 5.9. Effect of oxygen tension upon the uptake rate of ions (Hopkins 1956).

results. Rice (1974) has reviewed this work extensively, and leaves little doubt of the importance of the effect. Roots seem to contain fewer inhibitors of this type than leaves, though leachates and litter from the latter may reach the ground and exert their effects on or via roots. The wider effects are discussed in Chapter 7—here the possible direct effects on a single root are considered. A very wide variety of chemicals have been identified as allelopathic inhibitors, and the detailed mechanisms may be correspondingly complex and varied. Several of these block mitosis in root cells, and volatile terpenes (Muller 1965) also alter root cell elongation and dimensions. A variety of mechanisms at cell or sub-cellular level have been suggested, but our main interest here must be with direct effects on mineral nutrient uptake. Chambers & Holm (1965) found considerable differences in competitive effects of bean or weed on beans in uptake of placed ^{32}P, which could not be explained by simple competitive removal of phosphorus. Bucholz (1971) was able to show that a competitive effect of *Agropyron* on maize, which showed as a mineral deficiency, was completely overcome by supplying extra minerals to a small fraction of the root, and he concluded that the absorbing power of the individual roots could have been altered.

It appears clear from this that chemical root interactions cannot be ignored, and their absence or presence in solution culture experiments may need special attention. The high uptake rates found in dilute flowing culture solution have been noted above and it seems possible that the rapid removal of such inhibitors may be involved. If different individuals of one species can inhibit each other, one must ask whether a root can inhibit itself.

5.5.7   Transpiration and water stress

Several mechanisms are confounded in the effects of water stress on the nutrient uptake of a plant. The growth rate and sink strength of the plant may be decreased (Chapter 7), or roots grown in water-stressed conditions may have different properties to roots grown with ample water. However, the points of immediate interest here are firstly, that increased water demand will cause a greater flow of water through the roots; secondly, depending upon the availability of water, this may or may not induce a lower water potential in the roots.

The effects of transpiration rate on uptake by plants were reviewed by Russell & Barber (1960), Viets (1972) and Hsiao (1973). Essentially, they concluded that transpiration had little effect on uptake by roots in low external solution concentration, but was increased more by transpiration when the concentration was high (Table 5.7). This may be explained by assuming, not that there is simple mass flow of solute across the root, but that the cell membranes become more permeable in presence of high salt concentrations, and/or that an increased concentration of ions in the stele or xylem tends to decrease the rate of further uptake. Rapid transpiration

Table 5.7. Effect of transpiration rate on uptake of phosphorus by barley, in relation to external concentration (from Russell & Shorrocks 1959).

Ext. P conc.	0·1 ppm		31 ppm	
Transpiration g	9·3	0·7	8·4	0·8
Total P uptake, μg/4 plants				
Root	9·6	7·5	123	62·1
Shoot	11·5	11·7	291	79·8

will tend to move absorbed ions in the xylem from the root to the plant top, and hence keep the concentration in the root stele low (Fig. 5.10). By treating the root as an active leaky membrane to which the methods of irreversible thermodynamics may be applied, Dalton *et al.* (1975) have derived the following relation between the flux across the root surface of solute (F) and of water (v_a) which accords with the available experimental evidence, e.g. Fig. 5.11:

$$F = (1 - S) \, C_{la} v_a \qquad (5.4)$$

where S, the selectivity coefficient, is defined as

$$S = \frac{\sigma - F'/C_{la} v_a}{1 + i \, RT\omega/v_a}.$$

Here, σ is the reflection coefficient and ω the osmotic permeability of the root; F' is the active uptake component of the solute flux; and i is the Vant Hoff factor for dissociation of the solute.

Since α and v_a appear together in many important equations in subsequent chapters it is useful to derive a functional relation between them. It follows from equation 5.4 that

$$\alpha = F/C_{1a} = (1 - S)\, v_a$$

and, substituting for S

$$\alpha = \frac{(F'/C_{1a} + iRT\omega)\, v_a + (1 - \sigma)\, v_a^2}{v_a + iRT\omega}. \tag{5.5}$$

If $v_a \gg iRT\omega$

$$\alpha \simeq F'/C_{1a} + iRT\omega + (1 - \sigma)\, v_a. \tag{5.6}$$

It is evident that the effect of v_a on α is relatively greater when C_{1a} is large than when it is small.

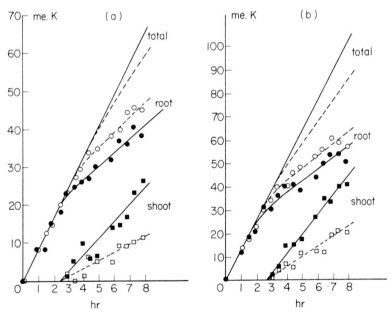

Fig. 5.10. Effect of transpiration on nutrient uptake (Hooymans 1969). (a) K absorption and transport from a 0·2 me l^{-1} K solution under normal ●■ and depressed ○□ transpiration rate (expressed per kg fresh weight of roots) (b) K absorption and transport from a 10 me l^{-1} K solution under normal ●■ and depressed ○□ transpiration rate (expressed per kg fresh weight of roots).

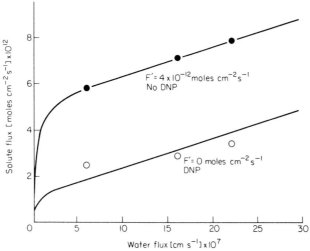

Fɪɢ. 5.11. Comparison between observed and calculated chloride uptake rate by barley as a function of water fiux.
$\sigma = 0.975$, $\omega = 0.5 \times 10^{-12}$ mol cm^{-2} s^{-1} bar^{-1},
$i = 2$, $T = 25°C$, $C_{1a} = 5 \times 10^{-5}$ mol cm^{-3} (after Dalton *et al.* 1975).

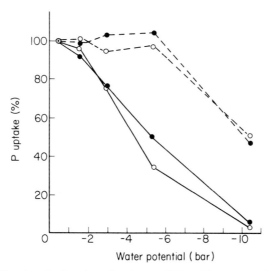

Fɪɢ. 5.12. Uptake of phosphate by intact 20-day-old tomato plants during 1 h, in relation to osmotic potential in solution medium. Results expressed as percent of uptake at -0.4 bar. ● dark; ○ light; — shoots; - - - - roots (after Greenway *et al.* 1969).

When plants are stressed, in the sense that their internal water potential is low, uptake of nutrients is usually decreased. Most experiments with roots in soil media are difficult to interpret because of the changing soil impedance to transport, but the work of Dunham & Nye (1974, 1976) (see Chapter 6) showed that α diminished as water potential decreased for chloride and phosphate uptake but not for potassium uptake. The work of Greenway *et al.* (1969) with osmotica in solution shows this effect clearly in a situation where effects on uptake must be a direct result of a change in root properties (Fig. 5.12).

5.5.8   Nutrient supply to plant

It is clear that there are feedback mechanisms which link the nutrient content of a plant with the activity of a particular root in ion uptake. An excellent example is given by Drew *et al.* (1969), in which the uptake of potassium over a defined length of 1 cm of root was greatly affected by whether the rest of the root was in soil or in sand, i.e. on the uptake of potassium by the rest of the root (see Fig. 6.8). α for a root may thus depend immediately upon the external concentration, and over a longer term also on how this latter affects the nutrient content of the whole plant. This is considered further in Chapter 7.

6

SOLUTE TRANSPORT IN THE SOIL
NEAR ROOT SURFACES

We discussed in Chapter 4 the movement of solute between small volumes of soil; and in Chapter 5 some properties of plant roots and associated hairs, in particular the relation between the rate of uptake at the root surface and the concentration of solute in the ambient solution. In the chapters to follow, we consider the plant root in contact with the soil, and deal with their association in increasingly complex situations. In this chapter we take the simplest situation that can be studied in detail, namely, a single intact root alone in a volume of soil so large that it can be considered infinite.

The essential transport processes occurring near the root surface are illustrated in Fig. 6.1. We have already examined in Chapter 3 the rapid

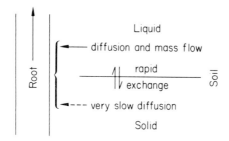

FIG. 6.1. Solute transport processes near an absorbing root.

dynamic equilibrium between solutes in the soil pore solution and those adsorbed on the immediately adjacent solid surfaces. These adsorbed solutes tend to buffer the soil solution against changes in concentration induced by the root. At the root surface solutes are absorbed at a rate related to their concentration in the soil pore solution at the boundary; and the ratio of flux to concentration is defined as the root absorbing power (α). Thus we may write (see p. 110)

$$I = 2 \pi a \alpha C_{1a} \qquad (6.1)$$

where

    I = inflow (rate of uptake per unit length),
    a = root radius,
    C_{1a} = concentration in solution at the root surface.

To calculate the inflow we have to know C_{la}, and the main topic of this chapter is the relation between C_{la} and the average pore solution concentration $\bar{C}_1$. The root also absorbs water at its surface in response to transpiration —as described in Chapter 2. This induces the soil solution to flow through the soil pores, thus carrying solutes to the root surface by mass flow (convection). Barber *et al.* (1962) calculated whether the nutrients in maize could be acquired solely by this process, by multiplying the composition of the soil solution by the amount of water the maize had transpired. Table 6.1 shows the results. The authors do not give figures for nitrogen, but assuming there is 20,000 ppm in the plant (2%) and the transpiration ratio is 500, the soil solution would need to contain 40 ppm N, a high value, to satisfy the requirement. Such calculations only give a rough idea of the importance of mass flow since in the early stages of growth the concentration of major nutrients in a plant is usually greater than it is later and the transpiration ratio may well be lower. In general more than sufficient calcium and magnesium (except in sodic or very acid soils), and sodium for the plants needs will be transported by mass flow to the root surface; but insufficient potassium and nitrogen, and quite insufficient phosphate. Similar calculations made for micronutrient elements in one soil suggest that mass flow can account for only a small part of the uptake of copper, iron, manganese and zinc, but more than half the uptake of boron (Oliver & Barber 1966a). Mass flow should be adequate to supply molybdenum to an 'average' plant on most soils, but not on those with less than about 4 ppb molybdenum in the soil solution (Lavy & Barber 1964).

Table 6.1. Supply of elements to maize by mass flow (after Barber *et al.* 1962).

Element	Composition of maize at harvest (ppm)	Conc. soil soln. needed if transpiration ratio = 500 (ppm)	Soil soln. of 145 mid-Western USA topsoils modal values (ppm)
Ca	2,200	4·4	33
Mg	1,800	3·6	28
K	20,000	40	4
P	2,000	4	·05

Some comparatively small organic solutes, such as many herbicides, are absorbed, relative to their concentration, at nearly the same rate as the water transpired—probably by passive uptake, as discussed in Chapter 5. Figure 6.2 shows the relation between transpiration of wheat seedlings and the total uptake of atrazine from two soils (Walker 1971). Variable amounts of linuron, atrazine and simazine are retained in the root depending on plant species, so that the composition of the xylem stream is less than the soil solution (Walker & Featherstone 1973; Shone & Wood 1972).

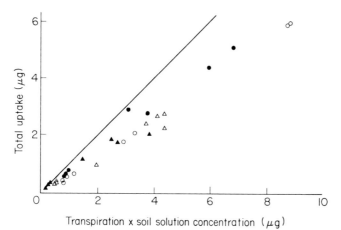

FIG. 6.2. Relation between transpiration and uptake of atrazine by wheat and turnip seedlings from two soils (after Walker 1971). The straight line represents the theoretical uptake if all the material supplied by mass-flow was taken up. ● wheat, Soakwaters; ○ turnip, Soakwaters; ▲ wheat, Little Cherry; △ turnip, Little Cherry.

If the solute is absorbed at a relatively greater rate than the water, as with phosphate and potassium, then the concentration in the soil solution at the root surface must fall. In response, ions may be released from the solid, tending to buffer the concentration. Nevertheless, there will be some lowering of the concentration at the root surface, and this will induce ions to diffuse towards the root. The development of the resulting zone of depletion is sketched in Fig. 6.3(a), and is revealed by an autoradiograph of a root growing in soil labelled with ^{33}P in Fig. 6.3(b). Autoradiographs of soils labelled with ^{99}Mo (Lavy & Barber 1964), ^{65}Zn (Wilkinson *et al.* 1968a) and ^{86}Rb (Vasey & Barber 1963) have all shown narrow zones of depletion around roots. A similar but unexpected effect with ^{45}Ca (Wilkinson *et al.* 1968b) was probably caused by the method of watering, which would dilute the calcium in the soil solution severely.

If, on the other hand, water is taken up at a relatively greater rate than solute, the solute must accumulate at the root surface, and tend to diffuse away from the root. This accumulation zone is sketched in Fig. 6.3(a) and the autoradiograph of two roots of a rapidly transpiring plant growing in $^{35}SO_4$ labelled soil (Fig. 6.3(c)) shows this increase in concentration.

Thus there are two processes by which solutes move in response to the disturbance created by the presence of the root surface—mass flow and diffusion.

The proportion of solutes crossing the root surface that are initially in 'contact' with the root, in the sense proposed by Jenny & Overstreet (1939a)

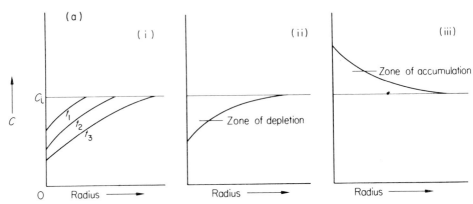

FIG. 6.3(a) Concentration of solute near a root surface (i) diffusion alone—increasing time t_1, t_2, t_3, (ii) water absorbed relatively slower than solute, $(v_a < \alpha)$ (iii) water absorbed relatively faster than solute $(v_a > \alpha)$.

(b)

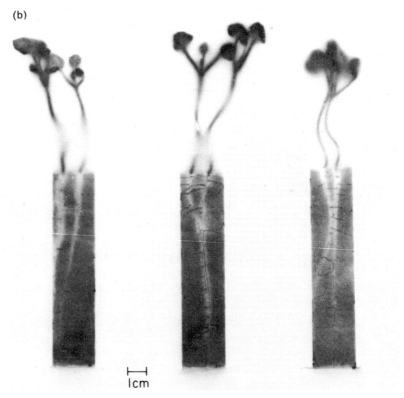

FIG. 6.3(b) Autoradiograph of roots of rape in a soil labelled with ^{33}P showing zone of depletion around roots, and accumulation within the axis and laterals (after Bhat & Nye 1973).

(c)

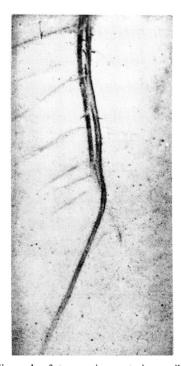

FIG. 6.3(c). Autoradiograph of two maize roots in a soil labelled with $^{35}SO_4$ showing accumulation near surfaces (after Barber *et al.* 1963).

(see Chapter 1) must be very small indeed, so that virtually all solutes absorbed undergo these two transport processes to the uptake sites on the root.

We consider later, in Chapter 7, how roots ramify through the soil; but it is necessary to be clear that they do not 'pick up' or 'intercept' or otherwise. ingest solutes as they do so, since the root cap is not an absorbing organ. They are rather to be pictured as following pores and channels, pushing soil solution and to some extent soil particles aside, whose solutes are then subjected to the transport processes of mass flow and diffusion induced by the new absorbing surfaces created. Much published work has treated root interception as a primary absorbing process; and the ideas of this approach may be found in Oliver & Barber (1966), and Halstead *et al.* (1968).

QUANTITATIVE TREATMENT OF MASS FLOW AND DIFFUSION NEAR A SINGLE ROOT

As we have postulated, diffusion and mass flow occur simultaneously, but quantitative experiments which investigate them together are few and difficult to devise. It is therefore convenient to consider first the theory of diffusion in the root zone; and describe experiments designed to test it, in which mass flow is negligible. There are so many possible complications in the zone of disturbance near roots that it is essential to establish how far theoretical concepts are verified in the simplest situations, before using them further to develop models of more complicated systems, or to use them in speculative interpretations of the results of experiments on solute uptake by whole plants. For these reasons also, experiments which are not complicated by root hairs will be discussed first.

6.1   Theory of diffusion in the root zone

If the rate of root elongation is rapid compared with the rate of spread of the disturbance zone, we may assume that movement of solute is normal to the root surface.

This assumption is reasonable if the spread of the diffusion zone, which varies as $(Dt)^{\frac{1}{2}}$, is small compared with the root's elongation in the same period, which varies roughly linearly with time. Hackett & Rose (1972) report that barley root axes extend 20 mm d^{-1}, first order laterals 4 mm d^{-1}, and second order laterals 2 mm d^{-1}. On the other hand the short stubby but long lived mycorrhizal roots of pine elongate much more slowly, and it may often be more correct to regard these as being at the centre of a spherical zone of depletion. Anderssen *et al.* (1969) have avoided the assumption that the direction of flow is normal to the root axis, but the resulting expressions seem too complex to be useful when there are so many practical uncertainties to be considered.

The appropriate continuity equation for cylindrical symmetry around the root has been given in Chapter 1

$$\frac{\partial C}{\partial t} = \frac{1}{r}\frac{\partial}{\partial r}\left(\frac{r\,D\,dC}{dr}\right). \tag{6.2}$$

The boundary conditions are: $t = 0, r > a, C = C_i$
$$t > 0, r = a, F = \alpha C_{la}.$$

6.1.1   Solution when α is constant

As we have seen in Chapter 5, at low concentrations α is fairly constant, and it helps the development of ideas to treat this as an ideal case, since equation 6.2 can then be solved analytically. The solution is given by Carslaw & Jaeger (1959, p. 337), and was first applied to the case of a root in soil by Bouldin (1961). The soil buffer power is also assumed constant.

In the following discussion we are particularly concerned with two questions: the lowering of the concentration of solute in solution at the root's surface; and the radial spread of the zone of depletion around it.

The root surface concentration ratio C_{1a}/C_{1i}

If diffusion to the root was very easy and there was abundant soil the initial soil solution concentration would be maintained and the flux would be αC_{1i}. When diffusion limits the supply to the root surface the concentration drops to C_{1a} and the flux is αC_{1a}. Thus if α is constant the ratio C_{1a}/C_{1i} is a measure of the extent to which diffusion is limiting the rate of uptake.

Figure 6.4 shows how this ratio varies with time for different values of $\alpha a/Db$. We note:

(a) αa is a measure of the root's demand for solute (see Chapter 5); and $Db \simeq D_1 \theta f_1$ (Chapter 4) is a measure of the ease of diffusion through the soil. Consequently when demand is high compared with ease of diffusion the surface concentration ratio is low. Conversely, when demand is low compared with the ease of diffusion the concentration at the root surface falls much more slowly from its initial value.

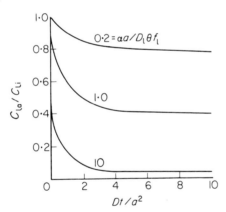

FIG. 6.4. Change in the root surface concentration ratio C_{1a}/C_{1i} with Dt/a^2 for different values of $\alpha a/D_1 \theta f_1$ (after Carslaw & Jaeger 1959).

(b)   There is an initial rapid fall in concentration at the root surface, particularly when $\alpha a/Db$ is high, followed by a much slower fall. Consequently, after the initial period the rate of uptake should decline only gradually—as shown in Fig. 6.5.

(c)   When $\alpha a/bD$ is small the rate of uptake increases in proportion to α; but as α increases so that $\alpha a/bD$ becomes large the rate of uptake reaches a maximum, as shown in Fig. 6.6. In this situation, which should often occur in practice when plant nutrients are deficient, the ratio C_{1a}/C_{1i} is very low, and diffusion through the soil limits the rate of uptake.

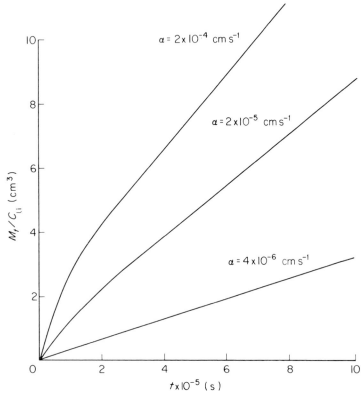

Fig. 6.5. Variation of uptake (M_t) per unit concentration in solution with time (after Nye 1966b). $a=0.05$ cm, $D=2.5 \times 10^{-8}$ cm² s⁻¹, $b=40$.

(*d*) The rate of uptake is not very sensitive to the buffer power of the soil, *b*. *D* does decrease as *b* increases (see Chapter 4), so that an increase in *b* tends to decrease the term Dt/a^2 on the *x*-axis of Fig. 6.4. Consequently, at a given time C_{la}/C_{li} is somewhat larger. This insensitivity to the buffer power arises because, as the zone of depletion spreads out radially from the root, an amount of solute increasing as the square of the radius becomes available in an infinite system. As will be shown later (Chapter 7), in a limited system when roots are competing for solutes the buffer power has a more important effect.

(*e*) A decrease in the soil moisture acts mainly to increase $\alpha a / Db \simeq \alpha a / D_1 \theta f_1$, since both θ and f_1 are reduced.

The numerator, αa, may also be reduced if appreciable moisture tension develops at the root surface. Lowering of moisture level also affects soil solution concentrations directly: for example, the concentration of an

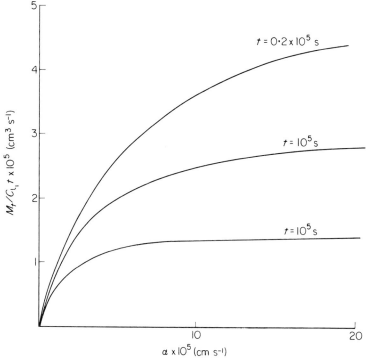

FIG. 6.6. Variation of the average flux per unit concentration

$$\frac{M_t}{t C_{li}}$$

with root absorbing power α $(a = 0.05$ cm) (after Nye 1966b).

unadsorbed ion-like nitrate will be increased, whereas that of an adsorbed ion-like phosphate may be reduced because the calcium ion concentration has been increased. For these reasons the effects of soil moisture changes will be discussed further in a later section.

There is no question that solutes diffuse to the root surface; but from these theoretical considerations we may draw some preliminary conclusions about the extent to which this process may limit the rate of uptake, as revealed by the balance between root demand and ease of solute movement $-\alpha a/D_1 \theta f_1$. For simple solutes $D_1 \simeq 10^{-5}$ cm^2 s^{-1}. In a moist loam at field capacity a reasonable value of θ is 0.25 and f_1 0.4. On the other hand, in a dry loam θ might be 0.1 and f_1 0.01. Thus the soil term in the denominator may range from 10^{-6} to 10^{-8} cm^2 s^{-1}. It is clear from Fig. 6.4 that diffusion seriously reduces the uptake rate when $\alpha a/D_1 \theta f_1 \simeq 1$, and is almost completely limiting when $\alpha a/D_1 \theta f_1$ exceeds 10. Thus we may predict that in a moist loam diffusion will be an important limiting process when αa exceeds 10^{-6} cm^2 s^{-1}; and in a drier soil it will be limiting when αa is considerably lower.

Values of αa in stirred culture solution have been given in Table 5.3, and for nitrate, phosphate and potassium at low concentration they exceed 10^{-6} cm^2 s^{-1}.

The spread of the disturbance zone

The development of the zone of depletion in time is shown in Fig. 6.7. There is no simple algebraic expression to describe the spread, and the concentration is in fact asymptotic to the distance axis. We note that the root-mean-square radial displacement of a molecule diffusing from a point source in two dimensions is $2\sqrt{Dt}$. Thus the extent of the initial disturbance created when the root begins to take up solute is given approximately by this expression, and the concentration approaches the original value very closely at

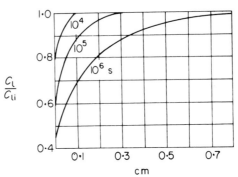

FIG. 6.7. The concentration profile near an absorbing root ($a = 0.05$ cm, $\alpha = 2 \times 10^{-7}$ cm s^{-1}, $D = 10^{-7}$ cm^2 s^{-1}, $dC/dC_1 = 0.2$) (after Nye & Marriott 1969).

this point. In practice, measurements of concentrations are not accurate enough to detect it, so we arbitrarily describe the distance of spread as that at which the concentration has departed from the initial concentration by 10% of the total concentration change. Since, for simple solutes at a given level of moisture, D varies as $1/b$, the disturbance limit varies as $\sqrt{t/b}$. Thus, for example, NO$_3$ with a buffer power 100 times less than phosphate will spread about 10 times as fast.

6.1.2 Solution when α varies with concentration

Although at low concentration α is fairly constant, at higher concentration it decreases—as described in Chapter 5. If the relation between α and concentration is known, equation 6.2 may be solved by numerical methods.

The relation between the flux across the root surface and the solution concentration there often approximates to the Michaelis–Menten form shown in Fig. 5.6(a). Kautsky *et al.* (1968) find that the assumption of a linear relation between flux and concentration rather than the Michaelis–Menten form causes only small discrepancies in the estimated uptake if C_{1i} is less than K_m; but if C_{1i} is greater than K_m the linear assumption underestimates the uptake, but overestimates the root surface concentration.

Without the trouble of a numerical solution, the flux may often be estimated approximately from Fig. 6.4 after the initial drop in concentration, by using a value of α that corresponds to the fairly steady low root surface concentration.

6.2 Experimental evidence for theory of diffusion near roots with restricted mass flow

6.2.1 Indirect methods

The model outlined in the previous section has been tested in several experiments. In the earlier work the value of α and hence the root surface concentration had to be derived indirectly by measuring the inflow through a portion of root, and the diffusion characteristics of the soil from which it was absorbing. The model was therefore tested only to the extent that 'reasonable' values of uptake were obtained. For example: (1) The observed uptake should not greatly exceed the amount that could diffuse to a root surface with a very high absorbing power. (2) Values of α derived from the experiments should be consistent with values obtained in experiments in stirred or flowing nutrient culture solutions. (3) Values of α should change with the concentration of nutrient or moisture status of the soil in a consistent manner.

Drew *et al.* (1969) and Drew & Nye (1970) measured the uptake of potassium and phosphate through a 1 cm length of intact onion root over periods up to 16 days, from soil with KH_2PO_4 added to range from very low to high potassium and phosphate status. They determined the diffusion characteristics of the soil independently. The onion has proved a convenient experimental plant because it may be grown to produce uniform cylindrical seminal roots, without root hairs, that develop lateral roots only after several days.

For potassium uptake they found: (a) Uptake through the 1 cm root length continued with little decline in rate for 16 days; indicating there was no great change in α with the age of the root element. The average root surface concentration ratio, and the root absorbing power they calculated are shown in Table 6.2. (b) When the remainder of the root was in sand the inflow from the segment was greater than when it was in soil (Fig. 6.8), presumably because the whole demand of the shoot was concentrated on the

Table 6.2. Root surface concentration ratio C_{la}/C_{li} and root absorbing power α of a single onion root in soil with graded additions of KH_2PO_4 (after Drew *et al.* 1969; Drew & Nye 1970).

Potassium			Phosphorus		
C_{li} $M \times 10^6$	C_{la}/C_{li}	α (cm s^{-1} $\times$ 10^6)	C_{li} $M \times 10^6$	C_{la}/C_{li}	α (cm s^{-1} $\times$ 10^6)
160	0·64	45			
200	0·39	77			
350	0·71	28	2	0·16	230
610	0·63	35	10	0·09	840
880	0·65	35	40	0·13	280
1300	0·65	24	87	0·35	76

segment. In a more extensive series of experiments using the same technique (except that the rest of the root was in moist air) on about 50 soils Bagshaw *et al.* (1969) found the relation between rate of uptake and concentration calculated at the root surface to have the form shown in Fig. 6.9(a). The ratio C_{la}/C_{li} is plotted against the initial concentration in Fig. 6.9(b). Diffusion was clearly a limiting process at initial solution concentrations below about 5×10^{-4} M. The apparently negative values calculated for the root surface concentration ratio at very low solution concentrations imply a greater uptake than the exchangeable potassium could supply by diffusion, and were possibly caused by release from non-exchangeable potassium.

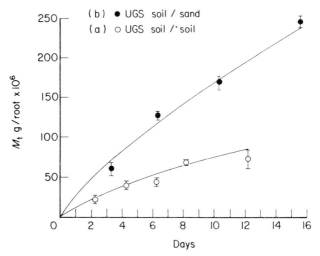

FIG. 6.8. Uptake of potassium through a cm length of onion root when rest of root is in (a) the same soil, (b) sand. Limits show standard errors of the means (after Drew *et al.* 1969).

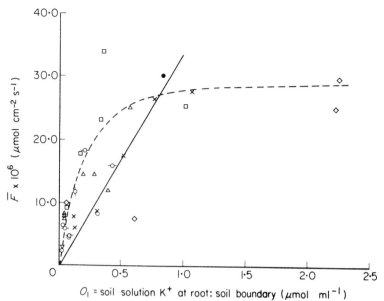

F<small>IG</small>. 6.9(a). Uptake of potassium through a centimetre length of onion root in a range of soils. Relation between the flux across the root surface and the estimated potassium concentration in solution there. The slope of the line fitting data of Drew *et al.* 1969 corresponds to $\alpha = 33 \times 10^{-6}$ cm s^{-1} (after Bagshaw *et al.* 1969) $\times$———— data of Drew *et al.* 1969. - - - - - data of Bagshaw *et al.* 1969.

In the experiments on soil ranging from low to high phosphate status the calculated values of α and C_{la}/C_{li} are also shown in Table 6.2. It appeared that except at the highest concentration of phosphate the root segment was approximating to a sink of infinite strength. In contrast to potassium, this great absorbing power for phosphorus was shown when the rest of the root was in soil as well as in sand. In the short period of 16 days it is unlikely that mycorrhizal effects would be observed (see p. 183).

The values of α for potassium and phosphate derived in these experiments may be compared with those obtained in stirred or flowing solution culture given in Table 5.3. For potassium, values of α are somewhat greater than for excised barley roots at comparable concentrations. The maximum flux greatly exceeds that of the grasses studied by Wild *et al.*—compare Fig. 6.9(a) with Fig. 5.6(a). As mentioned above, uptake was enhanced because it was restricted to a 1 cm portion of root (Fig. 6.8); no direct comparison of α values is therefore possible.

For phosphorus α values are of the same high order as those obtained for onion in solution culture at comparable concentrations.

For nitrate Clarke & Barley (1968) using wheat seedlings, grew a single root through a 5 mm soil layer in Wambi sand and Urbrae loam, each at two

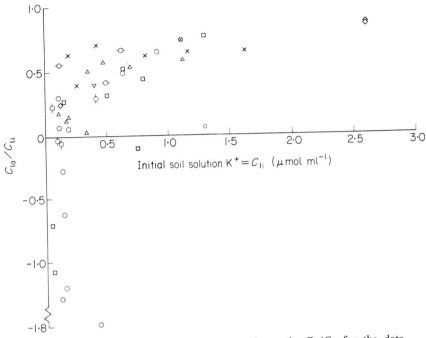

FIG. 6.9(b) Calculated root surface concentration ratio C_{1a}/C_{1i} for the data shown in Fig. 6.9(a) (after Bagshaw *et al.* 1969).

moisture levels and $C_{1i} = 200$ mg l^{-1} N as NO_3. Uptake over 75 h was satisfactorily predicted with $\alpha = 2.5 \times 10^{-6}$ cm s^{-1}. They also predicted satisfactorily uptake from Wambi sand by a planar root mat at four moisture levels, but uptake from Urrbrae loam was considerably less than predicted, due, it was suggested, to poor contact with the soil. In the same series of experiments they showed qualitatively that ammonium N, at the same equivalent concentration of N in the whole soil, was absorbed more slowly that NO_3 because, being adsorbed as an exchangeable cation, it has a lower diffusion coefficient, and a lower concentration in the soil solution.

Olsen & Watanabe (1970) sandwiched 10 maize roots, each 6 cm long, between blocks of three soils differing widely in texture and phosphate buffer power. Each soil was at three levels of phosphate. They showed there was a correlation between the phosphate absorbed during 24 h and the soil solution concentration, buffer power, and related diffusion coefficient, of the soils.

6.2.2 Direct methods

In all the work so far described, the actual concentration at the root surface and in the zone of depletion has to be calculated from the experimental data, so that there is no independent check that the model is correct. For more searching tests it is necessary to measure directly the concentration of solute. This has been achieved by two main methods.

(i) Autoradiography. This technique was initiated by Walker & Barber (1961) who grew plants in a box of soil uniformly labelled with radioisotope. One side of the box consisted of a mylar film, 1 μm thick, against which some of the roots grew. Autoradiographs of their contact with the soil were taken with X-ray film. The earlier work with this method was qualitative since (a) visual assessment of concentration from density on the film is very deceptive, and the contrast is greatly influenced by the time of exposure and development; (b) the isotopes used, e.g. ^{32}P, ^{86}Rb, ^{90}Sr had high maximum β energy, and resolution was correspondingly poor. For example, ^{32}P emits beta particles of maximum energy 1·71 Mev which can emerge from a depth of about 5 mm of soil.

Recently, more quantitative results have been achieved by strict calibration of film density against isotope concentration; scanning with a microdensitometer; and use of low energy emitters. For example ^{33}P with maximum beta energy of 0·25 Mev has a maximum depth of emergence of 0·3 mm, giving a resolution (defined as the distance at which the grain density is one half that observed directly over the source) of 0·25 mm (Bhat & Nye 1973). Autoradiographs have also been taken of sections across roots in soil. Sanders (1971) grew roots in a block of soil treated with ^{35}S. He impregnated

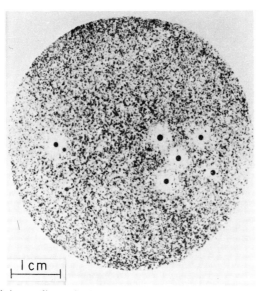

FIG. 6.10(a) Autoradiograph of section across onion roots in a soil labelled with ^{35}SO$_4$ (after Sanders 1971).

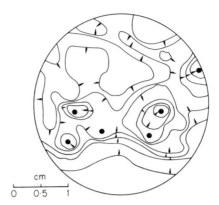

FIG 6.10(b) $^{35}SO_4$ concentration contour diagram. Contours are at equal intervals on the densitometer scale (after Sanders 1971).

the block with resin and cut sections across the roots. Figure 6.10(a) shows a typical autoradiograph and Fig. 6.10(b) the solute concentration 'contours' that may be derived by this method.

(ii)   The root plane technique. Farr *et al.* (1969) placed single seedling onion roots side by side to form a 'plane' of roots, which was then sandwiched

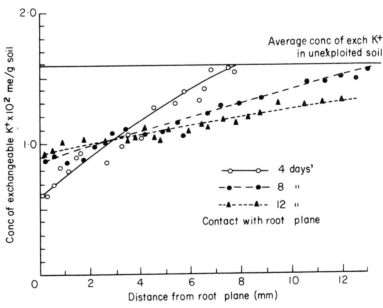

FIG. 6.11. Variation of exchangeable potassium in a block of soil in contact with a plane of onion roots (after Farr *et al.* 1969).

between two blocks of soil, and the uptake measured over several days. The blocks were then separated, frozen in liquid nitrogen, and sectioned parallel to the root plane in a freeze microtome—a technique developed by Brown *et al.* (1964) for measuring diffusion coefficients in soil. The sections were then assayed. Since sections of 20 μm can be cut concentrations very close to the root surface can be measured. An example of the results obtained is shown in Fig. 6.11.

With these methods the influence of a number of factors, represented in the theoretical treatment given earlier, have been measured and compared with predictions. It is well to recall that they measure the total concentration of diffusible solute in the soil. The concentration in the soil pore solution has to be deduced from an appropriate isotherm.

Diffusion of phosphate and potassium

Farr & Vaidyanathan (1972) measured the phosphorus concentration near a plane of onion roots in a highly-buffered clay soil. The concentration of phosphorus (exchangeable over 120 days) at the root surface was reduced by about 10%. The spread (as measured by a reduction of 10% in the total concentration change) was only about 1 mm after 12 days.

In a poorly-buffered sandy soil Bagshaw *et al.* (1972) obtained the concentration profiles shown in Fig. 6.12. The concentration of exchangeable phosphorus was reduced by nearly 50% at the root surface. The spread of the depletion zone was greater than that predicted from a simple isotherm. pH gradients were also measured, and it was found that the pH was reduced by at least 0·5 units near the root surface. In the soil used the concentration of phosphorus in the soil solution was increased as the pH was lowered, and this was suggested as a possible reason for the greater spread.

The concentrations of potassium in the same soil are shown in Fig. 6.11 (Farr *et al.* 1969). No comparison with predicted values was made, but the figure illustrates that the spread of potassium is greater than phosphorus at comparable times and on a similar soil.

In predicting concentration profiles for comparison with those observed experimentally, it is necessary to know the diffusion characteristics of the soil, and the value of α. The spread of the disturbance zone (see p. 136) depends mainly on D and t rather than α, and the comparison is largely a test of the correctness of the independently calculated diffusion coefficient when applied to the soil near the root. The value of α can in principle be determined independently from stirred or flowing solution culture experiments; though the extent to which solution and soil grown roots behave in the same way (even in absence of root hairs) is uncertain. It is also difficult

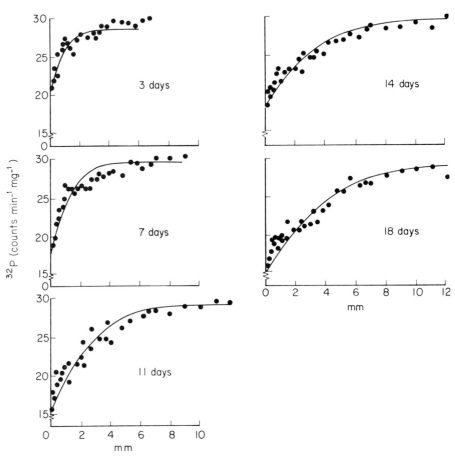

FIG. 6.12. Concentration profiles of ³²P exchangeable phosphate in a sandy soil in contact with an onion root plane (after Bagshaw *et al.* 1972). The experimental points are shown. The curves are those expected for a concentration independent diffusion coefficient and a constant flux across the root plane.

to reproduce in the solution culture the concentrations of all the other ions in the soil solution at the root surface.

When the concentration of phosphate in the soil solution at the root surface is less than about 10^{-6} M, the value of α for young roots is high, and their uptake rate depends very largely on the rate of diffusion through the soil. Using the autoradiograph scanning method Bhat and Nye (1974) found very good agreement between the observed and predicted phosphorus uptake, along the entire length (13 cm) of a single onion root. The concentration profiles of individual scans also agreed well with the predicted profiles (Fig. 6.13).

6.3   Roots with root hairs

As described in Chapter 5 most roots in soil bear a dense cluster of root hairs numbering 100–1000 per cm of root and in the range 0·1–1 mm long. The hairs appear a few millimetres behind the root apex, and in the main roots the hair zone usually extends for several centimetres towards the base.

Since the hairs are protuberances of 'piliferous' epidermal cells (Scott 1963) we may anticipate that they absorb nutrients through their walls like the remainder of the epidermal cells. As a demonstration of this, Barley & Rovira (1970) grew pea radicles through channels in clay soil into which root hairs penetrated, and in the same soil compacted so that no hairs penetrated it. Uptake of phosphate was 78% greater from the uncompacted soil.

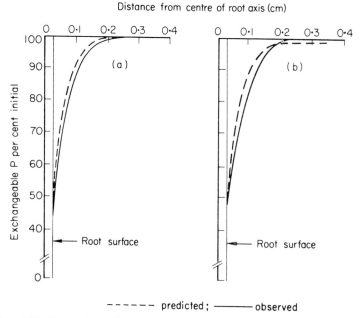

FIG. 6.13. Comparison of predicted and measured phosphate concentration profiles in soil around an onion root (a) 5 days, (b) 10 days old (after Bhat & Nye 1974).

Attention has often been called to the greatly increased absorbing area provided by root hairs (Dittmer 1940; Bouldin 1961). This must not however be taken to imply that the absorbing power of the root axes should increase in the same proportion, since the hairs are normally so crowded that they rapidly interfere with each others uptake. Hence the simple theoretical treatment of a root as a cylinder given on p. 132 must be modified for roots with hairs.

6.3.1 Theoretical treatment of roots with hairs

Since the hairs usually appear to be clustered within a fairly well-defined cylinder, Kramer & Coile (1940) and Wiersum (1961) have regarded this cylinder as an effective volume of soil being exploited by roots for water and nutrients. Passioura (1963) treated it as the effective surface to which nutrients moved by diffusion and mass flow. Nye (1966b) calculated that the hairs were usually sufficiently densely clustered that the concentration within the hair cylinder should rapidly become effectively uniform. The hairs should greatly increase the uptake of solutes with a low diffusion coefficient, especially during the first few days. He predicted the concentration profiles shown in Fig. 6.14 for a slowly-diffusing solute like phosphate, and a rapidly diffusing solute like nitrate. Clearly uptake of phosphate would be greatly enhanced, but that of nitrate little affected.

A more exact theoretical treatment (Bhat *et al.* 1976) is based on the solution of the continuity equation for diffusion around the root axis cylinder (equation 6.1), modified to include uptake by the root hairs in successive radial zones around the central axis. Competition between the hairs in each radial zone is treated by the method developed by Baldwin *et al.* (1973) for competition between whole roots in a soil volume (see Chapter 7).

6.3.2 Experimental observations on roots with root hairs

Drew & Nye (1969, 1970) measured the uptake of potassium and phosphorus from a 1 cm band of soil by single roots of rye-grass, which bore dense root hairs about 1 mm long. In the soil of lowest potassium status they calculated from Nye's theory (1966b) that the presence of the hairs enhanced uptake of potassium over 4 days by up to 77 %. In this soil the uptake of potassium was greater than could be accounted for by diffusion to the central root axis in the absence of hairs, so providing evidence that the root hairs were active in absorbing potassium. At higher levels of potassium the effect of hairs was less evident. On the other hand at high and low levels of phosphate it was calculated that the presence of the hairs had increased the efficiency of the central root 2 to 3 fold. The uptake could only be accounted for, if the root hairs were active.

Bole (1973), using strains of wheat that developed different root hair densities, found a slight, but not significant increase in phosphorus uptake per centimetre root length as the hair density increased to 50 per mm of root in a soil at both a low and high phosphorus level. There was no further increase in uptake at greater hair densities. The plants were harvested after 4

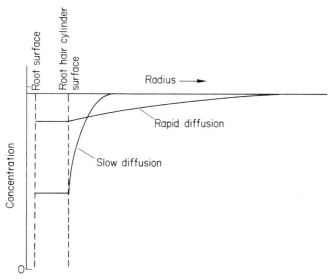

F<small>IG</small>. 6.14. Predicted effect of densely clustered root hairs on the concentration profile around a root axis for a nutrient with a low—or high diffusion coefficient (after Nye 1966b.)

weeks, and the data presented are presumably an average for roots of all ages and orders. Competition between individual roots, and a long period of uptake will tend to reduce the importance of the density of the root hairs.

More direct indications have been provided by autoradiography. Lewis & Quirk (1967) grew wheat roots in ^{32}P labelled Seddon gravelly loam containing 300 and 1000 ppm exchangeable P. Autoradiographs showed visually a pronounced diminution of intensity in a zone extending about 1 mm from the root axis, which corresponded to the approximate length of the root hairs. The zone developed in 3 days and remained for 17 days. If the hairs were inactive they calculated the spread of the disturbance zone from the main root axis should be only about 0·2 mm from the surface of the root axis.

Bhat & Nye (1973, 1974a,b), using the low energy isotope ^{33}P, and rape seedling roots, evaluated their autoradiographs quantitatively. A typical densitometer scan across the primary root axis is shown in Fig. 6.15.

It shows how the zone of intense depletion in the root hair zone merges into a more widespread zone of less intense depletion. The boundary between these two zones is not abrupt because the length of hairs per unit volume is greater near the axis, partly because of their radial distribution, and also because some extend further from the axis than others. At both low and high levels of phosphate the uptake was greater and the diffusion zone spread further than predicted. It appeared that rape had the ability to increase the concentration of phosphate in the soil solution—a conclusion supported by experiments with complete rape root systems on the same soils (see Chapter 7).

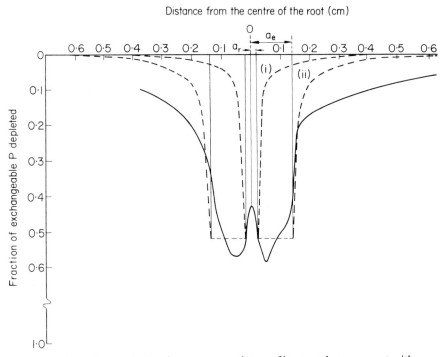

Distance from the centre of the root (cm)

Fraction of exchangeable P depleted

FIG. 6.15. Measured phosphate concentration profile around a rape root with dense root hairs (after Bhat & Nye 1973). a_r, radius of root axis; a_e radius of root hair cylinder. $----$ (i) calculated assuming the root hairs as inactive; $----$ (ii) calculated assuming intense root hair activity and uniform depletion from within the root hair cylinder; $-----$ observed in the experiment.

There was a marked difference between the primary tap root of rape and a single seedling root of onion in absorbing phosphate, particularly in the soil of low phosphate status, from which the rape root absorbed 10.5×10^{-7} mol P in 5 days, while the onion root absorbed only 0.25×10^{-7} mol P in 12 days, when it had attained a similar length of 14 cm.

In contrast to these findings Bole (1973) found rape and flax roots produced no root hairs in a soil of low phosphate status, yet their uptake of P per cm of root after 4 weeks of plant growth was about twice that of wheat with 45 hairs per mm of root. The average ages and dimensions of the roots were not given; and since the geometrical benefit of root hairs decreases with the time a section of root has been absorbing it is difficult to analyse these differences further.

6.4 Simultaneous diffusion and convection

Since roots normally absorb water as well as solutes, these solutes move in the neighbourhood of an absorbing root surface by both diffusion and by mass flow; and we have to consider how these two processes interact.

As we have seen, if diffusion acts alone a zone of depletion develops. If at the same time solute is transported into this zone of depletion its extent will be reduced. If the solute is transported towards the root faster than it is absorbed a zone of accumulation will develop (Fig. 6.3(a)).

6.4.1 Theoretical treatment

The continuity equation applicable to a cylinder of soil with a root at its centre, is, from equation 1.9;

$$\frac{\partial C}{\partial t} = \frac{1}{r} \frac{\partial}{\partial r} \left[r D \frac{dC}{dr} + r v C_1 \right] \tag{6.3}$$

where v is the water flux *towards* the root.

The boundary conditions when the volume of soil is infinite are:

$$t = 0, a < r < \infty, C_1 = C_{1i}$$
$$t > 0, r = a, F = \alpha C_1.$$

This equation cannot readily be solved analytically except for special cases, e.g. when $Db = rv$ (Geering 1967). Numerical solutions for different conditions at the root surface have been given by Passioura & Frere (1967) —for $\alpha = 0$, and by Nye & Marriott (1969)—for

$$\alpha = \frac{F_{max}}{K'_m + C_1}$$

the Michaelis–Menten condition, which includes

$$\alpha_{max} = \frac{F_{max}}{K'_m},$$

a constant, as a special case when $C_1 \ll K'_m$.

Nye & Marriott assumed in the first instance that F_{max} and α were independent of v, the water flux; and also that D was independent of v.

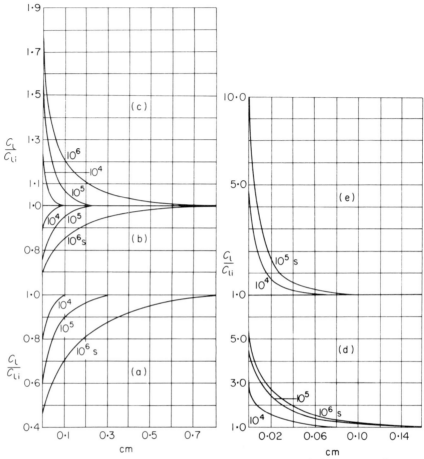

FIG. 6.16. The effect of the rate of soil solution flow on the relative concentration near a root surface ($\alpha = 2 \times 10^{-7}$ cm s^{-1}; $D = 10^{-7}$ cm^2 s^{-1}; $b = 0\cdot2$; F_{max} high; $a = \cdot05$ cm).

(a) $v_a = 0$ cm s^{-1}; i.e. diffusion alone. C_l/C_{li} continues to drop at the root surface and the zone of depletion to spread outwards at 10^6 s.

(b) $v_a = 10^{-7}$ cm s^{-1}. C_{la}/C_{li} has not reached the limiting value $v_a/\alpha = 0\cdot5$ after 10^6 s.

(c) $v_a = 4 \times 10^{-7}$ cm s^{-1}. C_{la}/C_{li} has nearly reached the limiting value $v_a/\alpha = 2$ after 10^6 s, but the zone of accumulation continues to spread outwards since $av_a/Db = 1$. For a steady state av_a/Db must exceed 2.

(d) $v_a = 10^{-6}$ cm s^{-1}. $av_a/Db = 2\cdot5$.

After 10^6 s C_{la}/C_{li} has almost reached the steady-state value $v_a/\alpha = 5$, and the zone of accumulation has ceased to spread outwards.

(e) $v_a = 2 \times 10^{-6}$ cm s^{-1}. $av_a/Db = 5$.

Both C_{la}/C_{li} and the zone of accumulation have nearly attained the steady-state after only 10^5 s. C_{la}/C_{li} is increased to $v_a/\alpha = 10$ but the zone of accumulation is compressed (after Nye & Marriott 1969).

As explained in Chapter 5, α may be expected to increase with v at high salt concentrations. It was also explained in Chapter 4 that D should strictly be replaced by a dispersion coefficient D^*. This numerical solution can readily accomodate such variations, provided the effect of v on F_{max}, α and D^* is known. The numerical solution will also allow for a diffusion coefficient that depends on concentration.

The effect, in theory, of an increasing water flux on concentration in the disturbance zone is illustrated in Fig. 6.16.

The effect of mass flow on solute uptake

The true mass flow at any point is the velocity multiplied by the solution concentration at that point. Hence the actual amount passing across the root surface by mass flow may be very different from that deduced in the earlier concepts of Barber (1962) as the product of the transpiration and bulk solution concentration. The processes of mass flow and diffusion occur together and the resulting concentration profile is not the result of two independent processes. Hence it is not possible to state that a given proportion of the total solute absorbed has 'arrived' or is 'taken up' by mass flow and the remaining proportion by diffusion—an approach used by many authors. It is however correct to state that a given amount of solute has been brought into the zone of disturbance around the root by mass flow, assuming the radial inflow of solution from outside it does not vary with distance from the root. This has been called the 'apparent mass flow' contribution by Brewster & Tinker (1970). Because of the interaction between the two processes, Marriott & Nye (1968) considered by how much transpiration might increase the flux across the root above that which would occur by diffusion in any case. From an approximate solution introduced by Passioura (1963)

$$F = (C_{1i} - C_{1a}) \frac{bD\gamma}{a} + v_a C_{1i} \qquad (6.3a)$$

they showed the formula

$$F = \alpha C_{1i} \left(\frac{1 + v_a\, a/bD\gamma}{1 + \alpha a/bD\gamma} \right) \qquad (6.4)$$

was correct within 10% when $v_a < \alpha$ (γ is a term tabulated by Jaeger & Clark (1942), depending on Dt/a^2. Provided $1000 > Dt/a^2 > 1$, $\gamma = 0.5$ within a factor of 2). The important point is that the flux increases linearly with v_a; and further that transpiration should increase the flux of any solute by

$$\frac{100\, a\, v_a}{b D \gamma}\, \%,$$

independently of the root absorbing power.

Since $bD \simeq D_1\, \theta f_1$ (see Chapter 4) and $D_1 \simeq 10^{-5}$ cm² s⁻¹ for simple solutes the increase for a given $v_a a$ will depend greatly on the moisture content θ. For example at a high transpiration rate, $v_a a = 10^{-7}$ cm² s⁻¹, in a moist soil ($\theta = 0.4$, $f_1 = 0.52$), the increase in flux is calculated as only 12%; but in dry soil ($\theta = 0.16$, $f_1 = 0.09$) the increase is 120%, provided the same transpiration rate is maintained, which in practice is unlikely (see Chapter 2). These conclusions assume that α is independent of v_a (see Chapter 5). Accordingly, they do not apply to solutes passively swept into the plant in the transpiration stream, e.g. many herbicides, in which case there is no depletion around the root, but there could be accumulation if water enters faster than the solute.

6.4.2 Experimental evidence for predictions from the theory of diffusion with mass flow

Qualitative experiments

We have seen that the condition for an increase in concentration at the root surface due to convection is that $v_a > \alpha$ (Fig. 6.3(a)). Even if this is so the increase in moist soil should be very small, because

$$\frac{v_a a}{b D \gamma}$$

is small. Conditions for a large increase are therefore high transpiration; low moisture; low root absorbing power. Such conditions are most likely to be found in saline soils under hot dry climates (Section 7.4.6).

Here it must be noted that accumulations of cations, specially of calcium, which is the cation usually dominant in the soil solution near the root, are not necessarily caused by mass flow. Since roots usually take up more anions than cations they normally excrete bicarbonate ions which react with the soil adjoining the root to make it less acid, as explained in section 6.7.1. The degree of cation saturation of the soil exchange complex necessarily increases, and the extra cations accompanying the anions move by salt diffusion to this soil without mass flow.

More recent direct measurements of accumulations have been given by Riley & Barber (1970), who found much greater salt concentrations in the 'rhizosphere' and 'rhizocylinder' volumes around soybean roots than in the bulk soil. They state that their results agree with the calculations of Passioura & Frere (1967), but it is hard to accept this, since the relative increase in concentration in the 'rhizocylinder' was larger (about 12 times) with low salt concentrations in the soil and low transpiration, than with high salt concentration and high transpiration (about 6 times). In the former case little accumulation would be expected since the stated ratio of apparent mass flow to plant uptake was only 1.2.

Quantitative experiments

It is easier to devise an experimental system for measuring concentration profiles near a root surface when the geometry is planar rather than cylindrical. Such a system has the additional advantage that there is a convenient analytical solution of the continuity equation in one dimension (equation 1.6) Nye (1966c), for the appropriate boundary conditions:

$$t = 0, \; x \geqslant 0, \; C_1 = C_{1i}$$
$$t \geqslant 0, \; x = 0, \; F = \alpha C_1$$

the solution is

$$C_1/C_{1i} = 1 - \tfrac{1}{2} \left\{ \text{erfc} \; \frac{x + vt/b}{2\sqrt{Dt}} + \frac{\alpha - v}{\alpha} \left(\exp \frac{-vx}{bD} \right) \text{erfc} \; \frac{x - vt/b}{2\sqrt{Dt}} \right.$$

$$\left. + \tfrac{1}{2} \frac{2\alpha - v}{\alpha} \exp \left(\frac{(\alpha - v)(x + \alpha t/b)}{bD} \right) \text{erfc} \; \frac{x + (2\alpha - v) t/b}{2\sqrt{Dt}} . \right. \tag{6.5}$$

At large times the distribution of solute near the root tends to a steady state, expressed by the simple relation

$$C_1/C_{1i} = 1 + \left(\frac{v}{\alpha} - 1 \right) e^{-vx/bD} \tag{6.6}$$

The final concentration at the root surface ($x = 0$) thus depends only upon v/α, and the distribution with distance from the root depends only on v/bD.

To test the predictions of this equation, Wray & Tinker (1969b) grew a single onion root through a rectangular block of very moist soil only 1 mm thick, so that movement of solutes to the root was effectively linear. The soil solution contained SO_4 labelled with [35]S. They were able to control the rate of transpiration by passing a stream of dry air over the onion shoot, and measured the rate by collecting and weighing the water transpired. Hence they determined v. The soil surface was traversed with a moving collimated slit 0·2 mm wide and 15 mm long, set close to the surface and parallel to a straight portion of root. The β radiation emitted from the soil and passing through the slit was recorded directly. Details of the apparatus have been given by Wray & Tinker (1969a).

Values of the parameters b and D in equation 6.5 were determined in independent experiments, and α was measured over the appropriate range of concentration in stirred solution culture.

The accumulation of sulphate near the root induced by a high rate of transpiration is shown in Fig. 6.17(a). Note the approximately exponential form of the concentration-distance curve after 18 h. When the transpiration rate was again severely restricted the accumulation rapidly dispersed.

The time course of the accumulation at the surface followed approximately the rate predicted by theory, as shown in Fig. 6.17(b).

The values of α and D determined in all the scanning experiments not spoiled by experimental hazards are compared with those measured independently in Table 6.3.

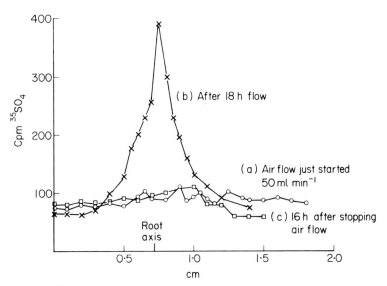

FIG. 6.17(a). Concentration profile of [35]SO_4 in moist soil near onion root (a) air flow over shoot just started; (b) after 18 h airflow; (c) 16 h after stopping airflow (after Wray 1971).

The agreement in α is unexpectedly good bearing in mind that the onions in solution culture were growing in different circumstances. The agreement

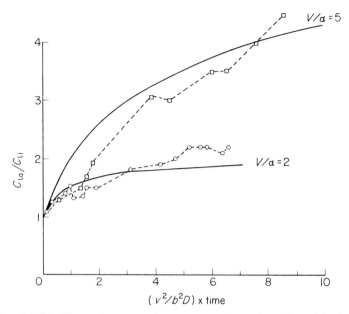

FIG. 6.17(b). Change in root surface concentration ratio C_{1a}/C_{1i} with time at two transpiration rates. Solid lines are theoretical curves (after Wray 1971).

in D is also very satisfactory. It is however noticeable that the D derived from the scanning experiments tends to increase with v. In *natural* soils the known values of hydrodynamic dispersion (see p. 88) would have been more than

Table 6.3. Diffusion and convection of sulphate to an onion root. Comparison of values of α and D derived from concentration profiles with values measured independently (after Wray 1971).

Time of observation hours	Scanning experiments			Independent measurements	
	v cm s^{-1} $\times 10^6$	α cm s^{-1} $\times 10^6$	D cm^2 s^{-1} $\times 10^6$	α cm s^{-1} $\times 10^6$	D cm^2 s^{-1} $\times 10^6$
22	3·1	0·9	(1·2)	0·8	not det.
108	3·4	1·8	2·0	1·2	1·4
70	4·4	1·0	2·7	0·8	1·4
230	2·2	1·1	1·7	1·0	1·4

sufficient to account for this; though if the soil in these experiments was indeed homogeneously packed, the effect would not have been sufficient. In general though, these experiments confirm theory, at any rate in a very moist homogeneous soil; and they indicate that the same concepts may be applied to cylindrical systems with some confidence.

6.5   The effect of soil moisture level on solute absorption by single roots

The soil moisture level has a marked effect on the absorption of solutes by whole plants. The detailed interpretation of these effects is extremely complicated because it involves both the transport of soil solutes, and complex plant physiological responses. Here we consider the effect of reducing a favourable moisture level on solute absorption by a single root. We deal with the effect on a whole plant in Chapter 7.

Some of the more obvious effects on a single root are as follows:

6.5.1   Plant effects

(a)   The root absorbing power, α, may be reduced by decreased water potential within the plant, affecting many features of plant growth.
(b)   Contact between root and soil may be reduced by shrinkage of soil or root.

6.5.2   Soil effects

(c)   The diffusion coefficient will decrease because θ and f_1 are reduced.
(d)   Convection to a root may be reduced by decrease in rate of transpiration.
(e)   In drier soil the diffusion coefficient near a root may decrease sharply because the water level decreases sharply near the surface if transpiration is still appreciable (Chapter 2).
(f)   The solution concentration of non-adsorbed solutes, e.g. nitrate and chloride will increase.
(g)   The concentration in solution of exchangeable cations will increase. Since the anion concentration rises, the concentration of cations will rise. If calcium and magnesium are the dominant cations, their concentration will increase approximately directly with the total anion concentration. The monovalent cations, e.g. potassium, will increase so that the reduced activity ratio is maintained:

$$\text{i.e.} \quad \frac{(\text{K})^2 \text{ dry}}{(\text{K})^2 \text{ wet}} \simeq \frac{(\text{Ca}) \text{ dry}}{(\text{Ca}) \text{ wet}}$$

(h) The concentration of adsorbed anions, e.g. phosphate, will tend to decrease since the activity product $(Ca)(H_2PO_4)^2$ tends to be constant and (Ca) increases according to (g).

A number of experiments at varying moisture levels have been made in which transpiration has been minimized so that transport is solely by diffusion,

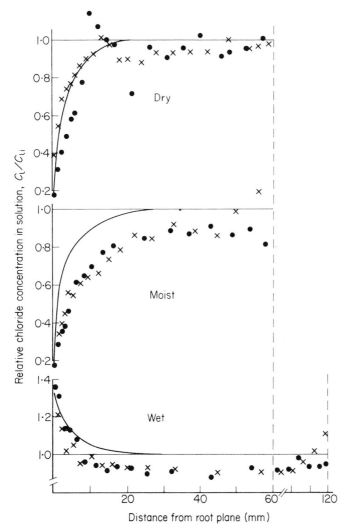

FIG. 6.18. Relative chloride concentration after 6 days uptake by onion root plane from a dry, moist and wet soil. Curves are derived from equation 1.6 with α selected to give the best-fit at the root surface (after Dunham & Nye 1974). Measurements on opposite soil blocks are shown by the symbols $\times$ and $\bullet$.

and in which uptake over short periods (72 h at the most) has been measured. The first batch of work reviewed indicates that reduced moisture reduces uptake in spite of increased concentration of ions in the soil solution.

Wiersum (1958) found that ^{86}Rb uptake from sand by buried excised roots decreased greatly with decreased water level, when the total rubidium concentration in solution was low—even though the solution concentration increased in inverse proportion to the moisture level. Likewise Danielson & Russell (1957) found that rubidium uptake by 2-day-old corn seedlings decreased linearly with decreasing soil water content. Place & Barber (1964) measured uptake of ^{86}Rb by a short length of root of a corn seedling in contact with soil of variable moisture level, with the rest of the roots growing in moist sand. They found a linear relation between uptake of rubidium and the self-diffusion coefficient of rubidium in the same soil.

Paul (1965) found the uptake of chloride by 3-day-old wheat seedlings decreased slightly with decreasing moisture level in the soil in spite of the fact that the chloride concentration in the soil solution increased. He attributed the effect to reduction in the diffusion coefficient of chloride in the soil.

The above work suggests that the part played by diffusion in reducing uptake as a soil dries is important, but does not give a detailed quantitative explanation. As Clarke & Barley (1968) have pointed out, there is no obvious reason why uptake should be linearly related to the diffusion coefficient: the relationship will depend on a variety of factors.

An attempt to identify all the factors involved has been made by Dunham & Nye (1973, 1974) using the root plane technique. They measured the uptake of ^{36}Cl by a plane of onion roots from a block of wet, moist or dry soil, the remainder of the root system being in unlabelled soil of similar moisture content. The initial soil solution concentration, C_{1i}, was 1·0 mM for each soil moisture level. In the same experiments they measured the transpiration rate, the chloride concentration profiles near the root surfaces, and the soil moisture profiles. Since the chloride concentration at the root surfaces was measured they were able to measure the effect of water potential on the root-absorbing power. The root absorbing power was also measured in stirred solution culture at various chloride concentrations.

The profiles of soil moisture have been shown in Fig. 2.8. Figure 6.18 shows the profiles of chloride concentration at the different soil moistures. The solid lines are those predicted from a numerical solution of equation 1.6 with values of α selected to give the best fit at the root surface.

The main findings are shown in Table 6.4. There is a marked drop in v in the moist, and a further drop in the dry soil, in which the water potential at the root surface falls to -25 bar. The root surface concentration ratio C_1/C_{1i} at the end of the experiment approaches v/α; there is a marked drop in the dry and moist soils; but in the wet soil C_1/C_{1i} although reduced to less than 1·0 at $2\frac{1}{2}$d, rises to 1·4 at $6d$ because v becomes greater than α. In the wet soil the value of α agrees fairly well with that determined in solution culture. In the moist soil the value of α is somewhat lower and in the dry soil much lower than that expected from solution culture.

In the dry soil, decreased mass flow, reduced root absorbing power, and reduced chloride diffusivity contribute to the lowered uptake, but since they are not additive, it is meaningless to say by how much each contributes. The importance of the experiment is rather that using values of α determined from the experiment (in the wet soil from solution culture), but otherwise independently determined diffusion parameters, it was possible to predict satisfactorily the concentration profiles of a non-adsorbed ion from a diffusion-convection model.

The geometry of this system is linear, and the results do not represent in detail the effects to be expected around an isolated root—for example the moisture gradients around a single root would not be so steep. However, now that the plane model has been verified experimentally, one can have confidence in predictions about behaviour around an isolated root, based on a cylindrical model that uses the same ideas and parameters.

In a similar single experiment (Dunham & Nye 1976) uptake of adsorbed ions, phosphate, potassium, calcium and magnesium was measured from blocks of wet, moist and dry soil. The rest of the roots were not in soil, but in contact with moist filter paper. In interpreting the experimental data, we recall that in each case the exchangeable ion was measured. The concentration in pore solution, which was required for the estimate of α (column 5 of Table 6.4), was inferred from a desorption isotherm determined in a separate experiment.

The main effects are:

Phosphate

In the wet and moist soils the root surface concentration ratio C_l/C_{li} was low, showing there was considerable diffusive resistance. In the dry soil at a water potential of $-3\cdot3$ bar the root absorbing power was markedly lowered; this was the dominant effect in limiting uptake and the supply by diffusion and mass flow were well able to satisfy the greatly reduced plant demand.

Potassium

There was considerable diffusive resistance at all moisture levels and the root absorbing power remained high in the dry soil. The roots had been cultured in calcium nitrate and hence were low in both phosphorus and potassium. It is clear that the uptake of phosphate and potassium are differently affected by moisture stress in the root.

Table 6.4. Effect of soil water potential at the root surface on uptake, root absorbing power and surface concentration ratio of a plane of onion roots (after Dunham & Nye 1974, 1976).

Initial soil moisture θ	Duration days	Average water flux cm³ cm² s⁻¹ × 10⁶	C_l/C_{li} at root surface	α cm s⁻¹ × 10⁶ soil experiment	α cm s⁻¹ × 10⁶ stirred* solution	Final water potential at root surface bar	Uptake μmol measured	Uptake μmol predicted
Chloride
0·20	6·0	0·25	0·20	1·0	7·0	−25	0·84	0·51
0·27	6·0	0·63	0·20	2·0	7·0	−6·5	2·7	1·7
0·45	6·0	2·09	1·40	1·2	1·0	> −0·05	5·7	4·3
Phosphate
0·14	4·5	0·12	1·0	0·4		−3·3	0·004	0·003
0·18	4·5	0·32	0·06	10	> 100	−0·15	0·015	0·020
0·36	4·5	0·70	0·06	20	> 100	−0·10	0·043	0·050
Potassium
0·14	4·5	0·12	0·004	(100)		−3·3	5·7	1·3
0·18	4·5	0·32	0·077	10		−0·15	3·7	1·4
0·36	4·5	0·70	0·036	50		−0·10	7·3	4·5

* The root absorbing power for uptake from a free solution having the same concentration as the final concentration at the root plane surface.

The high uptake of potassium from the dry soil proves that the low uptake of phosphate was not due to lack of contact between the roots and the soil.

Calcium and Magnesium

Under all the conditions the relative solution concentrations of these ions was between 1·0 and 1·5, indicating that transport through the soil did not limit uptake.

6·5·3   The prediction of ion uptake

The measured uptakes of chloride, phosphate and potassium in these two root plane experiments, exemplified in Table 6.4, covered a range—in mols— of about 10,000 fold. The theoretical uptakes, based on numerical solutions of equation 1·6, use independently determined parameters, with the exception that the values of α were unknown in roots under moisture stress and were chosen to give the best-fit of the theoretical concentrations at the root surfaces to the measured ones. When the relative concentration at the root surface, C_l/C_{li}, is below 0·2 large changes in α make little difference to the spread of the depletion zones or the total uptake, so that it is not a serious 'adjustment factor' in these experiments. The theoretical uptakes agree with the measured uptakes of phosphate within a factor of 1·5, and systematically underestimate chloride and potassium uptakes by factors up to 2·0 and up to 4·5 respectively. These discrepancies could be accounted for by experimental errors, notably in measurement of desorption isotherms and impedance factors; and by the possibility that the diffusion coefficient should be increased to allow for dispersion. Considering the very wide range of uptakes, the agreement between measured and theoretical values indicates that for simplified soil–root systems the theory presented here is substantially correct. We have now to consider some of the complications that arise in more natural root–soil situations.

ROOT MODIFICATION OF THE RHIZOSPHERE

In discussing ways a root may affect the adjacent soil it is well to bear in mind the spread of the zone being exploited for a particular solute: if this is wide there may be no point in emphasizing effects close to the root; but if it is narrow predictions based on the behaviour of the bulk soil may be wide of the mark. In a moist loam after 10 days a simple non-adsorbed solute moves about 1 cm, but a strongly adsorbed solute will move about 1 mm. In a dry soil the spread may be an order of magnitude less. The modifications to the soil in the rhizosphere may be physical, chemical or microbiological.

6.6   Physical Effects

Roots tend to follow pores and channels that are not much less in diameter than their own. If channels are much larger than the root there is the problem of contact between root and soil discussed below. If smaller, the roots do not grow freely unless some soil is displaced as the root advances. This may have the effect of compaction, and reorientation of soil at the root surface. For example, for winter wheat Low (1972) cites minimum pore sizes of 390–450 μm for primary seminal roots, 320–370 μm for primary laterals, 300–350 μm for secondary laterals, and 8–12 μm for root hairs. The soil density change adjacent to roots has been studied by Greacen. Wheat roots penetrating a uniform fine sand, only increased its density from 1·4 to 1·5 close to the root; (1968, private communication) and a pea radicle raised the density of a loam from 1·5 to 1·55 (Greacen *et al.* 1968). A pea radicle penetrating clay decreased the voids ratio from 1·19 to 1·10 within 1 mm of the surface (Cockroft *et al.* 1969). These changes in density are unlikely to have much effect on soil diffusion characteristcs (see Chapter 4).

A layer of clay oriented parallel to the root axis might well present a barrier to diffusion. To take an extreme case, diffusion of Na across the well-oriented flakes of montmorillonite was 300 times less than diffusion in the plane of the flakes (Mott 1967). However, in experiments already described in which single roots grow against a window in aggregated soil, orientation has not been observed; and in thin sections across cotton roots growing in soil Lund (1965) observed little orientation. Though he did note a slight packing effect close to the root, the soil structure within 1–10 mm of the root surface was typical of the whole soil. Photomicrographs of the tips of pea radicles penetrating uniform clay show slight reorientation adjacent to the surface (Cockroft *et al.* 1969). Where oriented clay near roots has been observed it is likely that it has been leached down channels of older roots after they have completed most of their life as absorbing organs.

In an aggregated soil diffusion may well differ within the aggregates from the bulk soil. For small mobile solutes there is no evidence yet that diffusion through aggregates is effectively delayed: e.g. there is no time dependence of the chloride diffusion coefficient (Rowell *et al.* 1967), and concentration–distance curves do not reveal a rapid diffusion through larger pores (see Chapter 4). Further, Green (1976) measured a chloride ion impedance factor $f_1 = 0·41$, similar to a bulk soil value, in saturated natural aggregates of porosity 0·38. Gunary (1963) perfused natural and artificial aggregates, about 4 mm diameter, with labelled phosphate solutions. Autoradiographs of sections showed the degree of penetration of ^{32}P increased with the concentration of the perfusing solution. When this was less than 0·1 ppm P only a thin surface layer had been labelled after 40 days. Though for immobile solutes there may be local differences on a microscale according to whether a portion of the root surface is adjacent to an intra-aggregate or an inter-aggregate region, use of a single diffusion or dispersion coefficient presupposes averaging over such microscale variations; and in practice longitudinal diffusion within the cortex will probably short-circuit irregularities. Also, high resolution radio-autographs (Bhat & Nye 1973) have not revealed significant irregularities in the phosphate depletion zones along roots. The fact that root hairs penetrate uncompacted crumbs contributes to this averaging process.

In drier soils only the intra-aggregate pore space contains appreciable water, and a diffusing solute has to pass through these aggregates—so increasing their significance.

Root-soil contact

Purely visual assessment in moist soil suggests that most of the root surface is in contact with solid particles or separated from them by no more than the normal range of soil pores, i.e. the interface is similar to an average cross-section through the soil. It may not be identical to it because the roots may not pass through dense aggregates and thus the proportion of their perimeter in contact with air exceeds the average soil air-filled porosity. Observations at the root laboratory, East Malling (Rogers 1939) show about 20–30% of main apple tree roots tend to follow worm holes. Contact with soil is about 40% in these cases. Such roots may later fill their own holes by secondary thickening—or may stay loose for years. Older roots may be loosened by sloughing off their cortex. In the remainder contact is good. Smaller roots about 1·5 mm diameter tend to compress the soil 1·5 mm on either side of them. The compression is less than 5%.

Lack of contact is reduced by the following effects: (a) Root hairs tend to proliferate in the humid air of large soil pores and channels. (b) Even though pores of diameter exceeding 10 μm are drained at -300 mbar water potential there is always a film of water on the root and root hair surfaces, linking them to the soil moisture continuum. (c) Roots tend to excrete a layer of mucilage up to 5 μm thick, but on average more like 1 μm thick, which will bridge small gaps (Jenny & Grossenbacher 1963; Greaves & Darbyshire 1972). The mucilage is probably a hydrophllic polyglucuronate and polygalacturonate polymer (Brams 1969) across which ions and uncharged solutes should be able to pass freely. Samtsevich (1968) describes on root caps of most agricultural crops in soil a colourless gel-like excretion of diameter 1·8–4·7 mm and length 2·2–14 times greater than the root caps that produced them; and Rogers (1939) has noted exudation drops

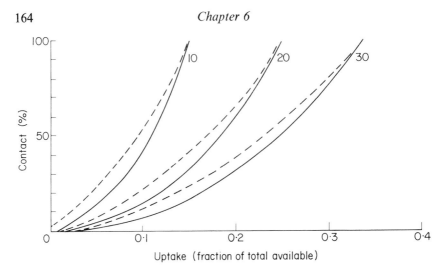

Fig. 6.19. The theoretical effect of % contact with soil around the root circumference on uptake of potassium after 10, 20 and 30 days (after Sanders 1971.)
————, cortical short;
– – – –, cortical block.

on root hairs. (d) Roots are usually linked to the soil by mycorrhizal
hyphae.

Roots may shrink at low water potentials and thus loose contact with
part of the wall of their channel, a possibility already discussed on p. 29.
Such an effect might occur in dry soil, though Dunham & Nye (1974) found
chloride ion was readily absorbed by hairless onion roots, sandwiched
between blocks of soil under gentle pressure, even though the water potential
of the soil at the interface was −25 bars. The extent to which lack of contact
should reduce the rate of uptake has been examined theoretically by Sanders
(1971) using an electrical resistance–capacitance network analogue (p. 223).
An example of his findings is shown in Fig. 6.19. The effect of poor contact
is increased if the root cortex does not transmit ions freely around its
circumference (cortical 'block' compared with cortical 'short-circuit'). If, as
is most likely, the cortex acts as a short-circuit, reduction of contact by 50%
leads to reduction in rate of uptake by about 25%.

6.7   Chemical effects

The root may alter simple predictions of the diffusion of solutes by:-release
of hydrogen or bicarbonate ions, evolution of CO_2 from respiration, creation
of changes in concentration of other ions and solutes that may affect the
ion of interest, and excretion of organic substances.

6.7.1 Excretion of hydrogen and bicarbonate ions

As explained in the historical introduction in Chapter 1, it has for long been supposed that roots excrete H^+ ions at their surfaces and render nutrients available, or exchange them for cations. It is only recently that this view has been challenged as a general rule; though it has for long been known from work in nutrient culture solution that the pH tended to rise if the source of nitrogen was NO_3^-, but to fall if it was NH_4^+. In 1960, Walker concluded from the data then available that plants absorbed more anions than cations, and would be expected to excrete more OH^- or HCO_3^- ions than H^+ ions into the soil. Later work has confirmed his predictions. In 1964, Cunningham surveyed the cation and anion contents of 62 common plant species grown in soil and showed that if the nitrogen was absorbed as NO_3^- then as a rule plants absorbed more anions than cations. Median compositions were cations 250 me/100 g (oven dry) shoots, anions 360 me/100 g oven dry shoots, i.e. an excess of 110 me of anions/100 g. There were wide variations and a few species absorbed more cations than anions. In order to maintain electrical neutrality across the root surfaces, there must in general be an excretion of HCO_3^- rather than H^+ ion and the soil near roots should become more alkaline instead of more acid. A neglected paper of Metzger (1928) had shown that the concentration of bicarbonate ion in soil sampled near the roots of a number of crops was 10–20% greater than at a distance. This effect must occur whatever means the plant adopts internally to control its pH (see Kirkby 1969 for a discussion of this problem).

6.7.2 Evolution of CO_2

In addition to excretion of bicarbonate ion, carbon dioxide is liberated at the root surface by respiration; and this has—usually wrongly—been assumed to have a local acidifying effect. The carbon dioxide respired by roots in the production of 100 g of shoots may be deduced from Parker's (1924) observation that roughly the same weight of CO_2 is produced as dry matter of shoots, i.e. 2.26 mol CO_2 per 100 g dry shoots. Thus a working median figure for the ratio

$$\frac{\text{mol } CO_2 \text{ respired}}{\text{mol } HCO_3^- \text{ excreted}} = \frac{2\cdot26}{0\cdot11} \simeq 20.$$

However, under aerobic conditions CO_2 diffuses rapidly away from the root through the air-filled pore space, so its effect is widely diffused over the whole soil. On the other hand, the bicarbonate ion is confined to the soil solution where its mobility in some 10,000 times lower than gaseous carbon dioxide; or it splits to carbon dioxide and hydroxyl ion, and the latter then behaves as bicarbonate. It reacts with acidic groups on the soil adjacent to the root and accordingly raises the soil pH close to the surface.

There is now plenty of experimental evidence for such changes. Riley & Barber (1969) grew soybean for 3 weeks in a silt loam soil, carefully picked out the roots, and shook them gently. The 'rhizoplane' soil, which adhered to the roots (c. 0–2 mm from the surface of the axis) had a pH of 6·9 compared with 6·2 in the pot soil, and a corresponding increase in the concentration of bicarbonate ion in the soil solution. The 'rhizosphere' soil, which fell off the roots on shaking (c. 1–4 mm from the surface of the axis), showed only a slight increase in pH, suggesting the effect was restricted to a narrow band round the root. At a lower level of NO_3 in the soil the increase in pH was also lower.

pH profiles near onion root planes were observed by Bagshaw *et al.* (1972). The pH fell from 6·0 to 5·6 at the surface because the roots absorbed potassium rapidly. When the soil was treated with $CaNO_3$ to reduce potassium and increase NO_3 uptake the pH rose towards the root from 6·4 to 6·8.

As an example of the consequences of such pH changes, Riley & Barber (1971) and Miller *et al.* (1970) have noted that supply of ammonium rather than nitrate ion increases phosphorus uptake from neutral soils. Absorption of the ammonium ion tends to lower the pH in the rhizosphere, and in the soils studied there was a corresponding increase in the concentration of phosphate in the soil solution.

In plants that fix a large proportion of their nitrogen symbiotically it will cross the root nodule—soil interface as uncharged N_2. It is to be expected that in such plants more cations than anions will cross the root soil boundary, and consequently hydrogen ions will be excreted to maintain the charge balance.

Clearly the pH change at the root surface depends upon the rate of release of bicarbonate or hydrogen ion, the soil pH buffer power and other factors. A quantitative model has been developed by Ramzan (1971).

6.7.3   Concentration changes of other ions

Change in pH and associated bicarbonate ion concentration is only one example of the disturbance near the root which may influence diffusion of other ions. In general we are dealing with multiple ion diffusion and an accurate solution has not been attempted. We may however discuss some guiding principles.

It is well to consider first the unadsorbed anions, since as pointed out in Chapter 3 their concentration controls the overall solution strength. Anions such as nitrate and chloride move to the root from a distance. They are accompanied by cations, hence the appropriate diffusion coefficient approximates to that of the calcium salt since calcium is the dominant cation in normal soil. Accurate knowledge of the diffusion coefficient of an ion requires a knowledge of the concentrations and concentration gradients of all the other ions present (see Nye 1966a equation (10)). One approach is by a method of successive approximation. For the first approximation over the first time step each ion is assumed to move independently of the remainder. Excess charge carried across the root surface is calculated and a compensating export of H^+ or HCO_3^- assumed. For the next approximation the concentrations of each ion is recalculated using the accurate expression for their diffusion coefficients in which the concentration and concentration gradients of the other ions obtained from the first approximation are used. If there are many ions the computation is tedious, and a more approximate approach based on physicochemical insight into the problem will indicate whether accurate solutions are justified.

For the cations the calculation of b and hence D is simplified if the overall solution strength may be regarded as constant. Then for an adsorbed cation present as a minority component e.g. K^+ the value of b_{K^+} is obtained from the buffer curve of $(K)/(Ca + Mg)^{\frac{1}{2}}$ in which (Ca plus Mg) may be taken as constant. The concentration gradient of the total anion concentration in solution is often negligible. We have seen that soil solution concentrations are usually greater than $10^{-3}M$ in agricultural topsoils, and they are less than $10^{-3}M$ only in very poor natural soils. The total anion inflow rarely exceeds 10^{-11} mol cm^{-1} s^{-1} in a fertile soil. These values lead to a value of $\alpha a \ll 10^{-6}$ cm^2 s^{-1} for the anions as a whole, though individual anions like nitrate may well have much higher values. At the same time we have seen that $v_a a$ lies in the range $10^{-8} - 10^{-7}$ cm^2 s^{-1}. Since all the anions have approximately the same diffusion coefficient in solution ($D_1 \simeq 10^{-5}$ cm^2 s^{-1}) they may for this purpose be treated as a common species. In moist soil Db ($\simeq D_1\theta f_1$) is c. 10^{-6} cm^2 s^{-1}. Hence both $\alpha a/Db$ and $v_a a/Db$ will be much less than 1. Hence, whether there is a fall in concentration to the root surface or a rise it will be shallow (Nye & Marriott 1969, Figs 2 and 3 where vr is 10^{-7}). An example of this behaviour is to be found in Brewster & Tinker's (1970) study of ion uptake by leeks from soil. In dry soils the change in concentration may be greater: though both αa and $v_a a$ are reduced, the product θf_1 is reduced more. If there is a significant gradient in the overall concentration of the soil solution then a numerical solution must be used. This involves calculating the solution strength at each time step and then using the appropriate value of b—adjusting (Ca plus Mg).

Another complication arises when, due to transpiration, the soil moisture level changes during the uptake period. If the hydraulic conductivity of the soils is sufficiently high compared with the rate of water inflow, there will not be appreciable water gradients to the roots, i.e. the soil will dry out uniformly, see Chapter 2. In these circumstances a numerical solution is readily achieved by adjusting θ and dependent parameters accordingly before calculating the ion concentration at each time step. Examples are provided by Dunham and Nye (1974, 1976).

6.7.4 Release of non-diffusible nutrients

It has been explained in Chapter 3 that lowering of the concentration of potassium in the soil solution often leads to the slow release of potassium not exchangeable to ammonium acetate. The rate of release tends to obey first-order kinetics. The very marked reduction in concentration of potassium near root surfaces should stimulate release there, though this effect has, as yet, only been shown with exchange resins. Kauffman & Bouldin (1967) measured concentration gradients in soil near H-saturated cation exchange resins. When the exchangeable potassium at the soil–resin boundary was lowered from 100 to 80 μg K per g soil, the resin absorbed 14 μg of non-exchangeable K in addition to 54 μg of exchangeable K in 74 h.

6.7.5 Organic exudates

It has been known for many years that organic compounds are exuded from healthy, undamaged plant roots (see Rovira (1962) for a review of early work). It is normally considered that 'exudates' are soluble, low molecular weight compounds; but other materials are also provided by the root. The mucilage excreted by the root has been referred to earlier, and there is a steady supply of decomposed and abraded root cap, dead root hairs and epidermal cells. It is by no means obvious whether this is included in published measurements of exudate; but where this is recovered from a solid growth medium, and especially when the total recovery in the soil of ^{14}C, from ^{14}C labelled plants is measured, then it is likely that much of the 'exudate' is in these solid forms. In a non-sterile environment, where there is a continuous turnover of organic materials, the distinction is particularly difficult to make; and it is likely that the higher values reported for solution culture also include substantial amounts of material which was initially part of the plant structure. The presence of these materials is a major characteristic of the rhizosphere, and we are particularly interested in two of the consequences: their direct effect on the chemistry of the soil around the root, and the development of a large microorganic population which utilizes these compounds as substrate. It is difficult to separate these aspects completely, since the microorganisms will alter both the composition and quantity of material in the rhizosphere by their metabolism, but in the first instance we consider what effects the exudate alone could have. A third possibility, that exudates affect neighbouring plant roots, will be discussed in Chapter 7.

The importance of these effects will depend upon the amounts of material exuded, but it is extraordinarily difficult to decide how large this is. Results have been reported in many different forms, which makes comparison difficult , but for our purpose the most useful way is undoubtedly as a fraction of the total dry matter production of the plant. Reported results cover a very wide range, from 0·1% (Riviere 1960), though a range of values up to 25% quoted by Krasilnikov (1958). Much of this variation may result from changes in growing conditions, since Virtanen & Tornaien (1940) showed that stress can greatly increase exudation. This has been confirmed repeatedly, and Hale *et al.* (1971), and Rovira & Davey (1974) in their reviews of exudation list the factors affecting it as plant species, plant age, plant nutrition, light, temperature, soil moisture, microorganisms and supporting medium.

The last two are particularly interesting, since microorganisms utilize exudates, and also appear to encourage their production. The effect of rooting medium is important because of the number of investigations made in solution culture; later work indicates that exudation is several times greater in a solid medium such as sand or soil (Boulter *et al.* 1966; Barber & Gunn 1974). Hale *et al.* (1973) found several hundred times as much amino-N exuded in soil as in solution, though the ratio may be overestimated due to mechanical damage to roots and root hairs. It is of considerable interest that Shay & Hale (1973) found that four times as much sugar was exuded (220 μg/plant) in a solution of 10 mM Ca as in 25 mM Ca, since these are comparable to the calcium levels often found in the soil solution.

It is obvious from this that no single value can define the quantity of total 'exudate', but it is important to decide on some reasonable figure so that exudate effects can be discussed. Rovira (1969) reviewed the literature and suggested that 0·4% of 'total synthesized carbon' was exuded; if the latter includes root respired carbon dioxide, it corresponds to perhaps 0·6% of plant dry weight. Bowen & Rovira (1973) also concluded that 1–2% of total root carbon was released, which corresponds well with the earlier figure; but higher values have also been quoted, e.g. Shamoot *et al.* (1968) found that the total recovery of ^{14}C compounds from soil corresponded to 11·2% of plant dry weight for fescue and 5·6% for lucerne though these figures include root hair and fine root material. Barber & Gunn (1974) found a comparable value for barley, again growing in soil, of 3·7% of total plant dry weight; and Martin & Barber (1976) reported that about 20% of plant dry matter was lost into non-sterile soil, excluding CO_2 from root respiration. For purposes of discussions, we choose 2% and 20% of total plant dry weight as possible figures for exudate, excluding respired carbon dioxide.

The composition of the material is equally uncertain. A very long list of compounds have been detected (Rovira 1969; Rovira & Davey 1974), the main groups being organic acids, sugars and amino acids. Most work has been with seedling plants, hence it may be important that Smith (1970) found that mature trees released much larger quantities of organic acids per g of roots than seedlings, and in different ratios.

Effect of exudates on nutrient uptake

At present there is no direct and unambiguous evidence that exudates affect mineral nutrient uptake. It has however been suggested that they can alter the chemical environment of the rhizosphere in ways beneficial to the plant; and as pointed out on p. 147, there are occasional serious discrepancies between diffusion theory prediction and measured nutrient uptake, which suggests that all alternative or additional mechanisms should be carefully considered. The main possibilities are that organic solutes displace phosphate ions from adsorption sites on the soil, or form soluble chelates with metal trace elements, and so increase the soil solution concentration of these elements. Such organic compounds could be direct exudates, or bacterial products resulting from the metabolism of exudates.

The most likely compounds to cause phosphate desorption are the poly-carboxylic or hydroxy-carboxylic acids (e.g. citric acid), which are known to sorb competitively with phosphate (Nagarajah *et al.* 1970). This process has not been considered quantitatively in terms of the total amount of complex-forming exudate, the amount of phosphate which could be desorbed, and the fate of the latter. Analyses of exudate (Riviere 1959; Smith 1969) suggest that polybasic acids usually form much less than half of the total (Table 6.5). Microbial action may produce more, but this would involve a loss of total weight, so it seems that the amount of such acids produced may be less— probably very much less—than 1 % of plant weight. Tinker & Sanders (1975) have suggested that such acids, because they must be strongly sorbed to displace phosphate (Talibudeen 1957), will diffuse very slowly out into the soil from the root, and the situation will probably approximate to a narrow cylinder in which most adsorption sites are occupied, with very little movement beyond this. Long root hairs may be particularly effective in delivering exudate away from the root axis (Bhat *et al.* 1976). There is little information

Table 6.5. Material released from tree roots, in mg g^{-1} dry root weight in 10 days (after Smith 1969).

Compound	*Pinus* species Monterey (*radiata*)	Sugar (*lambertiana*)	Black locust *Robinia pseudoacacia*
Acetic acid	31	34	15·4
Glycolic acid	1·6		
Malonic acid		1	1·4
Oxalic acid	56	21	1·4
Succinic acid		0·3	1·4
Carbohydrates	3·6	3·5	2
Amino acids and amides	41	31	10
Root length (cm mg^{-1})	1·1	0·6	1·2

on the sorption of organic acids which is relevant to this situation, but iso-therms determined by Nagarajah (1969) implied that kaolinite sorbed about 10 μmol g^{-1} and sesquioxides 40–60 μmol g^{-1} of citrate before the solution concentration increased appreciably. If we take 10 μmol g^{-1} as the minimum requirement for whole soil this gives some 1 mg citrate per g of soil. If, for the sake of argument, this displaces 0·2 mg P per g of soil, 1 g of citrate from a 100 g dry weight plant could in total displace 200 mg P, which is near to the phosphorus required by the plant. However, this assumes that all phosphorus displaced reaches and enters the root, whereas it can equally well diffuse away from the root and be re-adsorbed, though the presence of dense root hairs would make uptake of such displaced phosphate ions more likely. In brief, it is possible to show that useful quantities of phosphorus could be absorbed in this way, but only by a series of assumptions which in total appear unlikely, if the lower figure for exudate quantity is correct. If the higher value of 20 % is used, and any significant fraction appears as chelating acids, then the process becomes likely.

Chelation effects

The possibility of metal chelate formation by exudates is not proved, though the amino-acids in particular would tend to complex copper very strongly. There is much evidence that organic chelating compounds can hold trace element metals in solution and thus accelerate their transport to the root, where they may be absorbed as such, or dissociate so that the metal ion alone is absorbed (Hill-Cottingham & Lloyd-Jones 1965). Lindsay (1974) reviewed the subject and showed clearly that the equilibrium level of, for example, ionic iron in the soil solution would be insufficient for plant require-ments except at very low pH values. The effects of added synthetic iron and zinc chelates on plant uptake and growth were also in good agreement with their known stability relationships. Increased organic matter in the soil solution has been found to correlate with the total copper and zinc in the soil solution of a series of calcareous soils (Hodgson *et al.* 1966). It therefore seems extremely likely that the increased quantities of soluble organic matter in the rhizosphere will cause increased soluble metal chelates there, and that the latter will aid plant nutrition, but we are not aware of any direct work on this yet. However, decreases in manganese uptake have been related to the presence of Mn-oxidizing bacteria in the rhizosphere forming MnO_2, rather than to the direct action of exudates.

To predict theoretically the uptake of the metal ion by roots one needs to know the relative proportions of metal and metal-complex in the soil solution and the absorbing power of the root for each form. The fate of the complexing agent at the root absorbing sites is also important: if the metal complex dissociates when the metal is absorbed free complexing agent may accumulate at the root surface. Hodgson (1968, 1969) has explored these possibilities theoretically, using iron and copper chelates as examples, and has identified some of the parameters that need to be determined. His model predicts that when a steady state is attained the concentration of dissociated iron falls sharply towards the root, but because of its low overall concentration little iron should move in this form. On the other hand the concentration of the chelated iron falls gradually towards the root, but since its overall concentration is large most of the iron moves in this form. The liberated chelate may accumulate near the root unless other ions such as calcium form complexes with it—a process that depends upon the pH. If the free chelate accumulates it may limit further uptake by competing with the iron absorbing sites in the root. Clearly a great deal of experimental work is needed to verify these stimulating predictions.

6.8   Microbiological effects

All normal roots, in natural conditions, support a large population of microorganisms on their external surfaces (the rhizoplane) and in a thin sheath of soil immediately adjacent to the root surface (the rhizosphere) (Clark 1949; Clark & Paul 1970; Rovira & McDougall 1956). The species composition of the latter does not differ greatly from that of the general soil population, but their numbers are very much larger. The relative increase in microorganism numbers is expressed by the $R : S$ ratio, R and S being respectively the numbers per gram of soil taken from the rhizosphere (soil adhering to the root) and the bulk soil.

The method of counting can greatly influence the number of microorganisms found, and it is necessarily imprecise. In this terminology, the

Table 6.6. Microorganism populations at different distances from the root, for lupin seedlings by plate count (after Papavizas & Davey 1961).

Distance from root (mm)	Microorganisms (1000's per g oven-dried soil)			Fungi (per g oven-dried soil)		
	Bacteria	Strepto-mycetes	Fungi	*Aspergillus ustus*	*Cylindro-carpon radicicola*	*Paecilo-myces marquandii*
0*	159,000	46,700	355	5,650	4,940	9,000
0–3	49,000	15,500	176	3,360	0	2,800
3–6	38,000	11,400	170	2,920	0	1,600
9–12	37,400	11,800	130	2,880	0	1,500
15–18	34,170	10,100	117	2,270	0	0
80†	27,300	9,100	91	1,000	0	0

*   Rhizoplane.
†   Control soil.

rhizosphere usually includes the rhizoplane, or root surface proper. The numbers obtained by the usual dilution and plate count technique are themselves suspect, since direct counting techniques always show larger populations and smaller $R : S$ ratios (Trolldenier 1967). Louw & Webley (1959) found that the ratios of direct count to plate count results varied with time and with the distance of the sample from the root, being within the range of 1·2–11·3. The absolute values quoted in the literature are therefore uncertain, but a typical set is in Table 6.6. Fungi may not increase to the same extent as bacteria, but both symbiotic (Harley 1969) and non-symbiotic (Parkinson 1967) fungi are extremely frequent. Pathogens which cause specific diseases by invading the roots are excluded from this discussion, which is largely confined to the effects of the organisms on the nutrient supply to apparently normal plants.

Such effects might occur by:

(a)　Changes in the morphology or properties of the root system or of individual roots;

(b)　Changes in the phase equilibria of soil nutrients, such that they become more easily absorbed by plants, and/or more readily transported to the roots (e.g. pH or redox changes, or complex formation);

(c)　Changes in chemical composition of the soil, with similar results (e.g. mineralization of organic matter, or irreversible decomposition of soil minerals);

(d)　Symbiotic processes in which nutrients are transferred directly to the plant from other organisms;

(e)　Blocking of root surfaces, or competition by microorganisms.

Theories are much more numerous than agreed facts in this field, and we are obliged to state possibilities rather than firm conclusions in many cases.

6.8.1　Symbiotic bacteria

The nitrogen fixing bacteria which live in symbiosis with the *Leguminosae* and other plants and inhabit their roots are enormously important, but have little direct relevance to our subject, since there is usually no difficulty in transporting nitrogen gas to these bacterial nodules in soils where they are active. However, if much of the nitrogen requirement of the plant is supplied by this means instead of as nitrate or ammonium ions, the cation–anion uptake ratio into the roots will obviously be affected, which may have implications for other rhizosphere processes (see p. 165).

6.8.2　Non-symbiotic bacteria

It may be argued that if bacteria obtain their carbon requirements from the plant, and their presence facilitiates nutrient uptake, then an element of symbiosis is present. In this discussion we simply mean non-symbiotic to indicate that there is no direct physical link link between the plant and the microorganisms. A very long list of bacteria have been shown to develop large populations in the rhizosphere (Parkinson 1967; Gams 1967); but no clear conclusions have flowed from this work, and we do not distinguish between them here, except to note that motile Pseudomonads and Myco-bacteria, and in general, species which can reproduce rapidly on soluble substrates, appear frequently. The boundaries of the rhizosphere are in-definite, though Clark (1949) and Rovira & McDougall (1956) suggest dimensions of up to 1 cm (Table 6.6). This excess population can be at least $10^8 - 10^9$ bacteria per cm^3 of soil near the root. In the absence of certain evidence, we consider the possible effects of such populations on higher plants.

(i)   *Effects of bacteria on root uptake from solution culture*

It has been shown that the presence of microorganisms can alter the morph-ology of roots and root hairs (Bowen & Rovira 1961) (Fig. 6.20), probably by the production of growth hormones (Brown 1972), and some differences in nutrient uptake may result simply from this, though the information in published work does not allow this point to be decided. Alternatively, there

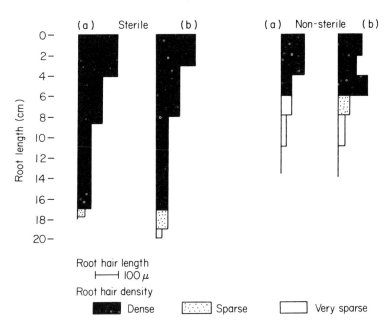

FIG. 6.20. Effects of rhizosphere microorganisms on root-hair growth and de-velopment of subterranean clover (a) sand (b) agar (after Bowen & Rovira 1961).

can be more direct effects of bacteria: in particular, they may compete for nutrient at low solution concentrations, or physically block access to root surfaces. The last suggestion seems less likely in view of recent findings (Rovira *et al.* 1974) that only about 7–13% of root surfaces in grassland species were covered by bacteria and fungi.

Direct competition is the likeliest explanation of results such as those of Barber (1969), in which the phosphate uptakes of sterile exceeded non-sterile plants at low concentrations. The largest fraction of phosphorus was transported to the shoot in sterile barley plants, and again the difference was largest at the lowest concentration. It is difficult to know how many of the rhizoplane bacteria were lost from the root, and how many remained on it and were therefore included in the analysis. Bowen & Rovira (1966) and Rovira & Bowen (1966), however, found that non-sterile tomatoes and clover both absorbed most phosphorus, and transported the largest fraction to the shoot. Barber & Rovira (1975) have recently reported the same effect in barley (Table 6.7). It is generally found that non-sterile plants show the largest incorporation of phosphorus into phospho-lipids and nucleic acids (see Rovira & Davey 1974). In all these experiments the uptake period was only a fraction of the root's life. At present it seems agreed that bacteria can have considerable short-term effects on phosphate uptake, but their origin and extent are still uncertain.

Table 6.7. Effect of microorganisms on the uptake of phosphate by barley seedlings of different ages, from 0·005 mM potassium phosphate (after Barber & Rovira 1975).

Age (days)	6		8		12	
Condition of plants	Sterile	Non-sterile	Sterile	Non-sterile	Sterile	Non-sterile
Roots						
Dry weight (mg)	6·0	5·5	7·5	6·7	10·4	8·8
Phosphate absorbed (pmole mg^{-1} wt)	559	1045	631	1004	584	908
Shoots						
Dry weight (mg)	13·2	15·1	17·5	19·6	27·0	26·9
Phosphate absorbed (pmole mg^{-1} dry wt)	19	40	21	31	15	12
Total phosphate absorbed (nmole)	3·64	6·34	5·06	7·23	6·51	8·33
Percentage in shoot	7·0	9·1	7·3	8·3	6·2	3·8

(*ii*) *Effects of rhizosphere bacteria in soil*

The most obvious possibility is that bacteria compete with plants for nutrients present in low concentrations, and this was the explanation of Benians &

Table 6.8. Effect of partial soil sterilization with 0·8 Mrad γ-radiation on growth of grass and clover in a brown earth (after Sparling 1976).

Soil treatment and plant species	Shoot DW (mg)	Root DW (mg)	Shoot P %
Limed soil			
Trifolium repens			
Control non-irradiated	12·3 a	4·36 a	0·288
Control & phosphate	406·5 b	153·5 b	0·317
Irradiated	7·8 a	3·72 a	0·111
Irradiated & phosphate	418·7 b	154·1 b	0·540 b
Unlimed soil			
Cynosurus cristatus			
Control non-irradiated	3·99 b	2·57 b	0·125
25 % irradiated soil	5·20 b	5·94 c	0·189
50 % irradiated soil	10·41 c	11·85 d	0·271
75 % irradiated soil	20·89 d	21·49 e	0·371 b
100 % irradiated soil	52·14 e	47·05 f	0·350 b

Figures in columns do not differ significantly when followed by the same letter (standard error of the difference, $p = 0.05$).

Barber (1974) for an improvement in growth and phosphorus content caused by γ-ray sterilization of soil. However, similar effects have been found by Sparling (Table 6.8) in grasses, though not in clovers, for only partial sterilization. There are two methods of considering this hypothesis: if a plant exudes only 2 % of its dry weight into the soil, and half of this is permanently converted to microbial tissue, which contains 1 % P, this represents 0·01 % P in the plant, which is trivial. Higher concentrations of phosphorus in bacteria seem unlikely in a phosphorus-competitive environment. Alternatively, we can say that if the rhizosphere, up to 2 mm from the root, contains 10^8 bacteria cm^{-3} more than in the bulk soil, with the large dry weight per cell of 10^{-12} g, and a phosphorus concentration of 1 %, then the content per cm of root is 4×10^{-9} mol P, which corresponds to about half a day's inflow to a root (see Chapter 5). Again this seems small to account for a large competitive effect. However, Halstead & McKercher (1975) suggest that some 25 μg g^{-1} of phosphorus in soil could be associated with microorganisms. If this is the average and the $R : S$ ratio is above 10, it suggests that the microbial phosphorus accumulation could be greater than appears from the above argument. This does however call for a very large total biomass of rhizosphere microorganisms, which could be acceptable if we take the highest value for exudate quantity. The mechanism whereby sterilization produces growth increases is therefore still uncertain.

It has been suggested that rhizosphere bacteria cause greater mineralization of soil organic matter (Thompson & Black 1970). There is some evidence that denitrification is rapid in the rhizosphere (Woldendorp 1963), but that the rate of nitrification is lower due to exudates reducing the numbers of nitrifying bacteria there (see Rovira & Davey 1974). There have been many suggestions that non-symbiotic bacteria in the rhizosphere can fix nitrogen (see Jurgensen & Davey 1970), and there is recent evidence for this with tropical grasses (Dobereiner & Day 1975).

Microorganisms can break down highly-insoluble sources of potassium, such as mica (Boyle *et al.* 1967), but it seems unlikely that this property is of much significance in the rhizosphere in normal soils.

However, there is a great literature on the 'solubilization' of soil phosphorus by bacteria, and in the USSR many claims were at one time made for bacterial fertilizer called 'phosphobacterin' (*Bacillus megatherium*) (Smith *et al.* 1962; Brown 1974). These claims seem to imply a dissolution of insoluble phosphate minerals or mineralization of soil organic phosphorus, but no detailed analysis of the requirements of such a process seems to have been given. A direct test of the 'bacterial solubilization' hypothesis is to grow plants in sterile and non-sterile media and compare their phosphorus uptake. Gerretsen (1948) made a classic experiment of this type, and found considerable growth increases in presence of bacteria. It is clear that bacteria may cause other effects than those on phosphorus uptake (e.g. the manganese deficiency found by Gerretsen), and Katznelson (1965) stated that he was not able to repeat Gerretsen's results. No fully substantiated claims seem to have been made, and the case cannot be regarded as proven. Swaby (1962) reviewed the use of 'bacterial fertilizers', and found little or no convincing evidence for their effectiveness, though work on them continues to appear. The essential proof of this theory would be the finding of clear and consistent differences in the L value* (Larsen 1967) of plants grown in sterile and non-sterile soil during fairly short periods, and that the non-sterile soil L value differed from the E-value. This does not appear to have been done.

In the absence of firm evidence, we discuss possible mechanisms whereby bacteria could alter the rate at which phosphorus compounds arrive at the plant root. Firstly, if the bacteria produce more organic acids in the rhizosphere than would be present in the root exudate alone, the mechanism described on p. 169 would be enhanced. Secondly, it is certain that many bacteria are able to utilize very poorly soluble phosphate sources, for their own growth. These bacteria have been isolated from soil, the rhizosphere (Swaby & Shurber 1958) and from seed coats of crop plants (Katznelson *et al.* 1962). Thus, Greaves & Webley (1969) have shown that bacteria can hydrolyse organic phosphates in culture, and others (see Parkinson 1967) have grown bacteria in media where the only phosphorus supply was hydroxyapatite. The latter type of experiment does not, of course, prove that bacteria

* The L value is the exchangeable soil P as determined by the specific activity of a plant grown in soil uniformly mixed with ^{32}P.

dissolve phosphates by some exceptional process; it could result from a large pH change in the medium, or chelation of calcium, or simply indicate that bacteria, with their large specific surface, can absorb effectively from the low solution concentrations (about 10^{-6} M P, depending upon pH) with which these compounds are in equilibrium. These processes might thus be of no significance in soil where pH changes are buffered, and calcium in large supply. It is in any case not sufficient to prove that bacteria can absorb phosphorus from sources which do not support plant growth; it must also be shown that this bacterial phosphorus can then be transferred to the plant. If bacterial phosphorus is released as phosphate ion, it will simply re-enter the phosphate equilibrium in the soil and be re-adsorbed. One direct experiment (Tinker, unpublished) showed that microorganisms in an artificial medium with an equilibrium phosphorus concentration of about 10^{-8} M could absorb ^{32}P freely, but that plants growing in it absorbed very little. It is therefore necessary to consider whether the bacteria can transfer nutrients to the root surface and make them available there. Griffin & Quail (1968) showed that the motile bacterium *Pseudomonas aeruginosa* moved little in soil at water potentials below -1 bar, but at higher potentials movements of 2 cm in 24 h were found, which is comparable to the root mean square displacement of a non-adsorbed diffusing ion in soil. It implies that a small amount of phosphate (and other nutrients), possibly of the order of several μg g^{-1} soil may have a diffusion coefficient similar to that in true solution. The bacterial phosphate is not, of course, immediately available to plants, though it could be released by root surface phosphatases (Woolhouse 1969); its concentration is not buffered in the same way as inorganic phosphate ion; and only some bacteria are motile; but it may be worth consideration.

Thirdly, much phosphorus may be incorporated in dead bacterial cells. Hannapel *et al.* (1964) have shown that up to 1 μg g^{-1} organic phosphorus can be in colloidal solution as bacterial debris in soil supplied with organic matter (as in the rhizosphere). The diffusion coefficient of such material may be low because of its large molecular weight (pp. 74, 79), but mass flow will operate upon it (Tinker & Sanders 1975). Given a transpiration ratio of 500, and a concentration of 1 μg g^{-1} solution, this mechanism could supply up to 0·05% P in dry matter, which is a useful contribution. Again, this mechanism depends upon the transported bacterial debris being broken down by or near the root, possibly by the root surface phosphatases.

6.8.3 Symbiotic fungi

(i) *Ectomycorrhizas*

The ectotrophic mycorrhizal fungi very clearly alter the root, even in its physical form. A mass of fungal hyphae form a sheath around the root

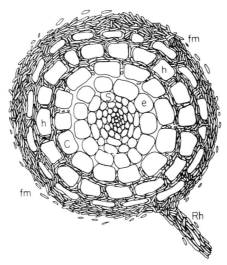

FIG. 6.21. Transverse section of a pine ectomycorrhiza showing (C) root cortex; (e) endodermis; (fm) fungal mantle; (h) Hartig net; (Rh) Rhizomorph—root-like hyphal structure; (S) stele (after Sanders & Tinker 1975b).

proper with hyphae extending from this into the soil, and into the inter-cellular spaces in the cortex (the Hartig net) (Fig. 6.21). The entire fine root system of plants may be mycorrhizal, so it is obvious that the mycorrhizal roots are actively absorbing, and plants which become infected in this way frequently grow very much better than if they do not have mycorrhizae. Thus there are a large number of reports of the failure of tree species intro-duced into new areas until mycorrhizal infection could be obtained (Mikola 1973), and quite startling increases in growth and phosphorus percentage have been obtained after infection (Table 6.9) (Harley 1969 p. 107). It is now generally accepted that mycorrhizae do aid nutrient absorption by the plant, particularly of phosphorus. There is little benefit from the inoculation of

Table 6.9. Weights and nutrient concentrations of naturally mycorrhizal and non-mycorrhizal seedlings of *Pinus virginiana* (after McComb 1938).

	Mycorrhizal seedlings	Non-mycorrhizal seedlings
Fresh-weight mg	1230	592
Dry-weight, mg	323	152
Nitrogen % dry-wt	1·78	1·88
Phosphorus % dry-wt	0·185	0·097
Potassium % dry-wt	0·66	0·62

plants growing in nutrient-rich soils, and infection is often much more difficult in such situations. Most ions, including ammonium, are absorbed by mycorrhizas, but possibly not nitrate (Carrodus 1967). It has often been claimed that nitrogen is fixed by mycorrhizal roots, but the evidence appears inconclusive and conflicting (Harley 1969, p. 6).

A very large amount of work has been done on the uptake of phosphate by beech mycorrhizal roots by Harley and collaborators (Harley 1969, p. 105 *et seq.*). This has established that the uptake pattern is broadly similar to, but more rapid than, that of uninfected roots, and that the uptake is greatly diminished by metabolic inhibitors such as azide. The results were consistent with metabolic uptake into the fungus of the sheath, where a large fraction was accumulated, followed by a transfer of phosphorus into the root proper. Simultaneously phosphorus moved by diffusion through the sheath, and was absorbed metabolically in the root cortex. It was deduced that the uptake rate by the sheath would be so large that virtually no phosphorus would reach the root itself by diffusion in the very dilute solutions expected in soils. There is evidence that sheath tissue can absorb much more phosphate per unit weight than root tissue, whether or not they are excised and separately exposed to phosphate solution (Table 6.10). It is not clear whether this is a true physiological effect—the fungal sheath possibly having a higher potential α than any root—or a consequence of the larger surface of the hyphae of the excised mycorrhizal sheath than of the true root. There is a clear analogy here with the diffusion and active uptake of nutrients in the root cortex discussed in Chapter 5 (p. 117).

Table 6.10. Phosphate uptake by intact and sliced (0·5–1·5 mm) beech mycorrhizas, in μg P per 100 mg of dry mycorrhizal tissue (after Harley & McCready 1952).

| P conc. mM | 0·004 | | 0·032 | | 0·32 | | 3·23 | | 32·3 | |
Treatment	Intact	Sliced	Intact	Sliced	Intact	Sliced	Intact	Sliced	Intact	Sliced
P uptake	4·67	7·85	17·7	35·6	63·0	124·5	190	239	407	460
P in sheath	4·35	6·20	16·1	27·5	56·0	94·6	146	141	258	267
P in host	0·32	1·65	1·5	8·1	7·0	29·9	44	98	149	193
Per cent P in sheath	93·0	79·0	91·5	77·0	89·0	76·0	77·0	59·0	64·5	58·0

It seems certain that ectotrophic mycorrhizas can increase plant nutrient uptake rates, especially that of phosphate. The mechanisms which may account for this are:

(a) A change in root shape geometry or distribution, leading to a larger absorbing area;

(b) A change in the time for which a root remains active in uptake;

(c)   An increase in maximum uptake rate per unit surface, i.e. sink strength;

(d)   A shortening of the effective transport path from soil to root;

(e)   An ability to use other forms of soil phosphate;

(f)   Storage of phosphate in the sheath.

(a)   This certainly does occur. Mycorrhizal short roots are frequently more branched than uninfected roots, and they appear capable of developing for a longer time. However, there is no unambiguous data on the length and surface of roots formed by comparable infected and uninfected plants, so the point can only be made qualitatively. It is not clear that there is any difference in root area per unit total plant weight. The mycorrhizal roots are themselves thicker than uninfected ones, and the fungal sheath adds another 40–80 μm to the diameter. However, the treatment on p. 117, 132 *et seq.* shows how ineffective an increase in diameter of a root is when it approaches the condition of an infinitely strong sink, which is likely to occur in the phosphate-poor soils where mycorrhizae are advantageous.

(b)   There is strong evidence (Harley 1969, Chapter 3) that infection prolongs the effective life of short roots. Work by Bowen (1973) indicates that the sheath absorbs for longer than the root alone.

(c)   There is evidence for increased absorption rates caused by mycorrhizal infection from solution culture experiments (Harley 1969, Chapter 6). However, short-term uptake experiments in solution do not indicate that the same happens in long-term growth in soil, and differences in α are unlikely to be important in the latter situation (see pp. 117, 132 *et seq.*).

(d)   Melin & Nillson (1958) have shown that nutrients can be absorbed and transported through the external hyphae into the root. The mycorrhizal effect may arise by absorption by hyphae at a distance from the root. The importance of this fungal network is not so much in its additional surface area, which is often stressed, as in the fact that this surface is distributed well away from the root proper, and is thus in relatively undepleted soil. The great difficulty here is that very little is known about the distribution and total length of the dispersed hyphae, as distinct from the sheath, of ectotrophic mycorrhizas.

(e)   Fungal hyphae may be able to absorb phosphate in other forms than those immediately available to roots. These could be inorganic forms such as iron or calcium phosphates, where phosphorus could be brought into solution by chelating substances or pH changes. Secondly, organic phosphorus compounds could be utilized directly, or hydrolyzed by phosphatase enzymes on the hyphal surfaces, though the relatively small amount of material in *direct* contact with any surface in the soil must still be borne in mind.

(f)   The high concentration of phosphorus in the sheath has suggested (Harley 1969) that it may act as a storage organ for this element. The small weight of the sheath compared with the whole plant implies, however, that the total amount stored cannot be very large in relation to total plant demand.

So far it has not been possible to decide the relative importance of these mechanisms in the ectotrophic mycorrhizas.

(*ii*)   *Endomycorrhizas*

Most of the foregoing refers to ectomycorrhizas, since a major research effort has been in progress on these for longer than on the endotrophic

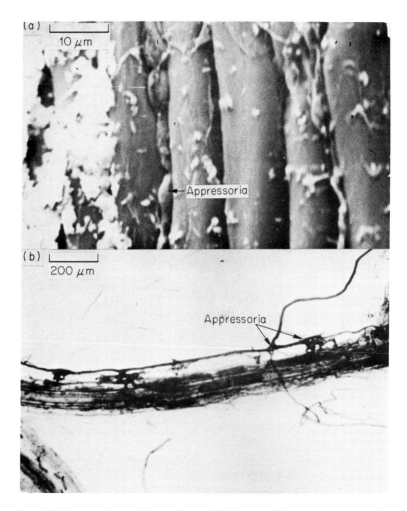

FIG. 6.22. Entry point and mycelium of V–A mycorrhizas; note appressoria (photograph Sparling & Sanders). (a) scanning electron micrograph of onion root surface, (b) stained preparation of onion root.

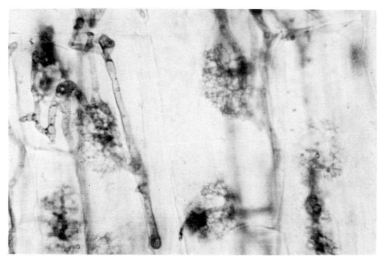

Fig. 6.23. Internal hyphae of V–A mycorrhizas in onion roots, with arbuscules in cells (photograph Sanders).

variety. The latter form no definite sheath around the root, and are consequently much less easily observed and investigated. The most important group is known as the vesicular–arbuscular endomycorrhizas. The fungi are poorly defined, and there is still uncertainty about their final taxonomy (Gerdemann & Trappe 1975), though most workers (until recently) have placed them in the *Endogonaceae*. Their importance springs from the very wide range of hosts which they associate with, which includes the majority of crop plants apart from the *Cruciferae* and the *Chenopodiaceae*; in fact it is becoming clear that by far the greater part of the worlds' vegetation is infected by them (Gerdemann 1968; Mosse 1973; Tinker 1975a).

The form of the association varies in detail with the host, but the essential stages are, in almost all cases, (a) initial infection, from a spore or external hyphae, with development at an entry point (Fig. 6.22); (b) development of external and intercellular mycelium; (c) formation of many-branched structures called arbuscules (Fig. 6.23) and spherical vesicles within the root. The final situation is therefore of well-developed mycelium spreading between and within the cells of the root cortex, and connected to arbuscules inside the cells, and external hyphae ramifying up to several cm from the root surface in soil.

Many reports of yield increases in host plants following infection have been made (see Mosse 1973). In the great majority of these, the phosphorus nutrition of the host has been improved, with increases in phosphorus percentage and uptake. Other elements have also been implicated, but much more rarely. A similar range of explanations to those suggested for the ectomycorrhizas can be advanced here, except that no mechanism involving the sheath is possible. Attention is therefore concentrated upon the external mycelium, and this has allowed a clearer understanding of the process than has been possible for ectomycorrhizas.

Sanders & Tinker (1971, 1973) and Hayman & Mosse (1972) have measured the L value in soils with infected and uninfected onions and shown both to be similar. It is therefore unlikely that mycorrhizal plants are able to utilize any sources of phosphate not available to uninfected roots. Sanders & Tinker (*loc. cit*) also measured the phosphorus inflows into infected and uninfected onions (Table 6.11) and found them to differ by a factor of four.

Table 6.11. Growth, phosphorus concentration and inflow for onions with and without mycorrhizas at $7\frac{1}{2}$ weeks. Inflow averaged over period $4-7\frac{1}{2}$ weeks.

	Mycorrhizal	Control
Total plant fresh weight, g	13·05	2·22
P, % in dry matter	0·31	0·14
Mean inflow, moles cm^{-1} s^{-1}	17×10^{-14}	$3·6 \times 10^{-14}$
Root length infected, %	50	—
'Hyphal inflow', moles cm^{-1} s^{-1}	$17·6 \times 10^{-14}$	

They (Sanders *et al.* 1975; Tinker 1975b) estimated the maximum inflow to a root of infinite strength, and found it similar to the measured inflow of the uninfected roots. These results show that no explanation based on morphological changes in the root itself can be satisfactory, and the only acceptable mechanism is uptake by external hyphae and translocation into the internal mycelium, with later transfer into the host. If it is assumed that the inflow to the root cylinder itself is the same in both cases—which may be an overestimate for the infected roots, since the hyphae will absorb phosphate in competition with them—then the extra inflow must come via the hyphae. Since only half the roots were infected on average, the 'hyphal inflow' would be about 18×10^{-14} mol cm^{-1} s^{-1} P, or about six times the rate to the single root. In effect, the hyphae act as a bypass for phosphate through the depleted zone immediately around the root, in exactly the same way as root hairs (p. 145). There are certain differences in detail; hyphae may persist for longer, or be less regular in distribution than root hairs, and may extend much further out into the soil. The competition between individual hyphae may therefore be less than between adjacent root hairs (Tinker 1975b). Sanders & Tinker (1973) found, by careful sieving of soil, about 80 cm of hyphae per cm of infected roots in onions, which is comparable with the total lengths of root hairs in other species, and Baylis (1970, 1975) has suggested that mycorrhiza formation is only beneficial to a host plant when it has few or no root hairs. Sanders found that the amount of mycelium per cm of root was remarkably constant during the increase in size of onions by 30-fold, in a growth chamber environment (Fig. 6.24). Application of equations for diffusion of phosphate to roots (p. 132) to the hyphae indicates that the absorbed uptake via the hyphae could easily be accounted for if half the hyphal length were active; but it is of course less reasonable to assume the narrow hyphae are traversing a random cross section of aggregated soil, than it was for the much thicker roots.

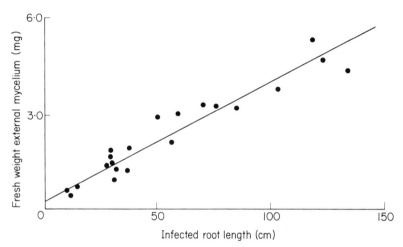

FIG. 6.24. Mycelium weight with different lengths of mycorrhizal onion roots (after Tinker 1975a).

The interpretation given here regards the hyphae as an alternative transport pathway to the soil, and a good deal of interest has been shown in their uptake rates and mechanism for translocation of phosphorus. Internal fluxes have been estimated at 3.8×10^{-8} mol cm^{-2} s^{-1} by indirect (Sanders & Tinker 1973), and up to at least 1×10^{-9} mol cm^{-2} s^{-1} by direct means (Pearson & Tinker 1975). These values probably imply transport by protoplasmic streaming, possibly aided by the formation of polyphosphate granules in the fungal vacuoles (see Tinker 1975b).

The effect of environmental changes on the mycorrhizas is not established, except in very general terms. It is likely that increased levels of phosphate (see Mosse 1973) and nitrogen (see Hayman 1975) in the soil diminish the amount of infection, and it is also lessened at low temperatures (Furlan & Fortin 1973) and very low light intensities (Hayman 1974). Precise experimentation in controlled environments on the effect of soil properties is still awaited. It is however evident that no studies on plant nutrition in soil, especially if it involves phosphorus, should be undertaken unless the presence and effects of mycorrhizas are defined and understood. In particular, it has been shown (Daft & Nicolson 1966) that plants appear able to utilize low-solubility sources of phosphate more efficiently when they are infected with mycorrhizas; this could arise simply from the smaller mean distance between phosphate mineral particles and absorbing surfaces when plants are mycor-rhizal, but more complex mechanisms cannot be excluded (Tinker 1975b). A typical set of results is in Table 6.12.

Table 6.12. Dry matter yield, and phosphorus concentration of mycorrhizal and non-mycorrhizal maize grown in sterilized soil with phosphate sources of different availability (after Murdoch *et al.* 1967).

Treatment	DM (g)	P (%)
Check	7·54	0·061
Check + inoculum	12·41	0·065
Tricalcium phosphate	8·24	0·067
Tricalcium phosphate + inoculum	23·10	0·088
Monocalcium phosphate	40·43	0·075
Monocalcium phosphate + inoculum	42·03	0·080

6.9   Conclusion

This chapter has outlined a consistent and quantitative model of the supply of solutes to a root by physicochemical processes, and for most ions this appears to be correct and sufficient. In the last part of the chapter alternative, less well-defined mechanisms involving biological modification of the rhizo-sphere have been discussed. Some of these are beyond question, such as pH modification and the activities of mycorrhizal fungi. Other processes are uncertain and speculative, such as the effects of root exudate and the non-symbiotic bacteria which live on it. The difficulty of accounting for phosphate uptake by rape plants (Bhat *et al.* 1976; Brewster *et al.* 1976) by known mechanisms suggests that some of the speculative processes, or others yet unknown, are operating, particularly on ions at very low concentrations in the soil solution.

7

THE MINERAL NUTRITION OF SINGLE
PLANTS IN SOIL

INTRODUCTION

Earlier chapters have dealt with the various components of soil and plant systems. Here we attempt to synthesize them into a unified treatment of a single whole plant, growing in soil. Certain parts are still simplified—for example we tend to consider homogeneous soils, and constant growing conditions. This is necessary to deal in any depth with so complicated a system, but we do not believe that it leads to omissions of any essential principles. Some of the complications are dealt with in Chapter 8. Though solute uptake is our main subject, it is essential to discuss this in relation to the growth of the plant as a whole, since this provides the developing sink for the absorbed solutes and the expanding root system through which they enter. Much attention has recently been given to the geometry of the root system, which greatly affects its efficiency.

7.1 Root system morphology and measurement

The individual root has been described in Chapter 5. Here we discuss the form, description and measurement of root systems belonging to single whole plants. Two quantitative methods of description are possible. The classical one is plant based, and defines all roots in terms of their branching system. This method can be used in soil or water culture, and is certainly easier in the latter case. The second method is soil based, and describes the system in terms of root densities (length per unit volume of soil, L_V) at particular points in the soil, in relation to the plant base. It does not consider the relationships of different roots within the system, and it is obviously useless in solution culture. Each method answers certain purposes, and the choice must depend upon the aim of the work.

7.1.1 Root system descriptions

Many authors have recorded the morphology of root systems in pictorial form (Fig. 7.1), usually with expenditure of considerable effort. (Weaver 1926; Muller 1946; Kutschera 1960; Rogers & Head 1969). It is difficult to see that great benefit has arisen from this labour. Certain generalizations can be made (Cannon 1949), for example, that particular species have tap roots, with regular first order laterals which penetrate horizontally (Fig. 7.1(a)), whereas others put forth a heavy crop of main axes near the soil surface (Fig. 7.1(b)). However one suspects that often the detail of the distribution depends far more upon soil properties than the genetic composition of the plant (see p. 195) (Fig. 7.2). The methods used in such studies (see Schuurmann & Goedewaagen 1971 for details) may often fail to detect the very finest roots, nor can these often be reproduced in drawings. These fine roots may be a large fraction of the total length, and hence of greatest interest for nutrient uptake studies (Fig. 7.3).

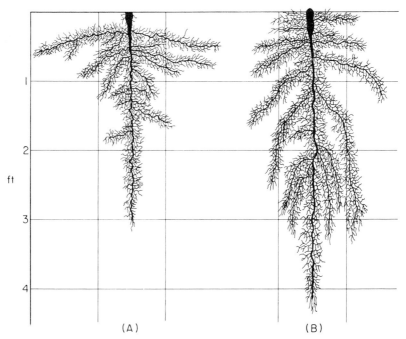

(A)   (B)

Fɪɢ. 7.1 (a). Root systems of sugar beets about 3 months old, A without, B with irrigation. Note tap root (after Weaver 1926).

Work such as that of May *et al.* (1965) and Hackett (1968) in describing the distribution of the different branching orders of a root system in exact quantitative terms is valuable, in that it gives a precise idea of which orders contribute the majority of the root length or weight (Fig. 7.3) (Table 7.1). However, there is not necessarily any greater uniformity of root properties (except for radius) within a single order than within the system as a whole.

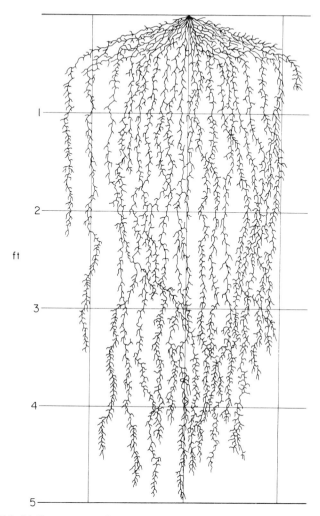

Fɪɢ. 7.1 (b). Root system of rye grown in dry sandy soil (after Weaver 1926).

Unless we can be certain that, for example, first order laterals behave in a recognizably and regularly different way to second order laterals, there is little advantage in discussing them in these terms. The same point applies to the distinction between seminal roots and nodal roots in cereals.

Occasionally root systems are highly distinctive in form. Thus many of the *Proteaceae* of the Western Australian flora have roots which are in the main uniform, long and poorly branched, but carry a few short zones with intense branching of short laterals of up to a centimetre in length (Jeffrey

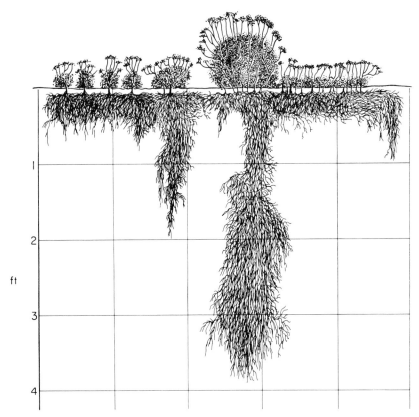

Fɪɢ. 7.2. One-year-old guayule plants. A highly-developed hard layer of soil impeded rooting; note different root habit where the layer was penetrated by tap root (after Muller 1946).

1967; Lamont 1972). These are called 'proteoid' roots; both the factors which cause them to form, and any possible value to the plant, are still obscure, though there have often been suggestions that they are particularly effective for uptake on their native phosphate-poor soils.

7.1.2   Root system measurement

To avoid extensive overlap and repetition, techniques used in field crop root sampling have also been included here, though they may be more relevant to the topics of Chapter 8.

The most useful data for soil-grown plants are those specifying the quantity and dimensions of roots at each point around the plant base. Information of this type can be obtained in a variety of ways (Table 7.2).

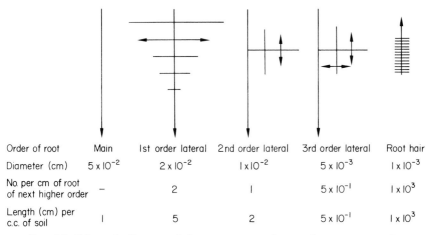

Order of root	Main	1st order lateral	2nd order lateral	3rd order lateral	Root hair
Diameter (cm)	5×10^{-2}	2×10^{-2}	1×10^{-2}	5×10^{-3}	1×10^{-3}
No. per cm of root of next higher order	–	2	1	5×10^{-1}	1×10^{3}
Length (cm) per c.c. of soil	1	5	2	5×10^{-1}	1×10^{3}

FIG. 7.3. Schematic diagram of the components of a cereal root system (after Barley 1970).

Direct sampling

Cores can be taken, and the roots washed out, weighed or the length measured. Schuurmann & Goedewaagen (1971) and Troughton (1957) have discussed the best procedures, but root washing is notoriously tedious and sometimes inaccurate, particularly with heavy soils. The measurement of weight is easy in principle, but surface area is difficult to measure directly; Carley & Watson (1966) have suggested that it can be obtained by measuring calcium nitrate retained on the root surfaces after immersing the roots in a solution of this salt, and then centrifuging them. In view of the amount of liquid that could be held between fine roots, the method appears open to error. It has been used by Raper & Barber (1970a).

The measurement of root length has been greatly speeded by Newman's (1966) 'intercept' method of counting intersections of the roots, laid out in a suitable way, with random lines. The number of intersects per unit length of random line N, is then related very simply to the mean length of root per unit area, L, by

$$L = \frac{\pi N}{2}.$$

The method has been modified by Marsh (1971), and the same principle is employed in length measurements with the Quantimet Image Analysing Computer (Baldwin *et al.* 1971) and in automatic apparatus for measurement of root length (Rowse & Phillips 1974). Other methods have included direct measurement with a ruler or a map-measuring wheel, and measurement of mean cross-section with a travelling microscope and consequent derivation

Table 7.1. Lengths (cm) of different orders of root in two varieties of barley, per plant and per root, as affected by potassium and phosphorus deficiencies (after Hackett 1968.)

	Complete nutrient		Potassium deficient		Phosphorus deficient		LSD
	Maris Badger	Proctor	Maris Badger	Proctor	Maris Badger	Proctor	(P = 0·05)
Total length	5720 (3·75)	3820 (3·58)	1410 (3·14)	1890 (3·27)	2650 (3·42)	3060 (3·48)	(0·13)
Axes							
Seminal	418	309	339	297	343	254	80
Early nodal	141	111	74	61	80	82	40
Late nodal	243	205	129	121	86	99	54
Primary laterals							
Seminal	3050	1920	630	1420	1640	1760	450
Early nodal	1230 (3·08)	1000 (2·99)	190 (2·12)	30 (0·79)	220 (2·25)	170 (2·20)	(0·71)
Late nodal	330 (2·40)	230 (2·26)	20 (1·08)	20 (0·98)	110 (1·94)	80 (1·92)	(0·51)
Secondary laterals							
Seminal	290	70	0	0	160	570	356
Mean length							
Axes							
Seminal	64·0	54·7	57·7	48·6	58·5	43·0	8·9
Early nodal	56·1	52·4	50·2	44·7	49·0	50·0	9·3
Late nodal	14·7	14·6	13·9	17·6	10·1	12·5	3·7
Primary laterals							
Seminal	3·50	3·55	1·14	2·16	2·17	2·36	0·87
Early nodal	3·47	3·68	1·02	0·72	1·24	1·21	0·68
Late nodal	1·18	1·21	0·23	0·23	0·90	0·69	0·37

transformed data in parentheses.

of root length from the fresh weight. This assumes that the specific gravity of roots is 1, which appears to be correct for young non-woody roots. Comparisons of direct and intercept methods have shown slight bias, with 8–10% over-estimate (Reicosky *et al.* 1970) or under-estimate (Brewster & Tinker 1970). The cause of this is not wholly clear, but is possibly connected with non-random distribution of roots in relation to the outer boundary of the area in which they are distributed.

Isotopic labelling of plants

Rennie and co-workers (Racz *et al.* 1964) tested the possibility of injecting ^{32}P into plants, and allowing it to be translocated into the roots. Core samples were then taken and assayed for ^{32}P, which indicated the quantity of roots present, but the figures obtained need calibrating against direct measurements of root length or weight. This principle has been developed by Russell and co-workers (Russell & Ellis 1968). They used ^{86}Rb, which is easily translocated throughout the root system and is a γ emitter, allowing the radiation to be measured with minimum self absorption. A modified technique has been published by Gerwitz & Page (1973). A comparison of various methods of determining root distribution has been reported (Newbould *et al.* 1969), and shows good correlation between results of direct counting of weighing, and of the ^{86}Rb method.

Ellis & Barnes (1973) also compared the ^{86}Rb plant injection method (see p. 244) with direct measurement of root weight and found good agreement with barley at 9 weeks from sowing. The discrepancies at crop maturity were ascribed to the death of part of the root system. Similar comparisons were also made for rye grass; in this case the ^{86}Rb method underestimated the roots near the surface.

Isotopic labelling of the soil

The converse method to the above is to inject labelled nutrient ions into the soil, and detect them in the plant. These methods are not strictly capable of measuring root quantity, but only nutrient uptake from defined areas, for which purpose they are very suitable. However, conclusions about root distribution have often been drawn from such data; Hall *et al.* (1953) originally suggested that ^{32}P would function as a measure of root activity at a given point. Good examples of this method are in work by Nye & Foster (1961), Russell & Newbould (1969), Boggie *et al.* (1958) and Lipps & Fox (1964), and it has been discussed by Danielson (1967). Soileau (1973) has used strontium as a tracer in this way, and Broeshart & Nethsinghe (1972) have shown that when used for trees, the variance between replicates was greatly reduced by double labelling with ^{32}P and ^{33}P. A minor development of this for field work has been to place the isotope in gelatine capsules, which dissolve in the soil solution (Bassett *et al.* 1970). One problem is that the labelled isotope (most usually ^{32}P) is diluted by nutrient already in the soil, and measurement of the isotope only gives the uptake from the labelled zone if this dilution is determined by estimation of the labile or exchangeable pool of nutrient. Other problems arise from uncertainty whether the injection causes a local change in root density, and—unless whole plants are assayed— whether the isotope is localized in particular parts of the plant. Russell & Newbould (1969) pointed out that phosphorus and strontium isotopes were absorbed in differing ratios from different horizons, which implies that certainly one, and possibly both, are not closely related to root quantity. Oberlander & Zeller (1964) found good correlation between ^{32}P uptake and measured weights of lucerne root, but this was in boxes filled with uniform soil. Any attempt to infer the amount of root from the measured uptake of nutrient in the labelled zone can only be made relative to other soil zones, and assumes that the uptake rate per unit amount of root is the same. Such conclusions are therefore to be interpreted with caution.

Root intersections on a soil face

An interesting principle—related to that used by Newman—was employed by Melhuish & Lang (1968, 1971). They cut out soil blocks containing roots, embedded the soil in resin, ground the surfaces and counted the cut root ends on these faces under the microscope. Three-dimensional geometry (Kendall & Moran 1963) gives a simple relationship between the number of root ends per unit surface, N, and mean length, L_V, per unit volume: $L_V = 2N$. This assumes roots to be randomly oriented in space. It is applicable to anisotropic root distribution, if root density is measured in three mutually perpendicular planes (Baldwin *et al.* 1971; Marriott 1972).

The idea is attractive, but the embedding and counting procedure too tedious for regular use. Baldwin *et al.* (1971) showed that, for onion, the roots could be labelled with ^{32}P, and the cut soil face placed against photographic film, the position of the roots being indicated by black spots when the film was developed. The use of ^{35}S and ^{33}P, and the possibility of distinguishing two root systems by separate labelling, was investigated by Baldwin & Tinker (1972). The method is successful for fairly coarse roots in the laboratory, and was successfully used in the field with onions (Baldwin *et al.* 1971), but the method needs testing with finer-rooted species before it can be generally acceptable.

Table 7.2 lists the available methods. Given careful sampling and washing techniques, there is little doubt that the first method is the most reliable, and gives most information on root dimensions. The second method does, however, yield the distribution of the roots in space (see pp. 195, 221) which is not otherwise obtainable.

Table 7.2. Summary of methods of root measurement, and the data obtainable from them.

General method	Data
Washing out from cores	Length, area or weight of roots in selected soil volume
Soil block or autoradiography method	Root length and distribution and orientation in space
^{32}P or ^{86}Rb injection into plant	Root volume in selected volumes
^{32}P injection into soil	Root 'activity' at specified point in the soil

It has been suggested (Pearson 1974) that the pattern of water extraction can be used to determine root distribution in the field, though the uncertainties concerning water transfer within the soil and evaporation from the surface are noted. The method is clearly most useful where the crop utilizes stored soil moisture rather than continuing rainfall during its growth; in a humid climate the pattern could become extremely confusing, and it can probably not be relied upon for much more than a general indication of rooting depth.

7.1.3 Root observation chambers and rhizotrons

A relatively simple way—given the original investment—of observing the growth and quantity of roots is to allow them to proliferate against a glass plate behind which the soil has been removed. A good example of this type of installation has been described by Rogers & Head (1969). More recent quantitative work has been reported by Atkinson (1973); and Taylor *et al.* (1970) used a similar system to show that maize produced three times as large a root density as tomatoes in similar growing conditions.

The weakness of the method is that it is essentially descriptive. Even for this type of use, it must be assumed that the conditions adjacent to the glass window are identical to those in the bulk soil. The diversion of roots which strike the glass, and then run along its surface, may give an exaggerated idea of the true root density in the bulk soil.

7.1.4 Root system patterns and orientation

Certain orientations are often favoured in the soil, due to geotropism, lamellar or columnar soil structure, or orientated fissures (see Fig. 7.2). This characteristic can be measured by orientating the planes of measurement in

Melhuish and Lang's method (block surfaces) and in Baldwin's method (photographic film surfaces) along different axes. It is in fact essential to do so if one direction is strongly preferred, since the formula for root length then breaks down, and a correction is necessary (Baldwin *et al.* 1971; Marriott 1972; Melhuish & Lang 1971).

Distribution of the cut root ends determines the 'pattern' of roots; this can be either contagious, random or regular ('over dispersed' or 'under dispersed' have rather ambiguous meanings, see Greig-Smith (1964)). Various mechanisms can be postulated which could cause departure from randomness; for example, a coarse-structured cloddy soil is likely to produce contagion or clumping, whereas any mechanism which causes roots to repel each other will tend to produce regularity. This concept is clear so long as it is applied to a small part of a plant's root volume, which can be treated as homogeneous. Its application is more complex where a considerable part of the root system is considered. On this scale, it is clear that the genetic mechanisms tending to separate a root system's major axes become important and a whole root system developing in an unrestricted way cannot be regarded as a wholly random assemblage of roots. The major changes in root density, as observed in moving outwards from a point below a single plant's stem, or downward under a crop, can also be regarded as aspects of pattern. These points are dealt with in more detail later.

7.2   Factors affecting root system form and distribution in soil

7.2.1   Genetic effects

The genetic constitution of a plant only expresses itself clearly in root distribution when it is growing in reasonably homogeneous and favourable growth media. If soil conditions are extreme—for example, if there is a plough pan—this may completely alter the root distribution from that which is expected (see Pearson 1974). Different species tend to have different root patterns, and often respond in different ways to environmental effects. Troughton & Whittington (1969) reviewed the whole topic, and listed the following root characteristics which may be genetically determined: root weight and size, pigmentation, nutrient uptake, exchange capacity and tolerance of heavy metals. Different varieties of a given species often have different amounts of root (Weihing 1935), though there are no clear reports of differences in distribution. Hackett (1968) (Table 7.1) found differences of up to 50% in root length of two barley vareties in solution culture, though the difference was reversed if the plants were potassium or phosphorus deficient. Raper & Barber (1970a) found considerable differences in the root weight, and up to 100% difference in the root surface area, of two soyabean varieties.

There is a clear possibility of breeding new varieties of crops with the correct amount of root for different environments, and it is of some interest to discover what effect breeding programmes have had on the root system. The short-stem cereal varieties could form a good subject, since less extension above ground might be matched by less root growth below. However, Welbank *et al.* (1974) found little difference between roots of normal and semi-dwarf wheat varieties.

7.2.2 Time and growth stage

A root system changes quite considerably during a plant's lifetime. The change in cotton roots, as seen in a root observation chamber (Fig. 7.4), shows that root density does not always increase with age regularly down the profile. Mitchell & Russell (1971) noted that soyabean laterals tended to grow out horizontally for 40–50 days, then turned downwards to considerable depths. They suggested three growth phases. (1) Vegetative shoot growth, downward tap root growth and horizontal lateral growth. (2) Flowering and pod setting on shoot, with root development down to about 75 cm, and the filling out of the spaces between the laterals in the topsoil with secondary and tertiary branching laterals. (3) Seed maturing on shoot, and deep penetrations of main laterals at the end of their horizontal spread. Beever & Woolhouse (1975) found with *Perilla frutescens* that root weight increased less rapidly

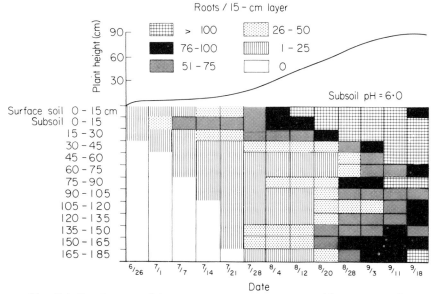

FIG. 7.4. Development of the cotton root pattern as measured by root count in a root observation chamber (after Pearson 1974).

after flowering commenced, but a simultaneous change in the branching pattern and mean diameter delayed the equivalent change in root length for 12 days. Welbank *et al.* (1974) found that the total root weight of barley decreased from about 75–90 days after sowing (Fig. 7.5), and that the root : shoot ratio decreased steadily throughout the growth period, from 0·28 to 0·064, in the plots receiving 100 kg ha^{-1} of nitrogen (Table 7.4).

7.2.3   Chemical effects

Effects of acidity

Soil acidity effects are due to hydrogen ion, aluminium, manganese and iron. The effect on root growth varies quite sharply with species or variety (Kerridge & Kronstad, 1968), arising from different aluminium tolerance. This may be physiological, but Foy *et al.* (1965) suggested that two wheat varieties differed in aluminium tolerance because of the different pH values induced around their roots. Aluminium toxicity is a particular problem in subsoils, because of the difficulty in introducing liming materials to that part of the profile (see Foy 1974); typical results for cotton and groundnuts grown in a subsoil are in Table 7.3. There is a very wide literature on the subject of acidity effects, including interactions with calcium and phosphorus availability, which are summarized by Pearson & Adams (1967), and specifically in relation to root growth by Foy (1974).

Table 7.3. Effect of pH and of aluminium concentration in the soil solution of a subsoil on the root growth of cotton and groundnuts (after Adams & Pearson 1970).

pH	Al in soil solution μM	Root extension in 48 h, mm Cotton	Groundnuts
5·0	33	50	85
5·5	0·4	97	81
6·4	0	149	95

Effects of nutrients

A general growth increase which results from applying plant nutrients will normally give a larger root system also. However, the shoot : root ratio almost invariably increases when nutrients are supplied, and it is possible that increased amounts of, for example, nitrogen could actually reduce root quantity at certain rates (Viets 1965). In the very comprehensive work by Welbank *et al.* (1974) nitrogen dressings increased the root weight of

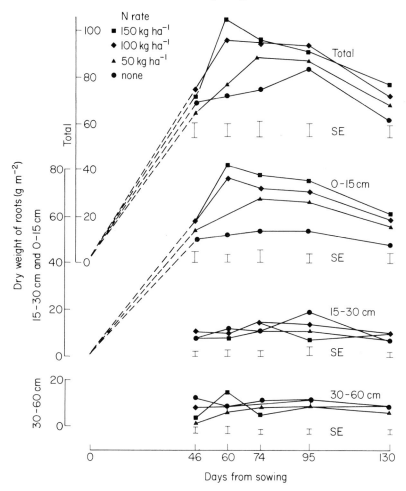

Fig. 7.5. Dry weights of root from different depths under a crop of barley receiving different dressings of nitrogen (after Welbank *et al.* 1974).

barley slightly but decreased the root : shoot ratio very sharply (Fig. 7.5; Table 7.4). In view of the common belief that phosphorus encourages root growth, it is interesting to note that there was no clear effect of either phosphorus or potassium fertilizers on root weight in this work. However, both nitrogen and potassium increased the root radius markedly, the extreme values for the specific length of dry root being 88 m g^{-1} with, and 226 m g^{-1} without these nutrients.

In addition to this general effect on the root system it has been known for a long time that locally-applied concentrations of nutrient ions promote branching of roots in these areas (Wiersum 1957). Much work on fertilizer

Table 7.4. Effect of fertilizer nitrogen on total root dry weight, as a fraction of crop dry weight (after Welbank *et al.* 1974).

Days from sowing	Fertilizer nitrogen (kg ha^{-1})				
	0	50	100	150	SE
46	0·417	0·362	0·280	0·300	0·010
60	0·208	0·195	0·171	0·169	0·014
74	0·188	0·151	0·124	0·120	0·010
95	0·133	0·101	0·079	0·081	0·011
130	0·081	0·080	0·064	0·067	0·003

Table 7.5. Increased growth and root demand coefficient in barley caused by local supply only of 50 μM phosphate solution (from Drew & Saker 1975b). Treatment 1, 50 μM phosphate to entire root system. Treatment 2, 50 μM phosphate to middle zone only, rest receiving no phosphate. Labelled phosphate supplied for 24 h to middle zones only.

Time since germination (days)	Treatment	Dry weight per plant (mg)		Relative growth rate of shoots (day^{-1})	Uptake of ^{32}P-labelled phosphate in 24 h by the middle zone (μmole)	
		Shoots	Roots (middle zone only)		per plant	per g root dry weight
5	1	21	0·24	0·76	0·018	74
	2	19	0·36	0·74	0·032	93
9	1	36	0·45	0·14	0·033	101
	2	31	0·91	0·12	0·281	343
12	1	74	2·4	0·24	0·224	97
	2	58	14	0·21	3·63	267
15	1	135	3·0	0·20	0·387	141
	2	93	20	0·16	6·50	253
21	1	340	17	0·15	2·74	163
	2	220	54	0·14	13·6	306
29	1	1390	7·9	0·18	1·15	110
	2	1130	160	0·20	38·3	240

placement has shown that roots eventually proliferate particularly strongly within the fertilized zones (Cooke 1954; Duncan & Ohlrogge 1958). A clear distinction must be made between the effect of increased supply of nutrients throughout the plant rooting volume, and to restricted parts of it. In the former case, root growth may be encouraged, but always less so than the shoot. If the nutrients are applied to a small part of the root zone, growth in that part is encouraged but the total mass of roots may change little or not at all, and root development is in effect diverted to the enriched zones (Miller 1974). Nitrogen and phosphorus (Hackett 1972; Drew & Saker 1975a,b) are most effective in this way; work with the latter (Table 7.5) shows that the local application of 50 μM phosphate in solution, compared to that concentration on the whole root system, caused a slight reduction in growth rate of barley plants, but also a rapid and intense growth of roots in the supplied zone until those had 20 times as much root as in the equivalent zone of the control plants. This increased amount of root, which also had a greater $\overline{\alpha a}$ and hence uptake rate, almost compensated for the limited length receiving phosphate.

The biochemical and physiological mechanisms which cause this proliferation of roots have not been determined, nor are there yet any precise comparisons of the effect in different species. The process is of obvious value to plants, and is probably a major reason why such a large proportion of most root systems is found in the topsoil.

Apart from the effects of nutrients on the gross morphology of the root system, the anatomy of the individual root may be changed (Schnappinger *et al.* 1969). The diameters of both the taproot and the xylem vessels in alfalfa were increased when phosphorus and potassium levels in sand culture were increased from 0 to 2 and 0 to 4 mM respectively.

Effects of organic materials

The possible effects on root growth of organic compounds produced by fungi, bacteria or by other roots have been discussed for a long time (see Chapters 5 and 6) in addition to recent work on ethylene (Smith & Robertson 1971). The lack of growth of some root systems in soils which have carried specific crops (soil sickness) has sometimes been attributed to an organic compound toxic roots. A fuller discussion of root interactions is in Chapter 8, since it is more appropriate to the subject of interplant competition.

Effects of salinity

Large concentrations of salts—especially ammonium nitrate—are damaging to roots, causing deformation of root cells immediately behind the tips, enlargement and deformation of the tips themselves, and eventually death of the root (Isensee *et al.* 1966). The effect of salinity has been reviewed often (Richards 1954; Bernstein & Hayward 1958). It varies greatly with plant and salt, for example sodium chloride is known to damage many fruit crops though there is no clear evidence whether it has specific effects on roots. In general, it is osmotic pressure which causes ill effects, and the type of salt causing the pressure may not be critical. Most plants adjust to osmotic pressure up to 10–15 bars by increasing the osmotic pressure in the root cells, although this is accompanied by temporarily reduced growth (Magistad 1941). Xerophytes will of course tolerate much more, up to at least 50 bars.

Measurements of direct effects on roots are scanty. Younis & Hatata (1971) pointed out that root extension was often more sensitive to salt concentration than was germination, and that fertilizer damage to emergence of crops may be due more to the former. Table 7.6 gives data from Tiessen & Carolus (1963) on the rate of extension of tomato transplant roots in a sandy loam soil. The effects varied widely with the elemental composition of the fertilizer salts used, but increased osmotic pressure was detrimental above 0·17 bars. The osmotic pressures were estimated on a 2 : 1 water : soil suspension, and could therefore be multiplied by 6–10 times in the soil.

Table 7.6. Effects of fertilizer salinity on root elongation of tomatoes transplanted into a sandy loam (after Tiessen & Carolus 1963).

| Fertilizer, g per 15 cm pot. | | | Osmotic pressure | Root elongation |
N	P	K	(bars)	in 2 days (cm)
0	0	0	0·11	0·39
0·2	0·4	0·3	0·17	0·85
0·4	0·8	0·6	0·18	0·65
0·8	1·6	1·2	0·25	0·21

7.2.4 Water content of soil

This subject has often been reviewed (see Danielson 1967). Root : shoot ratios of plants normally increase in drier soils; for example, Struik & Bray (1970) found that the percentage of the total plant weight below ground for maize varied between 33% in dry soil and 12·9% in moist soil. This is also clear in results (Troughton 1957) with various grasses (Fig. 7.6). These also showed that the root : shoot ratio increased quite sharply when the soil water exceeded 85% of pore space, when the root zone was possibly becoming anaerobic.

Despite the lower root : shoot ratio, it is usually observed that roots extend most rapidly in moist soil and less rapidly in dry areas (Danielson 1967). However, it is often not clear whether this is a direct effect of soil water potential, or of the greater mechanical strength of the drier soil. Eavis (1972) found no evidence of direct effect of water potential above −3·5 bars on the penetration of pea roots in sand, and others (see Drew &

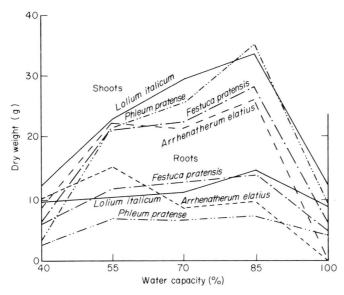

Fig. 7.6. Root and shoot weights of several grasses grown in soil at different moisture contents (after Troughton 1957).

Table 7.7. Root weight and length of tomatoes grown in soil at 3 moisture contents containing ^{32}P, when plants were supplied with adequate water by a split root system. Permanent wilting point: 13·7%; 1 bar suction: 22%. (after Thorup 1969).

Initial water content (%)	Root weight (mg)	Root length (cm)	P uptake (cpm)
4	300	20	37
12	880	90	408
22	1240	150	1189

Goss 1973) have found even less effect. The root cells appear able to adjust to low water potentials in the surrounding media with little effect on extension rate, and reported effects of different water contents are possibly due to secondary changes.

Nutrient uptake in dry soil will depend upon root extension, root properties and soil transport processes, and the separation of these effects is not easy (see Chapter 6.5). In experiments of Thorup (1969), tomato roots grew in soil at different moisture contents, while other roots were well supplied with water. Growth of roots in dry soil was reduced by a factor of 7 (Table 7.7), but uptake of ^{32}P was decreased by a factor of 30.

7.2.5   Mechanical resistance to roots

A general discussion of this important topic is given by Taylor (1974). Roots in solid media normally have to exert some pressure to compress or bodily remove particles which otherwise would prevent their extension, since the root tip diameter cannot be reduced significantly (Wiersum 1957). This external pressure arises from the expansion of the root tip cells, and the maximum which can be exerted has usually been found to be 9–13 bars. The turgor pressure of the cells is produced by the osmotic potential of the vacuoles, and is normally restrained by the cell wall. If the root presses against an external object, the pressure load is transferred to the latter. From this argument, it might appear that root extension rate would not be altered until the external medium was able to produce a counter-pressure greater than the turgor pressure, but this is not so. Drew & Goss (1973), in agreement with previous authors, found that roots growing in media composed of rigid spheres extended less rapidly when external pressures as low as fractions of a bar were applied (Fig. 7.7). The mechanism for this is unknown, but such low counter-pressures are easily produced in soil. Greacen & Oh (1972) have suggested that the rate of extension of any one root is proportional to the difference between the turgor pressure (assumed the same for all members of one root system) and the local yield pressure of the soil, but this seems unlikely in view of the curvilinear relation between extension rate and externally applied pressure in Fig. 7.7.

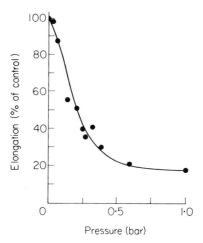

Fig. 7.7. Relationship between applied external pressure and relative elongation of barley roots growing in 1mm ballotini for 6 days (after Drew & Goss 1973).

The theory of the strength of soil under load is well developed (Barley & Greacen 1967), but it inevitably has to treat soil as a homogeneous body. This is acceptable when loads are applied to large areas, as by foundations or agricultural implements, but less so when the object applying the load is comparable in size to the individual soil aggregates.

If the soil is considered homogeneous, and the root as a smooth probe, the effects around the root (Fig. 7.8) may be divided into elastic (IV) and plastic deformations. The plastic changes may be divided into deformation

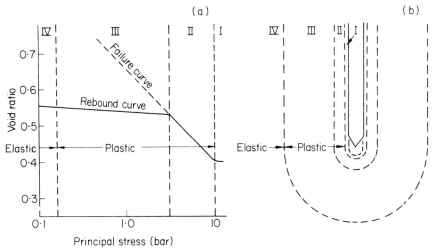

FIG. 7.8. Compression of soil around penetrometer or plant root tip, leading to mechanical resistance to root extension, and to changes in soil properties around root. (a) Compression curves, and (b) compression zones. See text for explanation (after Barley & Greacen 1967).

which is partly regained when this load is removed (III), deformation with failure (II), and complete failure with final compression to a minimum voids ratio (I); the different effects occur at different distances from the root surface. Large changes in voids ratio, and in θ, are therefore confined to zones I and II, which are of very limited radius. With this model they found good agreement with the force exerted on a probe in fine-textured uniform materials; the agreement was better when they rotated the probe to dissipate the frictional resistance on the surface of the probe. Some surface friction will occur with roots, since short lengths of root between the tip and the point of maximum extension is physically pushed through the soil, though mucilage from the root cap will minimize it. Direct comparisons of forces on extending roots and probes indicate that there is a fairly constant ratio between the two, but that root forces are always less, and never exceed the turgor pressure. Measurements with probes or rods thus give a good relative, but not absolute, indication of root elongation for any one species (Fig. 7.9).

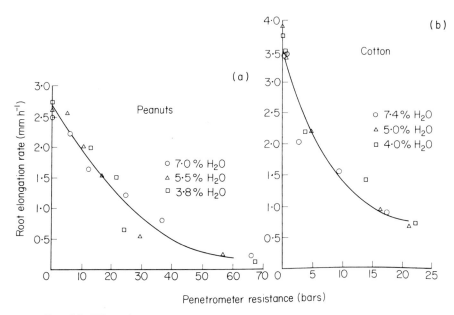

Fig. 7.9. Effect of penetrometer resistance and soil water content on (a) groundnut, and (b) cotton, root elongation rate (after Taylor & Ratliff 1969).

Pore size diminishes, and mechanical shear strength increases as soils are compacted, and the bulk density is often a useful guide to root impedance, a value of $1\cdot7$–$1\cdot8$ g cm^{-3} usually indicating little or no root penetration. The root then tends to thicken (which allows it to exert a larger force for the same turgor pressure), and may become distorted. There is also often additional branching and/or production of root hairs along the root; but there may be occasions when the hairs themselves are not able to penetrate a dense soil. Champion & Barley (1969) found that root hairs would only enter a uniform moist clay when the voids ratio was above $1\cdot0$ (i.e. a pore space of $0\cdot5$).

It proved difficult for a long time to separate the effects of mechanical strength alone from those of other factors often associated with it, such as water content or oxygen content, but the dominant effect appears to be soil strength (Taylor & Ratliff 1969; Eavis & Payne 1969). The net effect of varying soil strength on a developing root system is difficult to predict in detail. Roots will certainly tend to extend most rapidly in zones of low impedance, and the regular spiralling motion (nutation) of the extending root tip (see Rogers & Head 1969) allows the individual root to find the path of least resistance. Roots thus tend to follow cracks and channels in soil, and their extension may not be related to the bulk soil strength.

7.2.6 Oxygen concentration

A number of reviews deal with this problem (e.g. Greenwood 1969; Grable 1966; Luxmore *et al.* 1970). In soils water content, gas composition, pore size and soil strength frequently vary together, and the separation of their effects is difficult. Tackett & Pearson (1964) tested the effect of different O_2/CO_2 ratios with varying soil strength (expressed as soil density). The results (Fig. 7.10) show that there is a strong interaction between these factors.

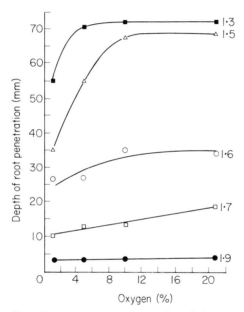

FIG. 7.10. The effect of oxygen concentration in the soil air and soil bulk density (numbers on curves) on depth of cotton root penetration into a compacted subsoil (after Tackett & Pearson 1964).

A common generalization is that root elongation is not affected until the gas phase contains less than 10% O_2, but root weight in some tree species has been reported to increase progressively with O_2 concentration right up to 20% (Bessel *et al.* 1967). Greenwood (1969) found that metabolism was only affected at very low oxygen pressure, and that the growth of seedling roots was not changed until the partial pressure of oxygen was below 0·05, while Huck (1970) using cotton and soyabean, reported reductions in growth rate to 50 and 10% of the original value when the oxygen partial pressure was reduced to respectively 0·03 and 0·01. The intimate connection between metabolism and growth was shown by the cessation of growth within 2–3 minutes of placing roots in pure nitrogen. It may be that the morphology of the roots themselves affects this question, and certainly oxygen can diffuse down from the shoot via the air-filled pores in the root if the medium external to the roots is oxygen-deficient (Greenwood 1969; Luxmoore & Stolzy 1972). This dual supply is clear from the work of Eavis *et al.* (1971), in which the oxygen pressures around root and shoot were independently varied (Fig. 7.11).

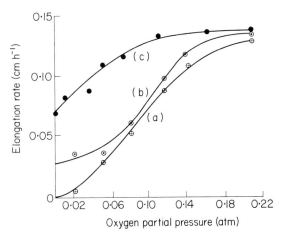

Fig. 7.11. Effect of varying oxygen partial pressure on pea seedling elongation rates when roots and shoots were treated separately. (a) The entire seedling was enclosed in the same partial pressure. (b) Cotyledon and shoots were at 0·21 atmosphere oxygen partial pressure, while roots were at different partial pressures. (c) Roots were in soil air with 0·21 atmosphere partial pressure oxygen while cotyledons and shoots were at different partial pressures (after Eavis *et al.* 1971).

7.2.7   Effect of temperature

Temperature affects all parts of plant growth, and the effects on roots are striking; recent reviews have been given by Cooper (1973) and Nielsen & Humphries (1966). The root : shoot ratio is largest at high and low temperatures, with a minimum in the range 20–25°C, even though root extension rate normally increases with temperature up to 25–30°C (Cooper 1973). Some species have very different extension rate maxima, e.g. pea roots extend most rapidly at 10°C. Generally roots are thicker and less branched at low temperatures, and cell size may be smaller (Nielsen & Humphries 1966), but the maximum number of laterals on a root system occurs at different temperatures depending upon species, from 20°C for maize to over 34°C for pine. The direction taken by roots is also temperature-sensitive, and Onder-donk & Ketcheson (1973) found the mean angle to the horizontal of maize roots was smallest (10°) at 17°C, but larger at both higher and lower temperatures. The maximum temperature in a daily cycle was most important.

The size and morphology of a root system is therefore strongly dependent upon temperature, and the uptake characteristics also change, due to both morphological and physiological changes (Chapter 5). The resistance to water uptake of unit amount of root certainly decreases with temperature, and this is affected not only by the temperature of the experiment, but also by the temperature at which the roots were grown (see Cooper 1973). Apple root growth in solution culture was greatly reduced at 35°C compared to lower temperatures (Gur & Shulman 1971), and this was associated with a sharp fall in the potassium content of the leaves. However, the clearest interaction of temperature and nutrition is probably for phosphorus (Power *et al.* 1963; Sutton 1969), which is absorbed much more slowly at low than at high temperatures. The origin of the effect is not clear; Sutton (1969) favoured physicochemical mechanisms in the soil, but it seems unlikely that this could account for the whole effect, and the remainder is probably due to a combination of slower root extension and lower α values.

7.2.8 Effects of defoliation, root cutting and shading

Root growth depends upon photosynthate from the shoots, and either ceases or is reduced after defoliation or cutting of grasses (for reviews see Troughton 1957; Milthorpe & Ivins 1966; Danielson 1967), the reduction being related to the frequency and intensity of defoliation. Successive defoliations may cause the death of part of the existing root system, and there is evidence (Soper 1958) that root hair numbers are greatly reduced. The respiration and nutrient uptake of roots are also reduced after defoliation, though it is not clear whether the latter is due to a change in root properties, e.g. by increased mean age following cessation of growth, or a lower demand from the much reduced shoot. Thus, Oswalt *et al.* (1959) found no uptake of ^{32}P from 15 to 25 cm below the soil surface until 19–30 days after defoliation of grass. If photosynthate supply is interrupted by any other means, uptake will also gradually stop. Clarkson *et al.* (1975) tested ion uptake by roots with 'take-all' (*Gaeumannomyces graminis*) lesions above the tested part, and found that both root elongation and uptake stopped over a period of 2–3 days after the fungus had disrupted the phloem. The root system itself has remarkably good powers of recuperation after it has been mutilated, and removal or desiccation of any part of the root system rapidly results in compensating growth elsewhere (Crossett *et al.* 1975).

7.2.9   Prediction of root system morphology

The remaining parts of this chapter discuss the mathematical modelling of the uptake of nutrients and water by single root systems. Ideally such methods would include the modelling of root system growth distribution, from a knowledge of soil and plant characteristics. It is obvious from the review of root responses to soil variables given here that it will be a difficult task, especially so in a natural environment. The information on the effects of these variables—especially where more than one is varying simultaneously—is often lacking, and the mechanisms by which they operate are largely unknown. As is seen later, it is possible to predict the *growth* of root by a plant when this is growing in constant conditions, but not the *distribution* of the roots. Lungley (1973) has published a mathematical model of root system extension, but this is based on the branching pattern of the root system, and it is not related to soil properties or to the growth parameters of the shoot. A more detailed model was proposed by Hackett & Rose (1972a,b), based upon several reports that the relation between numbers, length, surface area and volume of different root members remains constant during vegetative growth. From known extension rates of each class of root (axis, 1st, 2nd and 3rd order lateral), their density on the next superior class and the rate of extension of the region bearing laterals,the morphology of a single axis and its laterals could be reproduced fairly accurately. So far, however, no model of this type appears to have been tested on field soil-grown roots.

Detailed root branching and distribution patterns will probably have to be measured empirically for some time to come, but it does appear possible that the distribution of root density with depth under uniform crops may be predictable (Chapter 8, p. 252) from soil strength values and other parameters, or it may prove to be constant for specified crops and soil series. In uniform well-drained soils it may also be possible to predict the extension rates of taproots and main laterals, and hence the rooting volume.

7.3   Relationship of nutrient uptake and growth

Plant growth essentially consists of the ingestion and chemical elaboration of carbon dioxide from the air, and water and mineral nutrients from the growth medium. The balance between these processes is extremely complex, and many of them are still poorly understood, but the most basic aspect of the relationship can be stated easily, as follows (Nye & Tinker 1969; Brewster & Tinker 1972):

$$\frac{dU}{dt} = \frac{d(WX)}{dt} = \left(W \frac{dX}{dt} + X \frac{dW}{dt} \right) = \bar{I}L = \bar{F}A_{\mathrm{R}} = \bar{S}W_{\mathrm{RF}}. \qquad (7.1)$$

Here W is total plant dry weight, X plant composition of the nutrient in question, U the total nutrient content of the plant, t the time, L the root length, I the inflow, S the unit absorption rate, F the flux, A_{R} the root surface area and W_{RF} the fresh root weight. X will normally be relatively constant; at least, it rarely varies by more than a factor of 2–3 during a plant's growth, whereas W may change by a factor of 100 or more. If $dX/dt \sim 0$,

$$\frac{X}{W} \frac{dW}{dt} = X R_{\mathrm{w}} \sim \bar{I}(L/W) = \bar{F}(A_{\mathrm{R}}/W) = \bar{S}(W_{\mathrm{RF}}/W) \qquad (7.2)$$

where R_{w} is the relative growth rate. The growth of the plant is thus linked directly to the nutrient uptake process, in a way which indicates that a necessary condition for continual uptake is the increasing size of the plant as a nutrient sink.

7.3.1 Constant environmental conditions

There may be rapid changes in the nutrient concentration X in the plant in the earliest seedling stages, when it is dependent upon both seed reserves and root uptake. Later, if the environment remains constant, a plant normally grows in a reasonably well regulated and balanced manner characteristic of its genetic composition and the environment, so long as it remains in the vegetative phase. If R_{w} is approximately constant, a plant is said to be undergoing 'logarithmic growth' (Evans 1972), as for example in the results of Brewster & Tinker (1970) and Brewster *et al.* (1975a). It is probably rare for all the factors in equation 7.2 to be truly constant, since there is usually a slow drift of R_{w} and X, at least. Viets (1965) stated that the percentage of nitrogen is usually quite constant if the growth medium is continually re-supplied, but a study of rape growing in a constant environment in flowing culture solution showed clearly that the 10-fold decline in unit absorption rate at low external solution concentration ($5 \times 10^{-5}\mathrm{M}$ nitrate) was largely an inherent characteristic of the root system, which resulted in a decline in plant nitrogen concentration and relative growth rate (Table 5.3). At a higher level of nitrogen supply ($10^{-3}\mathrm{M}$ nitrate) the unit absorption rate and plant nitrogen concentration were maintained, and the decline in relative growth rate could be attributed to self shading. In onion at low external solution concentration there was only a 3-fold fall in unit absorption rate, which was associated with an increase in the proportion of root (decreasing W/W_{R}). At the higher concentration unit absorption rate was maintained. Flux and inflow showed similar trends. Percentage phosphorus often tends to decline

with plant age (e.g. Boken 1969; Mengel & Barber 1974a), and this change
may be greater in field-grown plants, since there is depletion of the nutrients
in the soil. The decline may be very sharp for nutrients which are in a
particularly labile form; for example, leaf nitrate concentration decreases
rapidly if soil mineral nitrogen is exhausted (Last & Tinker 1968). The
root : shoot ratio will also normally change with age (Aung 1974, and see
Table 7.4), but it is nevertheless likely that these parameters, including $\bar{I}$,
will change gradually or slightly during the early vegetative part of a
plant's life, after the early seedling stage, provided nutrient supplies are
maintained.

7.3.2 Changing environmental conditions

A change in the steady growth situation can be caused by many factors,
external and internal. Changes in temperature or light intensity will probably
alter R_w, and the root:shoot ratio. If R_w is low for reasons not connected
with nutrient supply, such as a low light level, we expect to have a relatively
high X and a low L/W, and $\bar{I}$ may, therefore, not be greatly altered (equation
7.2). If, however, the growth reduction results from a small supply of nutrients, then this will cause a small X and a large L/W value, and $\bar{I}$ should be
greatly lessened. The value of $\bar{I}$ may, therefore, prove to be a useful parameter for assessing the state of nutrition of a plant, in a more exact way than,
for example, the percentage content of the nutrient. The precise way in which
the plant adapts to any change cannot be predicted simply, because of the
extreme complexity of the interacting plant responses to any external change.
The modelling approach seems to offer the only hope of dealing with such a
system (see pp. 235, 243 *et seq.*), though attention so far has mainly been
concerned with modelling constant environment situations, and the deliberate
introduction of changes in environmental conditions during a simulation will
call for a considerably better understanding of the plant's physiology, and
more data on plant response to environment.

7.3.3 Plant growth phase

Changes in all factors in equations 7.1 and 7.2 will occur when the vegetative
phase changes to the reproductive phase in the plant's life, or as vegetative
storage organs develop. These may be relatively sharp, such as reduction of
root extension rate and root:shoot ratio at flower initiation (Beever &
Woolhouse 1975), or as implied by allometric relationships between the
weight of two plant organs. Figure 7.12 (Kato 1963) contains an example of
such a sharp change in root : shoot ratio which accompanied an increase in
the growth rate of the onion bulb. Near maturity the total nutrient content
of a crop may decrease quite markedly, e.g. Gasser (1961) found that wheat

lost 20% of its nitrogen before final harvest, and the analysis made above can no longer be applied directly, since it implies a negative $\bar{I}$.

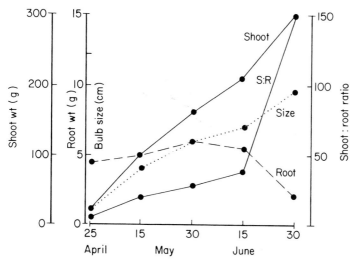

FIG. 7.12. Relationship of the shoot : root ratio and the increasing growth rate of the onion bulb (after Aung 1974).

7.3.4 Plant species

In a given environment, different plant species—or even varieties—will accommodate themselves to equation 7.2 in various ways. The differing root lengths of two barley varieties have been noted (p. 192). Rorison (1969) pointed out the low demand for nutrients of a slow-growing plant, and considered that the ability of *Deschampsia flexuosa* to grow at very low phosphate levels was due to its low relative growth rate and phosphate demand (Fig. 7.13). There are also major differences in root : shoot ratio between species, which allow some to grow rapidly with $\bar{I}$ or $\bar{F}$ values which for others would imply deficiency and reduced R_w. For example, Evans (1972, p. 430) noted that the ratios of root weight to total weight of *Helianthus annus* and *H. debilis* were respectively 0·27 and 0·18 when they had almost identical R_w values. Loneragan *et al.* (1968) determined the $\bar{S}$ values for calcium in several species, and showed how these related to their differing calcium demand. On the other hand, Wild *et al.* (1974) have found that the minimum value of $\bar{S}$ for potassium required to give optimum growth in several species growing in flowing solution culture was remarkably constant.

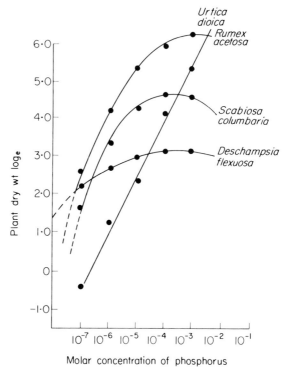

FIG. 7.13. Dry weights of various species (plotted as logarithm) produced in 6 weeks growth with different concentrations of phosphate in solution culture (after Rorison 1969).

7.3.5   Plant nutrient demand

We regard nutrient demand as arising directly from changes in the plant's weight and composition. A plant is a sink for two separate types of materials —carbohydrate and mineral nutrients—which are supplied from different sources, and the complexity of its growth arises because neither of these sources is simple and because both interact with the same sink, and hence with each other. The demand for nutrients is expressed at the root surface, but it is quite unrealistic to visualize this as some form of 'suction' for nutrients, since it is perfectly clear that the roots absorb them by an active process (excluding some solutes which are passively absorbed with the transpiration stream). On the other hand, it is obvious that the roots do not function as a simple 'pump' for nutrients, without regard for the conditions of the rest of the plant, since plants could otherwise have extremely large nutrient concentrations. Toxic concentrations of phosphorus in plants have been detected (Loneragan *et al.* 1966), but this seems to be a rather special case, possibly associated with high transpiration, and in general there must be a strong feedback control on root activity from the rest of the plant. The mechanism of this feedback is not known. It may simply be a matter of the total amount of carbohydrate or inorganic nutrient in the root or the shoot, or a hormonal mechanism (see, for discussion, Drew & Saker 1975a; Jungk & Barber 1974). For our purpose it is necessary to quantify this demand at the root surface. This problem was discussed in Chapter 5, where it was pointed out that a flow rate could be used to define the condition at the root surface, but the property of the root was best stated as the root demand coefficient, $\overline{\alpha a}$, or equivalent parameter. We therefore re-state equation 7.2 as

$$W R_{\mathrm{w}} X = \bar{I}L = 2\pi \, \overline{\alpha a} \, C_{\mathrm{la}} \, L = A_{\mathrm{R}} \, \bar{\alpha} \, C_{\mathrm{la}} = 2W_{\mathrm{RF}} \left(\frac{\alpha}{a}\right) C_{\mathrm{la}}. \qquad (7.3)$$

It follows from the definition above that $\bar{I}$, the mean inflow, is averaged over the whole root system of the plant. Some parts of a root system may be inactive, for example due to heavy suberization or loss of the cortex, and a reduced value of L may be selected to take account of this.

When applied to whole plants, as opposed to short pieces of single root, some care must be taken in the definition of the α parameter. If C_{la} is a constant over the entire root system, as in a well-stirred solution culture (see Chapter 5), then $\overline{\alpha a}$ is also an arithmetic average over the whole root system as for $\bar{I}$. If the root system is in soil, the value of C_{la} may vary, either due to soil heterogeneity, or because local variations in αa lead to different local depletions around the roots. In the latter case C_{la} is itself dependent upon αa at that point on the root (Nye & Tinker 1969). In consequence, if $\overline{\alpha a}$ is calculated from nutrient uptake for whole plants in soil, it must be regarded as a general indicator of plant demand rather than as an exact average of the real αa values at the root surface (Brewster & Tinker 1970). The precise meaning of $\overline{\alpha a}$ must be considered in each case.

The analogy with the absorption of CO_2 from the air is of interest here (Tinker 1969b). The net assimilation rate, E, is exactly analogous to $\bar{F}$ for nutrients:

$$R_{\mathrm{w}} = E \, \frac{A_{\mathrm{L}}}{W},$$

where A_{L} is leaf surface area. E is an average over the whole leaf surface, and the real net assimilation rate will vary widely from leaf to leaf. Despite this, E is found to be a useful parameter of plant growth (Evans 1972), and a similar conclusion should apply to $\bar{I}$, $\bar{F}$ and $\bar{S}$.

In Table 5.4 a series of values for $\bar{I}$, $\bar{S}$, $\bar{F}$ and $\overline{a\alpha}$ were given, obtained from whole plant experiments in solution culture. These experiments used young plants in rapid growth, and most of their roots would consequently be young. For ions where the uptake rate does not vary greatly with root age, these are a fair estimate of the real values at the root surface. Values derived from soil-grown plants must be less reliable, for reasons described above, but if the soil is homogeneous, the plants young, and the root density low, they are probably still acceptable.

7.3.6 Nutrient flow analysis

These concepts were applied to whole plants by Brewster & Tinker (1970; 1972). Their approach may be regarded as the simplest possible model (Table 7.9) of single whole plant nutrient uptake from soil. Leeks were grown in large containers in the field. The initial soil water content was sufficient for three months growth, rain being excluded, and transpiration was measured by repeated weighing. Plant weight, nutrient content and root length were measured at four harvests. The well-known formula for estimating net assimilation rate in plants was adapted to determine the inflow of nutrients (Williams 1946),

$$\bar{I} = \frac{(U_2 - U_1)(\ln L_2 - \ln L_1)}{(t_2 - t_1)(L_2 - L_1)} \tag{7.4}$$

where subscripts 2 and 1 refer to two consecutive harvests, in a way analogous to that used for determining R_w. The alternative would be to use more frequent but smaller harvests, and fit polynomial or other equations to these data (for references see Hunt 1973). The latter is the most reliable, in that no assumptions are involved in the method, whereas the derivation of the logarithmic formula assumes L and U are linearly related. In practice the choice may depend upon the convenience of the harvest sequences.

Brewster & Tinker found that $\bar{I}$ for all ions (see Table 7.8) were remarkably constant during the experimental period of one month. Probably this was because the plants were beyond the seedling stage, the logarithmic growth during this period caused the age structure of the root system to change little (most roots were less than 2 weeks old in 3-month-old plants), the mean environmental conditions were very constant during this time, in May, June and July, and there was a sufficiency of all nutrient ions.

These $\bar{I}$ values may be used to determine how much the concentration at the root surface differs from that in bulk soil. A modified version of the approximate Passioura (1963) equation (see Chapter 6, p. 151) gives:

$$\bar{I} = 2\pi \, \bar{C}_{la}\overline{\alpha a} = 2\pi \, (Db)(C_{li} - \bar{C}_{la}). \, \gamma + \bar{I}_w \, \bar{C}_{li} \tag{7.5}$$

where I_w is the mean water inflow per unit length of root. In a single short piece of root, γ, the flux parameter (Nye 1966b) will have a unique value at any time, but in a whole root system it will vary with root age. A final mean

Table 7.8. Measured data in nutrient flow analysis for leeks grown in soil. Calculated C_{1a} and α based on assumption of constant I over whole root system. (After Brewster & Tinker 1970; Brewster 1971.)

Inter-harvest period	Inflow I (mol cm^{-1}s^{-1} $\times 10^{-13}$)	App. mass flow M (mol cm^{-1} s^{-1} $\times 10^{-13}$)	C_{1a} (mol cm^{-3} $\times 10^{-7}$)	C_{1a}/C_{1i}	M/I	α (cm s^{-1} $\times 10^{-6}$)
			K$^+$			
1	14.8	1.04	1.10	0.26	0.07	71
2	11.8	0.99	1.80	0.42	0.084	35
3	10	1.04	2.00	0.46	0.104	30
			Ca^{++}			
1	4.4	36	177.5	1.18	8.2	0.13
2	4.8	35	176.4	1.17	7.4	0.14
3	4.4	36	179.7	1.19	8.2	0.15
			NO$_3$			
1	23	49.5	216.2	1.05	2.15	0.56
2	30	47.5	212.6	1.02	1.66	0.75
3	21	49.5	217.4	1.06	2.36	0.59
			SO$_4^=$			
1 2 3	2.3	6	27	1.09	2.6	0.45

Inflow I (mol cm^{-1} s^{-1} $\times 10^{-13}$)	App. mass flow M (mol cm^{-1} s^{-1} $\times 10^{-13}$)	$\bar{C}_{1a}$ (mol cm^{-3} $\times 10^{-7}$)	C_{1a}/C_{1i}	M/I	α (cm s^{-1} $\times 10^{-6}$)
		Na$^+$			
2.15	3.96	17.30	1.05	1.84	0.66
1.64	3.80	17.50	1.06	2.3	0.50
0.48	3.96	18.20	1.11	8.3	0.16
		Mg^{++}			
1.01	1.49	6.46	1.04	1.47	0.84
0.85	1.42	6.51	1.05	1.67	0.69
0.68	1.49	6.70	1.08	2.2	0.62
		Cl$^-$			
5.1	22	97.2	1.07	4.3	0.28
4.7	21	96.8	1.06	4.45	0.26
4.7	22	97.7	1.07	4.65	0.29
		H$_2$PO$_4$*			
1.2	0.050			0.04	
0.82	0.048			0.06	
0.94	0.050			0.05	

* It is probable that these plants were infected with endotrophic mycorrhizas, hence calculation of C_{1a}/C_{1i} as for other ions is not valid see p. 180). Calculations of the maximum I were made, with a numerical solution of the diffusion equation (Brewster 1971), with different assumptions about α and D, giving a best estimate of theoretical maximum I of about 0.67×10^{-13} mol cm^{-1} s^{-1}.

value G, was determined, weighted for the length of root in each age class. C_{1i} was directly measured at each harvest, and remained fairly constant throughout; Db was found by direct measurement. Equation 7·5 then yields results for $\bar{C}_{1a}$ and $\overline{\alpha a}$ (Table 7.8). The problems of applying this equation to a whole root system, rather than to a single short piece of root, have been touched upon earlier. If αa is constant, then $\overline{\alpha a}$ applies to all parts of the system, and $\bar{C}_{1a}$ is constant. If not, I will be largest where αa is largest and C_{1a} smallest. It is therefore certain that most of the total nutrient uptake enters at points on the root where the concentration is less than the calculated $\bar{C}_{1a}$. It is of course assumed in this model that there was no inter-root competition, which was reasonable in view of the smallest mean inter-root distance of about 1 cm. This complication is dealt with in the next section.

As expected from the work of Barber *et al.* (1962), Ca, Mg and Na were supplied or oversupplied by mass flow; for these ions C_{1a} is therefore greater than C_{1i}, and diffusion is in the reverse direction. The experiment showed that root surface depletion must have been serious for potassium at least. Severe depletions were also predicted for phosphate (Brewster 1971), but it is probable that the roots were infected with endotrophic mycorrhizas (see p. 182) and this result is therefore open to question.

It will be seen from equation 7.5 that when the mass flow contribution is negligible, as it is for potassium and phosphate (Table 7.8), the equation in effect is

$$\frac{\overline{\alpha a}}{Db} = \frac{(C_{1i} - C_{1a})}{C_{1a}} \gamma,$$

and γ varies relatively slowly with D and t. The decrease in the soil solution concentration, relative to that at the root surface, is therefore directly related to $\overline{\alpha a}/Db$. This term may be regarded as balancing the demand of the plant at the root surface $(\overline{\alpha a})$ with the ease with which ions diffuse to the root $(Db \simeq D_1 \theta f_1)$ in the soil solution, and therefore decides whether diffusion is a limiting factor in the uptake process, for that particular situation. In this book, a root-soil system with a large value of $\overline{\alpha a}/bD$ is referred to as 'a root with very high relative sink strength'.

7.4   Uptake by competing roots within a single root system

The step from the single-root models described earlier to a whole root system is a major one. The true difficulties may be ignored in suitable cases, as above, but these complications must be dealt with in a more precise model. The main complications are:

(a)   The depletion zones of individual roots overlap, producing a non-symmetrical concentration distribution around each root. Hence also the medium around each root cannot be regarded as infinite (Fig. 7.14). This is the main form of competition in uptake between individual roots.

(b)   General concentration gradients arise within the root system, produced by different root densities or demand coefficients in different zones. The 'concentration in the bulk soil' thus becomes an uncertain quantity. This problem has analogies with the discussion of transport of water to plant roots by Newman (1969): he distinguishes between a local impedance to flow around each root (rhizosphere resistance) and an impedance to movement over greater distances from soil volumes not yet exploited by roots (para-rhizal resistance).

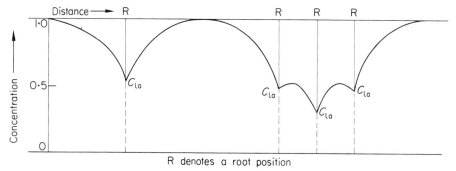

FIG. 7.14. Concentration–distance profile to show effect of overlap of diffusion zones on concentration at the root surface (after Baldwin 1972).

(c)   The continuing development of roots alters the size and geometry of the system. The age distribution of roots was dealt with in the Brewster–Tinker model, but without any reference to the spatial changes with time.

(d)   Soil nutrient concentrations will alter with time, which may affect the diffusion coefficient and buffer power.

Various authors have shown the general effects of root competition and uneven distribution on uptake. Cornforth (1968) grew plants in media of differing aggregate sizes, so that roots were confined to channels between them, and showed that uptake was least with the largest aggregates, where the roots were most concentrated and hence competed most strongly. Andrews & Newman (1970) used split pot systems in which the roots of plants of different species could intermingle, or be kept separate. They found that when the roots were intermixed, and hence competed, the uptake of nitrogen was reduced in the least vigorous plant, but the uptake of phosphate was not. This is what would be expected from the wide depletion zones for nitrate, and narrow zones for phosphate, thus making competition for nitrate much more likely. Methods of dealing with such effects in more detail are described in subsequent sections. They are summarized in Table 7.9.

Table 7.9. Summary of available models for interpreting the nutrient uptake of multiple root systems.

	Diffusion–depletion	Competition	Mass flow	Pattern	Time variation
1. Barber (1962)	—	—	√	—	—
Scope: Simple allocation to mass flow, diffusion or root interception.					
2 Lewis & Quirk (1967)	√	—	—	—	—
Scope: Diffusion to root surface with time-dependent concentration.					
3 Brewster & Tinker (1970)	√	—	√	—	√
Scope: Application of simple single-root model to whole plant growth. Mass flow by addition.					
4 Barley (1970)	√	√	—	√	—
Scope: Computer solution of diffusion equation for limited volume assigned to each root.					
5 Sanders *et al.* (1970) Baldwin *et al.* (1972)	√	√	√	√	√
Scope: Electrical analogue for accurate prediction of competition effects. Mass flow by addition, time variation by analogue change, or time averaging.					
6 Baldwin *et al.* (1973), Baldwin & Nye (1974)					
(a)	√	√	—	—	—
Scope: Steady state approximation for regular array of roots.					
(b)	√	√	√	—	—
As (a) Mass flow by addition.					
(c)	√	√	√	—	—
Steady state approximation, from Nye & Spiers (1964) accurate equation for mass flow.					
(d)	√	√	√	—	√
As (c), but using time-step method for calculation, with change in *L*, *b* or *D* with time.					
(e)	√	√	√	—	√
As (a) or (c) but using time-integral averaged values for the parameters.					
(f)	√	√	√	—	√
As (d), but allowing for concentration dependent *D*, and with improved accuracy at short times.					

7.4.1 Competition between groups of parallel roots

The simplest competing multiroot system consists of two similar parallel roots. The concentration contours around these will be as in Fig. 7.15. There is no useful analytical solution of the diffusion equation even in this simple situation, much less for groups of many parallel roots, and other means have to be used. It should be noted at once that all models presented up to the present have had to restrict themselves to parallel roots in a specified volume, so that each root has only two degrees of freedom in position. However, it has been shown (Marriott 1972) that the mean distance between any point chosen at random in the volume and the nearest root is the same whether the system is parallel or not. This gives reason for believing that results derived with parallel-root models are applicable in principle to root populations with non-parallel orientation (Baldwin *et al.* 1972). The complexities of dealing with groups of non-parallel roots seem very large in relation to the possible benefit which could be obtained. We therefore consider the uptake of nutrients by groups of parallel roots situated in volumes of soil that can be regarded initially as uniform. Parallel root populations in defined volumes can be described by the following small group of parameters.

(i) $\overline{\alpha a}$, the root demand coefficient, which is assumed similar in all roots. It introduces considerable complications if αa values differ, though they must do so to some extent in reality. This enforced uniformity in root properties is clearly a major simplification. However, there seems little point in considering a varying αa when so little is known about how this changes

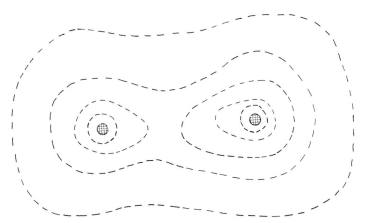

FIG. 7.15. Schematic drawing of concentration contours around two absorbing and competing roots in soil.

with position on a root system and soil properties. In particular, the effect of changes in αa on uptake is small when $\alpha a/Db$ is large (equation 7.13), and models of this type discussed here are most realistic where this is true, e.g. with phosphate and potassium ions absorbed by young actively growing plants.

(ii) The root radius, which has a significance additional to that expressed in αa (see p. 133, 227). This is assumed constant.

(iii) Root density, in roots per cm^2 on a surface normal to root direction, which equals the length of root per unit volume, L_V.

(iv) Pattern—the above three parameters completely define a regularly distributed root population in which all inter-root distances are the same. The roots can however be distributed in different ways and their arrangement, or pattern, may then be important. Pattern has frequently been discussed in quantitative plant ecology where essentially the same problems arise (Greig-Smith 1964; Kershaw 1973; Pielou 1969). A pattern of points may be regular (distances between any point and its nearest neighbour are similar), random, or contagious (where the nearest neighbour distances are on average short compared with random patterns) (Fig. 7.16). Pattern should be defined by a frequency distribution of distances between neighbours, but approximate single-valued parameters can be used, of which the most useful is the $V{:}M$ ratio:

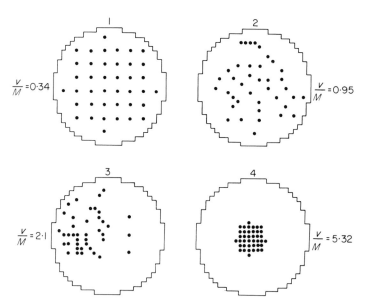

FIG. 7.16. Regular (1), random (2), clumped (3), and highly-clumped (4) patterns of points with their respective indices of dispersion (after Baldwin *et al.* 1972). (See text for symbols.)

$$\frac{\text{variance of numbers of roots in a volume}}{\text{mean number of roots in a volume}}.$$

This parameter is the same for the variance and mean number of points of intersection of the parallel roots with a surface normal to them. The same principle will apply to non-parallel roots, but the pattern may vary with the orientation of the intersecting surface.

Equivalent cylinder model

Barley (1970) made the ingenious suggestion that the problem of uptake by parallel but non-regular roots could be solved by assigning a 'Voronoi polygon' to each root on a plane at right angles to them (Fig. 7.17), and,

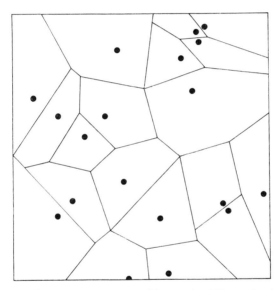

FIG. 7.17. A random pattern of points, with associated Voronoi polygons (after Barley 1970).

considering this as a cylinder of similar area. The problem would then be converted to uptake by roots from a series of soil cylinders of different radius, in which the frequency distribution of the radii would be a function of the root pattern. Youngs & Gardner (1963) have published an analytical solution for uptake from limited cylinders.

This model was used to test the effects of root distribution on uptake, which Barley (1970) found to be small. Baldwin *et al.* (1972), who found much larger effects, suggested that Barley's methods did not in fact produce a significant degree of contagion. Apart from the possible errors involved (Baldwin 1972), the method does not appear easy to apply in practice.

7.4.2 Electrical analogue of root systems

An exactly similar equation to the generalized equation for diffusion (equation 1.8) describes flow of electrical charge in a volume with electrical resistance and capacity. Diffusion problems can, therefore, be solved by electrical analogues (Karplus 1958), with suitable choice of scaling factors. Sanders *et al.* (1971) constructed such an analogue, based on a network of 525 units (nodes) each with the circuit diagram in Fig. 7.18. This represents a thin uniform slab of soil, and the analogue only simulates transfers in two dimensions; it is technically possible to make a 3-dimensional analogue, but very

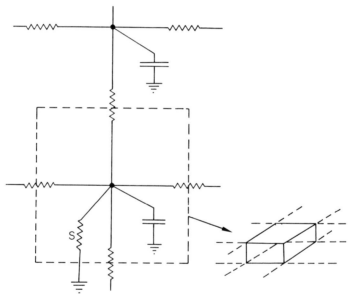

FIG. 7.18. Circuit diagram for two nodes of the analogue, showing sink resistor *S*, and equivalent portion of soil slab (after Sanders *et al.* 1971).

tedious. For roots in the soil, the 1st-order relationship for uptake, $F = \alpha C_{1a}$ (α constant), (p. 110) is easily simulated by connecting a resistor (S) representing a single root to the central point of the node and to earth, since Ohms Law is also 1st-order in form. Voltage on the analogue represents concentration in solution. After charging, the appropriate 'root' connections are made, and the charge (which represents quantity of diffusible solute) leaks out through the 'roots' to earth. The effect of different arrangements of 'roots' can easily and rapidly be tested on this analogue.

Apart from the solution of specific problems, e.g. see p. 231, the instrument lends itself to determining certain simple generalizations about root distributions. This is greatly aided by the fact that the uptake results may conveniently be represented in terms of three dimensionless variables only: $Dt\rho$ (where ρ is the root density in root ends per cm²), $\alpha a/Db$ for each root and soil, and a root pattern parameter.

Competition between two similar parallel roots in an infinite medium

In theory, competition starts instantly, since diffusion equations predict an instantaneous effect at all distances. However, the effect is vanishingly small at first. For convenience, we may define 'competition' as commencing at the point at which two roots of very high relative sink strength decrease each other's uptake by 10%. Results for this situation are in Fig. 7.19 (Baldwin *et al.* 1972); from this we may say that competition does not start until $\sqrt{Dt} =$ distance between the roots. Tests with the analogue showed that the competitive effects, in percentage terms, on one root due to two other roots in an infinite medium can be regarded as additive. If the roots absorb much less strongly ($\alpha a/bD$ is small) the competitive effect will be correspondingly less, and roots with very low values of $\alpha a/bD$ will not compete sensibly at any time, if they are in an infinite medium. If the medium is limited in size, the more rapid fall in mean concentration with two roots than with one is of course an effect of competition, but this does not depend upon root position.

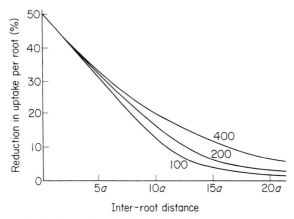

Fɪɢ. 7.19. Reduction in total uptake of a pair of similar roots, as percentage of uptake by isolated roots, plotted against inter-root distance relative to root radius *a*. Numbers on curves are Dt/a^2. $\alpha a/Db$ for roots is 15.9, which is a sink of high strength (after Baldwin *et al.* 1972).

The following generalization seems likely to be useful in practice: if the mean inter-root distance is greater than $\sqrt{Dt}$, and if the degree of contagion is small, a root system can be regarded as composed of isolated single roots.

Effects of pattern

Above we considered the situation of single roots at the beginning of effective competition. Once root competition becomes serious, it cannot effectively be dealt with on a single-root basis. Instead, we consider a large population of roots, distributed with uniform mean density. The analogue then simulates an average sample of the volume containing the roots, taken so that there is no solute transfer across the boundary. A constant number of roots (i.e. sink resistors) can be arranged on the analogue in different ways, varying from a regular distribution to a highly contagious one with roots grouped near the centre, and relative uptake rates determined. Baldwin *et al.* (1972) found that the variance : mean ($V : M$) ratio appeared the most useful pattern parameter, though a variety of different distributions may give the same $V : M$ ratio, and an exact relationship between $V : M$ and uptake is not to be expected (Fig. 7.20). These results are for high values of $\alpha a/bD$, i.e. for roots with high relative sink strength (see Chapter 6). The relative effects of different patterns will diminish as $\alpha a/bD$ decreases, and at very low values, the positioning of the roots is of little importance, any more than it is in stirred solution culture.

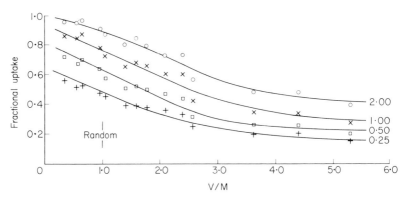

Fig. 7.20. Fractional uptake from a limited volume by roots of high sink strength, plotted against pattern parameter V/M. Numbers on curves are values of $Dt\rho$. (after Baldwin *et al.* 1972).

The results of this work showed that there is little difference (less than 20%) in uptake between regular and random distributions, but that highly-contagious distributions could be extremely inefficient, absorbing less than half as much as regular distributions at some stages of the uptake process. The degree of regularity or contagion in distribution of roots in field soils is at present unknown, though extreme clumping has often been observed qualitatively (Fig. 7.2). Heavy soils with large impenetrable clods may have almost all roots in the inter-ped spaces. Intense concentrations of nutrients, such as might be established around a fertilizer granule, will cause local proliferation of roots. Against this, roots will branch and grow less in zones already depleted of nutrients by other roots, so tending to a regular distribution. A similar effect would be produced by any toxic root exudate which limited the growth of nearby roots. Quantitative field measurements of this factor are necessary, though difficult to make.

7.4.3 Approximate steady-state models

Previous attempts have been made to describe a transient-state situation by a simple approximate steady-state model. A good example is that of Gardner (1960) for transport of water to roots, which assumes a constant flux of water at the root surface (see p. 24).

Baldwin *et al.* (1973) followed this principle in developing a series of root uptake models. They started from the assumption that the concentration profile around each root is given by the steady state equation

$$I = 2\pi\alpha a\, C_{la} = 2\pi r\, Db\, \frac{dC_1}{dr}$$

where r is the radius of a circular section in the soil with the root at its centre. Integration gives

$$C_{lr} = C_{la}\left(1 + \frac{\alpha a}{Db}\ln\frac{r}{a}\right). \tag{7.6}$$

With a regular parallel array of roots, of length L_V per cm^3, the cross-sectional area per root on a plane normal to them is

$$\frac{1}{L_V},$$

and they therefore assigned to each root a cylinder of soil around it of radius

$$x = \frac{1}{\sqrt{\pi L_V}}.$$

If roots deplete only their own cylinder, the perimeter at x is a zero-transfer boundary.

The mean concentration of liquid-phase solute $\bar{C}_1$ within such a cylinder is

$$\bar{C}_1 = \frac{1}{\pi(x^2 - a^2)} \int_a^x 2\pi r \, C_{1r} \, dr. \tag{7.7}$$

Substitution of the expression for C_{1r} from equation 7.6 and integrating gives

$$\bar{C}_1 = C_{1a} \left[1 - \tfrac{1}{2} \frac{\alpha a}{Db} + \frac{x^2 \left(\dfrac{\alpha a}{Db} \right)}{(x^2 - a^2)} \ln \frac{x}{a} \right]. \tag{7.8}$$

If $x^2 \gg a^2$, which will be true in the great majority of cases, this reduces to

$$\bar{C}_1 = C_{1a} \left(1 + \frac{\alpha a}{Db} \ln \frac{x}{1 \cdot 65a} \right). \tag{7.9}$$

Hence, from equation 7.6, $C_{1r} = \bar{C}_1$ at $r = x/1 \cdot 65$, which shows that the serious depletion zone around the root is not extensive.

The uptake of material may then be calculated by considering the change in concentration $\bar{C}_1$ with time. Equating rate of loss from soil with root uptake gives:

$$\frac{d\bar{C}_1}{dt} b = -2\pi a \alpha \, C_{1a} L_V \tag{7.10}$$

Eliminating C_{1a} between equations 7.8 and 7.10 and integrating from $t = 0$ when $\bar{C}_1 = C_{1i}$ gives

$$\frac{\bar{C}_1}{C_{1i}} = 1 - \frac{M_t}{M_\infty} = \exp \left\{ \frac{-2\pi \alpha a \, L_V \, t}{b \left[1 - \tfrac{1}{2} \dfrac{\alpha a}{bD} + \dfrac{x^2 \left(\dfrac{\alpha a}{bD} \right)}{x^2 - a^2} \ln \dfrac{x}{a} \right]} \right\} \tag{7.11}$$

where M_t and M_∞ are the amounts of solute absorbed at times t and ∞ respectively. This gives

$$\frac{M_t}{M_\infty} = 1 - \exp \frac{-2\pi \alpha a \, L_V \, t}{b \left(1 + \dfrac{\alpha a}{Db} \ln \dfrac{x}{1 \cdot 65a} \right)} \tag{7.12}$$

when $x^2 \gg a^2$.

This approximate method has been compared with a modified version of the accurate numerical solution of Nye & Marriott (1969) and gave good agreement, especially at values of

$$\frac{\alpha a}{b D} < 5.$$

This equation satisfied intuitive expectations; for example, L_V and αa are partly interchangeable in effect. If

$$\frac{\alpha a}{D b}$$

is very large, then the expression gives the largest possible uptake in that soil and with that root density:

$$\operatorname*{Lim}_{\frac{\alpha a}{b D} \to \infty} \quad \frac{M_t}{M_\infty} = 1 - \exp \frac{2\pi L_V t D}{\ln \dfrac{1 \cdot 65a}{x}}. \tag{7.13}$$

This approach has similarities with that of Barley (1970), though it is much simpler to apply. Any attempt to include non-regular patterns by varying x would meet the same difficulties found in his model, and it is easiest to deal with patterns by using the correction factors from Fig. 7.20.

Addition of mass flow

The method in the last section may be developed to include mass flow and transpiration, though models of the type developed here are most interesting when depletion zones are well marked, and mass flow is then unlikely to be important. It has been pointed out earlier that a high mass-flow component must mean a high soil solution concentration, hence the possibility of development of large gradients in C_1, and the relatively easy transfer by diffusion. Alternatively, a high soil solution concentration implies a low αa (see Table 5.4), and a low $\alpha a/Db$. Both arguments lead to the conclusion that when mass flow is important, only slight depletion zones develop (Chapter 6). Accumulations, with diffusion away from the root, may also be small in most real cases (Wray & Tinker 1969b). Nevertheless, it is perfectly possible to take mass flow into account using similar methods to those above (Baldwin et al. 1973). Nye & Spiers (1964) showed from the continuity equation 1.6 that in the steady state

$$\frac{C_{1r}}{C_{1a}} = \frac{\alpha}{v_a} + \left(1 - \frac{\alpha}{v_a}\right) \left(\frac{r}{a}\right)^{\frac{-a v_a}{b D}}. \tag{7.14}$$

By assuming av_a/bD to be small, we can approximate

$$\left(\frac{r}{a}\right)^{-\frac{av_a}{Db}} = 1 - \frac{av_a}{bD} \ln \frac{r}{a}$$

and hence

$$\frac{C_{1r}}{C_{1a}} = 1 + \frac{\alpha a}{Db} \ln \frac{r}{a} - \frac{av_a}{Db} \ln \frac{r}{a}. \tag{7.15}$$

Calculation of the $\bar{C}_1$ as before (equation 7.8) by integration gives

$$\frac{\bar{C}_1}{C_{1a}} = \frac{\alpha}{v_a} + \frac{\left(1 - \dfrac{\alpha}{v_a}\right) \dfrac{2}{2 - \dfrac{av_a}{Db}} \left[\left(\dfrac{x}{a}\right)^{2 - \frac{av_a}{Db}} - 1\right]}{\left(\dfrac{x}{a}\right)^2 - 1}. \tag{7.16}$$

If av_a/bD is small, it can be shown that equation 7.16 reduces to

$$\frac{C_{1a}}{\bar{C}_1} \sim \frac{1}{1 + \left(\dfrac{\alpha a - v_a a}{bD}\right) \ln \dfrac{x}{1 \cdot 65a}} \quad \text{or} \quad \frac{1 + \dfrac{v_a a}{bD} \ln \dfrac{x}{1 \cdot 65a}}{1 + \dfrac{\alpha a}{bD} \ln \dfrac{x}{1 \cdot 65a}} \tag{7.17}$$

i.e. uptake increases linearly with water inflow ($2\pi v_a a$) in the competitive situation, as in the infinite situation (see p. 151), provided α and bD remain independent of v_a.

Equation 7.16 can be used to calculate uptake from a given soil volume over a sequence of time steps. As before

$$x = \frac{1}{\sqrt{\pi L_V}},$$

and initially $\bar{C}_1 = C_{1i}$. Thus C_{1a} can be found by equation 7.16, and uptake over the period $\triangle t_1$ is given by

$$2\pi \, a\alpha \, L_V \, C_{1a} \, \triangle t_1.$$

The new $\bar{C}_1$, $\bar{C}_{12}$, is calculated from the uptake and b, and C_{1a2} calculated from $\bar{C}_{12}$ as before. By this procedure, any time period can be covered. The method lends itself well to programming for a computer (see Nye *et al.* 1975).

One of the great advantages of equation 7.16 is that it relates uptake to the average concentration within any chosen volume of soil, i.e. a soil compartment. Thus it may readily be incorporated in large scale models that divide the soil into a series of compartments—which are discussed later.

Parameters changing with time

This last model has the further advantage that the time step procedure, whether carried out by computer or manually, allows parameters to change during the simulated time. Where mass flow is occurring in a limited volume, it is particularly useful to be able to alter θ and f_1 as the soil is dried out. $Db \sim (D_1 \, \theta f_1)$ thus decreases during the run, and the relationship of f_1 to θ must be known. Where b alters greatly with C_1, as for phosphate, it is also important to be able to vary this parameter as $\bar{C}_1$ alters with time; av_a and αa may be changed, and any interaction of one on the other (see p. 274 *et seq.*) allowed for.

Baldwin *et al.* (1973) compared the results for uptake, using this approximate method, with the more accurate computer simulation method of Nye & Marriott (1969). The results were in reasonably good agreement, the largest errors (of around 12%) being for high α, short times and high v, as would be expected.

A major weakness of steady state approximations lies in the assumption that a depletion zone, extending to the outer limit of the equivalent cylinder, is established instantly. Baldwin & Nye (1974) found that the accuracy of the method, at short times, improved if the boundary of the depletion zone around each root is considered to spread outwards with time, until it coincides with the boundary of the equivalent cylinder at

$$x = \sqrt{\frac{1}{\pi L_V}}.$$

This is achieved by inserting in the programme the condition

$$x = 2\sqrt{Dt} + a \quad \text{until} \quad 2\sqrt{Dt} + a > \frac{1}{\sqrt{\pi L_V}}.$$

Root system of increasing density

The root system of real plants develops, i.e. L_V is time variable, and the time step procedure described above is appropriate to deal with a proliferating root system also. In this case L_V and x vary with time in a way determined by the growth rate of the root system. Since a steady-state profile is assumed around each root, there is no time effect for an individual root, apart from the general decrease in $\bar{C}_1$, and new roots behave like old ones as soon as they appear. The electrical analogue method does not make this approximation, but a comparison of results by these two methods showed excellent agreement (Fig. 7.21).

The time step method is useful, but may be too tedious for very rough calculation. A simple but approximate method is to use in equation 7.11 and 7.12 the time-integral average, for the root density, $\bar{L}_V$, which gives

$$\bar{x} = \frac{1}{\sqrt{\pi \bar{L}_V}},$$

by summing the area under the curve of L_V against t. A similar principle can be applied to other variables such as θ, f_1 or Db. The use of the time integrals was suggested originally for the electrical analogue (Baldwin *et al.* 1972), and was tested by Baldwin *et al.* (1973) with an exponential increase in L. Time averaging causes an underestimate of uptake compared with the accurate simulation of increasing root density on the electrical analogue. The maximum error was around 20 % when the root system density increased rapidly according to $L = L_0 e^{2.4t}$, where t is in days. The error arises because time averaging implies that one 'root-day' is equally useful at any point in the uptake period; this is not so, since many roots for a short period are more efficient at uptake than a few for a long period.

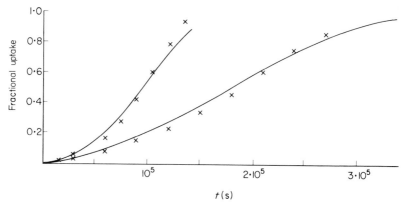

Fig. 7.21. Fractional uptake by a root in a limited soil cylinder, calculated by equation 7.12 and from the electrical analogue, showing good agreement (after Baldwin *et al.* 1973).

7.4.4 Models for whole root systems developing into fresh soil volumes— neglecting water uptake

If a boundary is drawn well outside the final extent of a developing root system (see Fig. 7.4) the uptake problem could be considered as one of pattern, with clumping being extreme at the outset of growth and becoming less with time. The larger the boundary drawn, the more extreme the clumping, but on the other hand, the larger M_∞, and the effect of changing the boundary on predicted uptake should be small. However, the problems in defining V/M in such cases may be serious. It is more effective to sub-divide the rooting volume into compartments, as is regularly done for computer simulation models of root uptake or simple solute transport in soils (e.g. Nye & Marriott 1969; de Wit & van Keulen 1972). Similar methods can easily be applied, with calculation of separate values of $\bar{C}_1$ in different compartments (Baldwin 1972). Transfer between compartments during each time step can be calculated simply from Ficks 1st law. If compartmentation is carried out in two or three dimensions, the size and complexity of the model becomes formidable, but the principles are unchanged, and simplified numerical methods, called 'locally one dimensional', of treating such problems are available (Mitchell 1969).

7.4.5 Testing and use of whole root system model

Any practical test of these models raises three questions:
(1) Can the model account for the measured uptake?
(2) Can it do so with an $\overline{\alpha a}$ which is of reasonable magnitude, in relation to measurement in solution culture?
(3) Is the 1st order boundary condition the best one to use?

For an adequate test of these methods very detailed sets of data are required, particularly of root length and distribution and soil solution concentration, and only experiments designed for these purposes will give them. Only two sets have been reported so far (Baldwin 1972; Brewster *et al.* 1975a; see p. 236). In the former the uptake of nitrate and potassium by rape plants growing in specially designed horizontally-sectioned pots, was measured. There were four compartments, separated by silicone grease membranes to prevent water transport. Root length and root distribution were determined by a radioautographic method and the soil solution concentration was determined. The measured and predicted value for potassium uptake (using equation 7.12) is in Fig. 7.22. In this particular case the exact value of αa was not very important, since it was so large that C_{1a} was very small. The value giving best fit (ca 2×10^{-6} cm² s⁻¹) is however quite compatible with data in Table 5.4. A set of model results for a constant flux boundary condition is shown, but this does not fit the experimental data so well as the 1st order boundary condition. For nitrate uptake the best fit was given by $\overline{\alpha a}$ of 1×10^{-7} cm² s⁻¹, which is also a reasonable value, but in this case a better fit would have been given by an $\overline{a\alpha}$ which decreased with time from 2 to about 0.5×10^{-7} cm² s⁻¹.

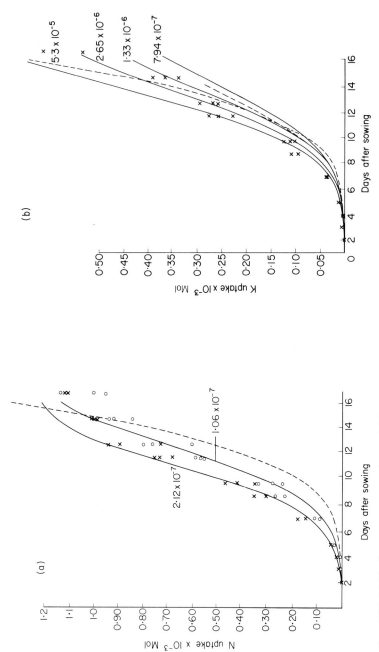

Fig. 7.22 (a) Single root system uptake model for nitrate, applied to rape plants grown in pots with four layers. Points show experimental data for uptake (x, plant analysis, o, soil analysis), full lines show model predictions for $\overline{\alpha a}$ values on curves. Dotted line is prediction for constant flux root surface property. (b) As above, for potassium. (After Baldwin 1972.)

The agreement of theory and experiment appears good enough to conclude that models of this type represent a simple developing root–soil system well. The main interest must centre upon whether they are flexible enough to represent crops in the field.

7.4.6 Model for whole root system simultaneously absorbing nutrients and water

The easiest way to visualize the nutrient and water uptake is by considering a block of soil cubes or compartments which form the volume affected by the plant root system. Each cube is small enough to be considered homogeneous in soil composition and root density. Nutrient uptake can be computed separately for each block, if L_V is known; but it must be assumed that αa and av_a are the same throughout the root system. The assumption about αa has been discussed above; that for av_a is more questionable, since it depends partly upon plant and partly upon soil. av_a, which is an inflow, not a root surface property analogous to αa, will certainly depend upon the value of θ in each compartment (Dunham & Nye 1973), in ways which are at present difficult to predict, but over short ranges of θ it appears reasonable to assume that uptake of water per unit length of root is constant (Taylor & Klepper 1975).

$\bar{C}_l$ and θ in each compartment vary at different rates, and at short intervals the transfer between compartments must be considered. While θ is still relatively large, the transfer for a mobile ion such as chloride or nitrate may be rapid. A water transfer, an associated mass flow of ions and a diffusive flow of ions must be calculated for each time-step.

All the various processes have been discussed in detail earlier, and models using them are described in Chapter 8, but one point requires further mention here. A mobile ion rarely suffers important depletion or accumulation involving a high-concentration gradient near a single root, except when the soil is dry. In the situation here, its transport from a distance may be more important than local transport around a root. Water moving in from distant compartments will carry such ions with it, and it is conceivable that these ions become concentrated in the zones where rooting is densest. For example, Jessup (1969) has found chloride to be concentrated round the base of salt bush in South Australia, which may in part be due to this process.

7.5 Whole plant models

Models dealt with earlier in this chapter refer solely to uptake by the root system, and the growth of the plant is, in effect, inserted into the model via the steady increase in L_V, which is a measured parameter. There have been a number of attempts to construct mathematical models of the growth of whole plants, from environmental and physiological data, and it is clearly useful if the root models described above could be joined to a plant growth model. There is then the possibility of representing mathematically all the major processes by which a plant increases in size.

Plant growth models are of widely varying complexity, ranging from extremely complex systems such as ELCROS (de Wit *et al.* 1969) to relatively simple sets of equations which aim to represent the most salient points in plant growth, such as that of Thornley (1972). Brouwer & de Wit (1969) included the root : shoot ratio as a variable, the root being regarded as a sink for photosynthate rather than as a source of nutrients. In general, all generate a value of the relative growth rate R_w from environmental variables and the inherent properties of the plant, as expressed in the equations of the model. The connection with root models can then be established from equation 7.2:

$$R_w \cdot X \cdot (W/L) = \bar{I} = 2\pi \, \overline{\alpha a} \, \bar{C}_{la} \, .$$

In addition, equations have to be formulated which express the change of R_w, W/L and $\overline{\alpha a}$ with X.

This view is probably oversimple, in that the mean percentage nutrient throughout the plant is a crude measure of its nutrient status, and R_w of its development. Reid & Bieleski (1970) showed the different effects of phosphorus deficiency on net assimilation rate E and on leaf expansion rate. Further, the percentage nutrient near meristems may be far more important than the plant average, and the two may not correlate perfectly. Avnimelech & Scherzer (1971) followed the growth of plants receiving varying rates of phosphorus in solution culture, and concluded that the growth rate at any moment was determined by the phosphorus content of the tissues during the previous history of the plant, rather than at that time. However, such refinements cannot be dealt with at present.

7.5.1 Whole-plant phosphate nutrition model—Nye–Brewster–Bhat model

Nye *et al.* (1975) and Brewster *et al.* (1975a,b) published a full model of this type, with the necessary solution culture experiments to determine $\overline{\alpha a}$, and a test of the model with soil-grown onion plants. The flow diagram for this model is in Fig. 7.23. In this, the central factor is the percentage of phosphorus in the shoot, which is taken to affect the various physiological and structural parameters determining the growth rate dW/dt, for given environmental conditions. These parameters are the net assimilation rate, the ratio of leaf surface area to shoot weight, and the ratio of shoot weight to root weight. In a more complex whole plant model, intended for use over a long period, the net assimilation rate would probably have to be amended by plant weight or time, to allow for self shading and leaf senescence.

The phosphorus percentage also modified $\bar{\alpha}$, the ratio of root surface area to weight, and the way in which absorbed phosphorus is partitioned between root and shoot. These variables, with solution concentration and with the shoot weight, then yield over a time step the new value for phosphorus percentage in the shoot, so completing the cycle. Such a scheme can be extended whenever it is believed that the processes governing any of the empirical relationships in it are so well understood that they can be formed into a sub-model. Until this happens, the model must only be used, strictly speaking, within the environmental and other limits within which the empirical relationships have been determined.

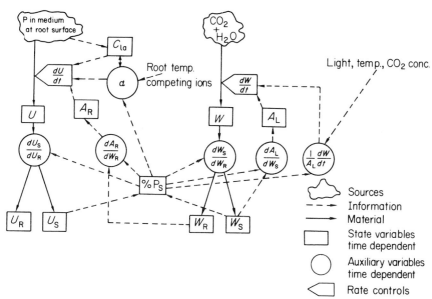

FIG. 7.23. Flow diagram for whole plant model of growth and nutrient content (after Nye *et al.* 1975). A_L = leaf area. A_R = root surface area. C_{la} = concentration in solution at root surface. $\%P_s$ = per cent P in shoot dry matter. U = Uptake of P. U_R = Root content of P. U_S = Shoot content of P. W = Weight (dry). W_R = Root weight (dry). W_S = Shoot weight (dry). α = root absorbing power.

The empirical relations needed were determined in solution culture experiments under a controlled environment (Brewster *et al.* 1975a), which gave the changes in plant parameters both at different levels of phosphate supply, and at different times of growth (Fig. 7.24a). The variation with time was occasionally greater than that which could be produced by different levels of phosphate, and it is clear that effect of growth stage needs careful attention.

With the onions used in this work, there seemed to be a physiological adjustment about day 16 of the growth period, with a shift in many parameters about this time, despite the constant conditions. There was occasional irregularity in the relationships between % P and the other parameters linked to it in Fig. 7.23, but despite this, it proved possible to simulate the uptake of phosphorus from *solution culture* satisfactorily by using the model and the measured parameters (Fig. 7.24 a & b).

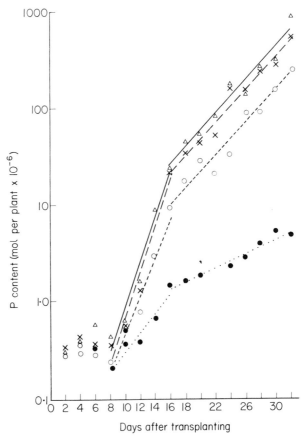

Fig. 7.24 (a). Measured content of phosphorus in onion plants grown in solution culture at different phosphorus concentrations (10^{-6}M. . . . . . ●; 10^{-5}M- - - - ○; 10^{-4}M — — — ×; 10^{-3}M————△). (After Brewster *et al.* 1975a.)

The essential part of this work lay in the test of the model on onions grown in soils of different phosphorus content (Brewster *et al.* 1975b), in a growth system in which water content and electrolyte concentration could be kept constant. Some growth parameters differed from equivalent ones in

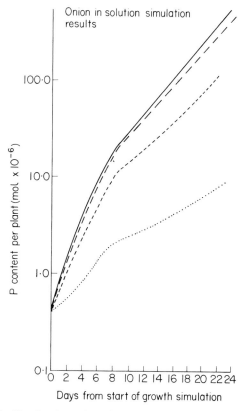

FIG. 7.24(b). Simulated uptake using model (after Brewster *et al.* 1975a).

solution culture: solution grown plants were able to achieve considerably larger shoot : root weight ratios and phosphorus percentages than any of the soil-grown plants, but in general the growth patterns were similar. The soil grown plants showed a steady drift in relative growth rate and associated parameters, rather than the sharp break found with the solution grown plants at day 16.

Simulation was carried out in two ways. First, the measured root lengths at different times were used as input into the model; second, the more interesting method was to use the plant parameters measured in solution culture to predict both growth and uptake (Fig. 7.25). In both cases in the soil with the lowest phosphate level the simulated uptake of phosphorus was less than the experimental; with increasing additions of phosphorus the simulated uptake became steadily larger than the experimental, i.e. the real plant growth and uptake was less influenced by the soil phosphate level than expected. Simulation was satisfactory for the intermediate levels of phosphate, but the processes taking place at low and high phosphorus levels were not being accurately modelled.

It is usually taken for granted that there is no competition for phosphorus between roots, for sparsely-rooted crops, but there was evidence that this may not be correct. Measurements on soil after harvest indicated that the effective buffer power in *structured* soil for plants might be as low as 2 in the high phosphorus soil, whereas that measured by equilibration of soil samples in solution gave values up to 20 times larger. Use of the smaller value, and equation 7.16 to calculate the concentration at the root surface, gave results closer to the measured values. The lower buffer power, and smaller soil solution concentration, explained an unexpectedly rapid decrease in mean flux with time.

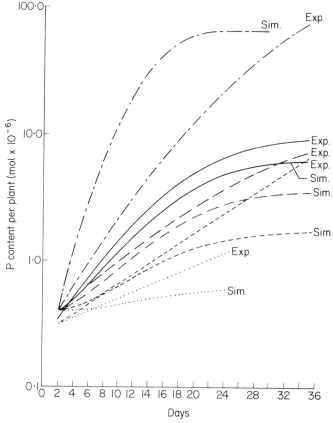

FIG. 7.25. Measured and simulated phosphorus content of onion plants growing in soil in pots, using full growth model, and taking buffer power equal to 2 (see text) (after Brewster *et al.* 1975b).

The general result of this modelling programme was satisfactory, but emphasized the problems still to be found in soil phosphate chemistry, and the need for more detailed understanding of the physiology of nutrient uptake and growth. Such clarification of the points where understanding is incomplete is one of the most useful aspects of the construction of mathematical models.

7.5.3 Scaife and Smith model

Scaife & Smith (1973) produced a model for phosphorus uptake by lettuce grown in pots of soil which was based on rather different ideas than those discussed here, though there were some underlying similarities.

Their assumptions were essentially that all phosphorus brought by mass-flow to the root is absorbed, and that a diffusion component is added to this. The latter was calculated by assuming a hypothetical concentration at the root surface, which was determined solely by the internal concentration of phosphorus in the plant. R_w was found, from the experiment, as a function of percentage phosphorus in the plant also.

The assumption about the concentration at the root surface is interesting. This is somewhat analogous to the methods given above, in that the plant root sink strength is determined by the phosphorus status in the plant, but the concept of a concentration determined solely by the plant (rather than by both plant and soil, via $\alpha a/bD$) is probably only acceptable for high relative sink strength. In fact, the method of determining this value (equivalent to C_{la} in our methods) depends upon a plot of observed phosphorus uptake rates per unit plant weight (closely related to $\bar{I}$) and soil solution phosphorus concentration (C_{li}). Essentially, therefore, the method is similar to the Passioura equation, determining ($D_1 \, \theta f_1 \, \gamma$) and C_{la} from the uptake data instead of from physical measurements (see p. 151). The main uptake equation was (using our notation)

$$\frac{dU}{dt} \bigg/ W = \text{COND} \times (C_{li} - C_{la})$$

where COND is a 'conductivity' or proportionality factor. The physical significance of COND is in effect

$$\frac{2\pi D_1 \, \theta \, f_1 \, \gamma L}{W}.$$

The method therefore assumes that L/W and γ remain constant in soils of different phosphorus content, which is unlikely. Despite this, the model was successful in predicting uptake of phosphorus by lettuce growing in two soils.

7.6 Conclusion

The main aspects of the uptake of nutrients from homogeneous soil by whole single plants have been considered, and certain generalizations are possible. Where the soil nutrient behaves in a relatively simple way, at least over a short time, such as nitrogen or potassium, our understanding of the soil processes is at least adequate. For phosphate, and probably for most micronutrients, the processes involved are so complex, and so open to modification by the plant itself, that we cannot regard the situation as satisfactory.

The knowledge of the required plant parameters is still less satisfactory, since most of the relationships are empirical, and probably oversimplified. For example, the mean nutrient percentage in the shoot is used as the main parameter, but it may well be the nutrient content in some metabolic pool (e.g. ATP) or in some defined part of the plant (e.g. shoot meristem) which is truly important, and the relationship of these may not be simple and direct. The vast quantity of data relating percent nutrient in plants or plant organs to growth is often useless for this purpose, since 'growth' is usually given as a single harvest, rather than a relative growth rate; leaf and root areas are unknown, and most of this type of work has not been done in controlled environments.

With so many problems remaining for accurate experiments in closely-controlled environments, it may seem vain to attempt to analyse field-grown crops in this detail yet. Various approaches have however been made, which are discussed in the next chapter.

8

MOVEMENT AND UPTAKE OF SOLUTES
IN FIELD CONDITIONS

INTRODUCTION

Previous chapters have discussed the principles of soil and plant behaviour which affect their interactions, and Chapter 7 presented a synthesis into an acceptable view of the nutrition of a single isolated plant. This chapter is arranged to discuss the processes that need to be incorporated in a model of a community growing in the field.

(a)   Shoot growth and associated root growth and pattern. Discussion of crop growth in relation to the aerial environment is beyond the scope of this book, and is covered by the many attempts currently being made to develop whole plant or community growth models (for references see Thornley (1976), Milthorpe & Moorby (1974), Wareing & Cooper (1971), Jeffers (1972)). In such models the section dealing with the root–soil system is usually much simplified, and we aim to contribute to the wider scheme by amplifying this aspect. We concentrate on the effects of agronomic practice and soil variation on root behaviour, though we also give examples (p. 273 & p. 282) in which detailed root–soil models are linked to crop growth models.

(b)   Water movement in the profile. This has been discussed in Chapter 2, and the principles are covered more fully in the works of Childs (1969), Kirkham & Powers (1972) and Hillel (1971). Methods of treating water uptake by crops from each part of the profile are not so widely known and we concentrate on this aspect.

(c)   Solute movement and balance in the profile. This has always received attention in the study of irrigation and saline soils, and more recently out of concern for eutrophication and pollution. The net intake of nutrients by a crop or community is part of the overall nutrient balance. Together with growth it determines the nutrient composition of the plants, which in turn is an important factor regulating their further growth. The division of the total intake among different parts of the root system is related to local movement of solutes in the adjacent soil. This has been little studied in situations where more than one plant species is involved, and raises the question of inter-specific root competition for nutrients, which we discuss at length.

We do not deal with overall solute balance sheets in agricultural and forest crop sequences or natural ecosystems, since full acounts are readily available

(Cooke 1967, 1969), and much more data should shortly be published as a result of the International Biological Programme and the Man and the Biosphere Programme (UNESCO 1971). It is planned that much of this data should be published within the framework of systems analysis. The wide scope for this approach in soil ecosystems is indicated by Russell (1975). He identifies the main limitation to the application of this method as the lack of quantitative information of soil and plant parameters, and we discuss how these may be supplied.

8.1   Detection and measurement of roots in the field

The principles of the available methods have been outlined in Chapter 7, and the discussion here is confined to the practical value of the various methods in field conditions, in yielding quantitative data on root quantities and dimensions.

The most dependable method, and one that allows root dimensions to be measured, is undoubtedly the physical extraction and measurement of roots; and a variety of ingenious pieces of engineering have been constructed for this purpose (Welbank & Williams 1968; Ellis & Barnes 1973; Welbank *et al.* 1974). On a suitable soil this type of equipment works well. The major delay is in washing out and measuring roots. For this reason there has been much interest in the alternative methods given in Chapter 7.

The method which has been most extensively used is the injection of ^{32}P into the soil (see p. 193; Russell & Newbould 1969; Boggie & Knight 1962) and measurement of the uptake into plants. It inevitably suffers from the disturbance caused in the soil if intensive injection is carried out, and from the uncertainty over the relation of uptake to amount of root in different soil horizons. If the soil moisture content varies seriously with depth, uptake of the ^{32}P will be decreased in the dry horizons. Basic theoretical problems associated with this method have been discussed in Chapter 7. The injection of isotopes—usually ^{86}Rb—into plants, followed by soil core sampling and isotope assay to determine root volume, has been used in a variety of experiments. This method has been brought to high technical efficiency (Ellis & Barnes 1973), but it will always be laborious, in the field. The coefficient of variation of ^{86}Rb count rate associated with any one core measurement under a 10-week-old barley crop (D.R. Hodgson, private communication) was found to be over 20 % in the topsoil, and numerous replicates are needed for accuracy. It is unsuitable for use in the earliest stages of plant development when root density is low, and cannot be used to measure root length. Regrettably, there appears to be no real alternative to excavation and washing methods for reliable measurement of this parameter, which is likely to be a most important one.

8.2 Root distribution and density in the field

For a whole crop, the root system may be described in terms of the amount and distribution with depth of root length and weight, if necessary in terms of size classes. These may be varied by almost any changes in the circumstances of the crop, and the effects are briefly described here.

8.2.1 Planting density

This affects the mean interplant distance, and hence the time when the individual root systems meet. It is also likely to determine the final size of the individuals within the crop. In general terms, we expect a greater planting density to cause a more rapid development of total root length and total root dry matter early in the life of the crop, and the earlier meeting of the extending roots will probably cause increased rooting depth. Later the crop density may have less effect on root dimensions, but high densities which greatly restrict the final size of each individual plant may cause a smaller final rooting depth. Since low densities may lead to a small rooting depth because of the lack of root competition in the topsoil (Fig. 8.1) we expect a maximum rooting depth and exploitation at intermediate plant densities. Kirby & Rackham (1971) examined the distribution of barley roots with depth, at planting densities of 50, 200 and 800 plants m^{-2}, on a light loam. They found that the middle density produced the greatest percentage and absolute weight of root at depth (Fig. 8.2) in agreement with this argument. The root distribution of apples planted at different densities, observed in a root observation chamber (Atkinson 1973), showed 92% of white root above 40 cm depth when trees were at 240 cm spacing, but only 11% when they were at 30 cm spacing. In the latter case 47% was below 100 cm.

8.2.2 Temperature and season effects for perennials

Most perennials, especially trees, gradually build up a net of main branches which persist for a long time, though grass roots may die back and be replaced annually (Stuckey 1941). The re-growth of tree–root systems in spring (and presumably at the beginning of the wet season where there is a monsoon climate) thus starts from a great number of points, instead of from the single one of the annual plant, but the principles governing uptake are exactly the same. The main roots of trees show secondary thickening, though there is evidence (Head 1968) that their rate of thickening is more irregular than that of trunks. The main branches of the network are unlikely to be active absorbers of water or nutrients, and the active fraction of the root system by weight will, therefore, always be much less than in young annuals. Thus Ruer (1967) showed that the percentage of roots assumed to absorb (tertiary and

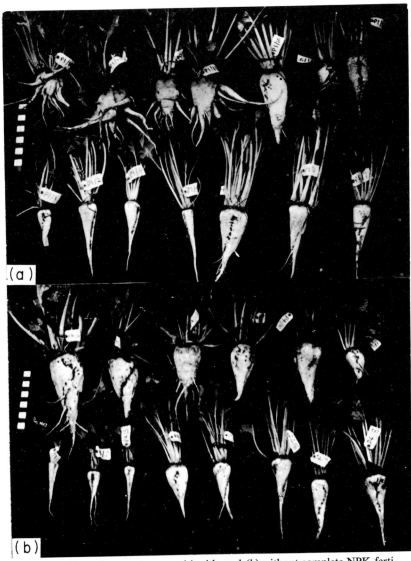

Fɪɢ. 8.1. Sugar beet roots grown (a) with, and (b) without complete NPK fertilizer, in a plant population trial. In both (a) and (b) the plant density increases
from top left round to lower left of picture. Note shallow, fangy root system
with low density and fertilizer. Increasing density has same effect on root shape
as less fertilizer.

quaternary) in oil palms formed less than 30% of the total weight of the root
system, and the weight of apple roots over 5 mm diameter formed over 75%

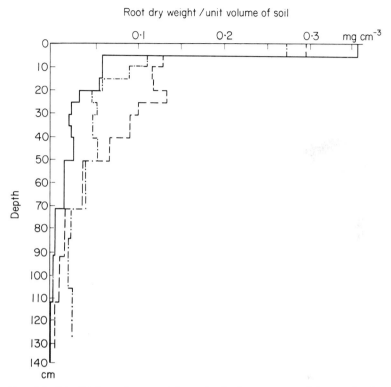

Fɪɢ. 8.2. Distribution of root weight under barley crops, at different planting densities. ————, fifty plants m⁻²; —.—.—.—, 200 plants m⁻²; ————, 800 plants m⁻²; (after Kirby & Rackham 1971).

of the total in cores taken 50 cm from the base of the trunk (Atkinson 1973, 1974). Rogers & Head (1969) give data for the amount of white (new) root on apples, showing a peak in spring, and a subsidiary peak in autumn (Fig. 8.3(b)); the sharp effect of pruning on root growth is also shown. Fine tree roots are continually replaced, possibly in bursts of growth in limited parts of the root system (Reynolds 1970).

Very good sets of data for grasses are available from the work of Troughton (1957), Garwood (1967) and Williams (1969). New grass roots are initiated at the plant base. The pattern of root initiation normally has a maximum in spring (Fig. 8.3(a)), and the elongation rate of the roots was greatest in May, August, and September. It is noteworthy that the greatest rates of extension were found in the soil below 24 inches depth. Many of the roots formed in spring die off relatively rapidly, and the mean life of roots appeared to be less than 12 months. Examples for trees and grasses quoted above are

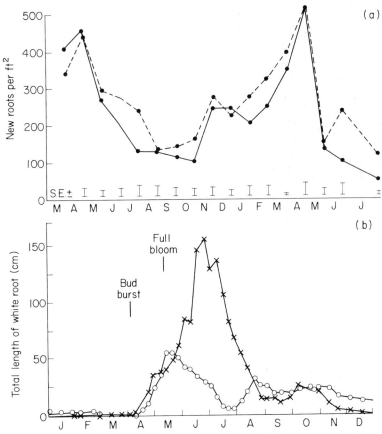

FIG. 8.3. (a) Initiation of new roots over 2·5 cm long on perennial ryegrass at different times of year, either (- - - - -) irrigated or (————) not irrigated (After Williams 1969.) (b) Amount of white root seen in root observation chambers, under Cox's Orange Pippin apple trees throughout season, o—o—o pruned; x—x—x—x unpruned. (After Rogers & Head 1969.)

for the temperate humid climate of Britain, and other weather patterns can produce different growth cycles (see Troughton 1957).

8.2.3   Crop nutrition

The general effects of mineral nutrients have been discussed earlier. The most obvious outcome of these processes in the field is the almost universal concentration of roots in the topsoil, which is rendered fertile by manuring or by natural return of litter. Placed fertilizer, especially nitrogen and phosphorus, always encourages the growth of roots locally; and in addition, nutrients may increase the size of plants, and hence whole root systems. Root distribution in the profile can also be altered by fertilizer, and Fehrenbacher

& Alexander (1955) found that the weight of maize roots was similar down to 100 cm both with and without a complete fertilizer, but roots extended for 25 cm further down in fertilized plots. Fertilizer placed at depth greatly increased rooting in the subsoil (de Roo 1969), though the increase in crop yield was much less dramatic.

8.2.4 Water relations

The development of the root system compensates for some of the variation in water supply in a profile during a year. If the rest of the profile is well watered, a water table is of little value, but if the rest of the profile dries out, the root system will tend to proliferate above the water table (Lipps *et al.* 1957) (Fig. 8.4). A static water table can thus be a very considerable benefit,

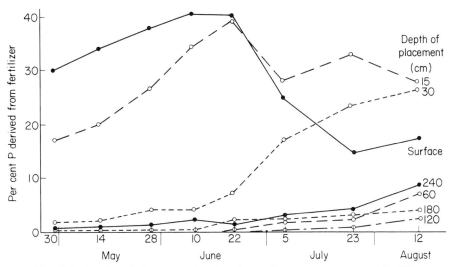

Fig. 8.4. Activity of a lucerne root system during the season, as measured by ^{32}P uptake, above a water table at 240 cm (after Lipps *et al.* 1957).

but an unstable one is very damaging if the root system has insufficient time to adjust itself to the varying conditions. There is a tendency for root systems to shift down the profile during the growing season. For annuals, this is partly the result of the general extension of roots, but for both annuals and perennials it may occur when the upper parts of the soil dry out, and the roots in the topsoil may become inactive, as in rye-grass sward (Garwood & Williams 1967) where no active roots were found in the top 6 inches in July. In areas of limited rainfall, root systems may be confined to the depth wetted during the rainy season. Thus Russell (1961) gives pictures of prairie grass-root systems, in which the plant size and root depth diminished with annual rainfall. However, in more humid conditions increased water favours shallow rooting. Bloodworth *et al.* (1958) examined the root distribution of several irrigated crops by a coring technique, and found uniformly high concentrations of roots in the topsoil, with a regular decrease in root density down the profile. Rowse (1974) compared roots of lettuce with and without irrigation, and found that irrigated plots always had more root in the topsoil, and less in the deeper subsoil, than non-irrigated. In both cases the vertical distribution could be expressed by an exponential equation (see p. 252). Root diameter was always least with irrigation, probably because of the smaller mechanical impedance in wet soil.

8.2.5 Soil structure and tillage

The most striking effect of this kind is shown by excessive soil compaction, usually caused by heavy traffic over soil at an unsuitable water content. Surface compaction may cause serious loss of crop yield (Flocker *et al.* 1960; see Taylor 1974), but it is normally cured easily by ploughing. Compacted layers below plough depth, or plough pans, are more insidious, but may be equally damaging (de Roo 1969). They rarely prevent all penetration of roots, but the lower part of the root system is restricted and its development delayed, which is particularly serious if the crop depends upon subsoil water to complete its natural growth cycle. Such situations are however the exception rather than the rule, and roots appear able to extend readily in most natural subsoils. Experiments with deep tillage (Russell 1956; de Roo 1969) have usually given no or small increases in yield on average, and it seems that direct evidence of unusual compaction—from root observation, soil bulk density or penetrometer resistance—is required before any benefits from deep cultivation can be expected.

   Zero-tillage or 'direct-drilling' methods, in which seeds are planted directly into undisturbed soil, have shown that a similar situation may hold in the plough layer also (Finney 1973; Connell & Finney 1973; Finney & Knight 1973). Their adoption usually leads to steadily increasing compaction in the topsoil at first, compared to ploughed areas, but a more or less constant level is reached which appears only rarely to be so large that root growth is seriously impeded. Seminal roots of wheat elongate more slowly (Finney 1973), but branch more, leading to a shallower root system than with ploughed soil (Table 8.1). The reverse, with deeper rooting on non-tilled soil, has however also been observed with spring barley (D.R. Hodgson, private communication).

8.2.6 Pests and diseases

By implication we have discussed healthy plants throughout this book, but in practice all field plants must suffer from some degree of infection or

Table 8.1. Root extension under barley crops grown without soil cultivation or after ploughing (after Finney 1973).

Cultivation treatment	Length of primary seminal root		Length of 1st pair of lateral seminal roots	
	22 October	Mean 1–13 November	22 October	Mean 1–13 November
Ploughed	2·21	14·1	0·89	12·2
Direct-drilled	2·05	10·6	0·88	9·2
Shallow cultivated	2·01	11·9	0·83	9·5
Deep cultivated	2·25	15·1	0·84	12·6
Standard error (single plot)	±0·25	± 1·4	±0·12	±1·0
Sig. diff. between treatment means $p = 0{\cdot}05/0{\cdot}01$	Not sig.	2·8/4·3	Not sig.	2·1/3·3

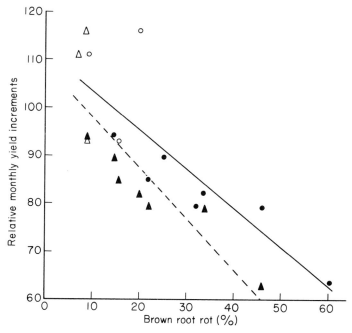

FIG. 8.5. Relation between the incidence of brown root rot and tomato yields. Data were derived from the third, fourth and fifth successive crops grown in untreated soil, the monthly yield increments being expressed as percentages of those from comparable plants growing in partially sterilized soil. Yields are plotted against the incidence of brown root rot either at the beginning (○, ●———) or one month before (△, ▲- - - -) the relevant month of picking. ○, △ = yields during the first months of picking in the different seasons; ●, ▲ = yields during succeeding months (after Last *et al.* 1969).

infestation with pathogens or pests, however minor. The precise way in which the plant is damaged varies widely, but we are particularly interested in organisms which damage roots directly. The complex effects of rhizosphere organisms in general on root form and function have been mentioned in Chapters 5 and 6, but here we are concerned with obvious effects such as loss of a large part of the root system (Fig. 8.5). This is caused by, for example, the 'take-all' fungus (*Gaumannomyces graminis*) which forms lesions cutting off large parts of the cereal root system. Many other pathogenic fungi (Chamberlain 1962) attack the finer roots, leading, for example, to 'damping off' of seedlings, caused by *Rhizoctonia* or *Pythium*. Nematode attack often has similar effects; free-living nematodes (for example *Longidorus* and *Trichodorus* species) may cause serious damage to root systems, leading to symptoms of magnesium and nitrogen deficiency on sandy soils (Whitehead *et al.* 1971). The root-knot nematodes (*Heterodora* species) have even more dramatic effects, so that attacked potatoes and sugar beet wilt readily because of their scanty root systems.

8.2.7   Profile distribution

It is almost axiomatic that root density decreases with depth, but a glance at Kutschera (1960) shows the variety of ways in which roots can be distributed. However, models of crop nutrient uptake require a function relating the distribution of the crop's roots with depth or a root extension model. At present this must be entirely empirical. Gerwitz & Page (1974) have found that a plot of the logarithm of root quantity in a soil layer of unit thickness against depth tends to a straight line. This reduces to

$$P_z = 100 \left(1 - e^{-fz}\right) \qquad (8.1$$

where P_z is the percentage of root between the surface and any depth z, and f is a constant. From this, $1/f$ equals the depth of soil containing 63% of the total root mass. Gerwitz & Page summarized data showing that this depth is less than 30 cm for the majority of crop plants investigated (Table 8.2), and considered that this function described the root distribution well for some 70% of the cases quoted.

8.2.8   Root length per unit surface area

In field work the most useful parameter of root quantity is usually the amount per unit land surface (L_A) since it is the latter which is related to the rate of transpiration, the rate of photosynthesis and (less directly) the rate of nutrient uptake in a crop with complete cover. The lengths vary quite widely, within limits for L_A of about 10–1000 cm cm^{-2} (Table 8.2) with grasses

Table 8.2. Collected data for root length per unit volume of soil and per unit surface area, for different species, depths and plant ages.*

Species	Plant age days	Depth cm	L_V cm^{-2}	L_A cm^{-1}	Source
GRAMINEAE					
Poa pratensis		0–15	55	840	Dittmer (1938)
Maize	79	0–15	4	145	Mengel & Barber
		15–30	2·1		(1974)
		30–45	1·4		
		45–60	1·2		
		60–75	1·0		
Maize*	37	0–15	1·0	30	Foth (1962)
		15–30	0·8		
		30–45	0·1		
	100	0–15	7·7	170	
		15–30	2·7		
		30–60	0·6		
		60–90	0·1		
Barley*	'almost	0–5	4·2	96	Kirby & Rackham
(200 plants m^{-2})	ripe'	5–10	1·6		(1971)
		10–20	1·1		
		20–60	0·7		
		60–140	0·3		
Spring wheat†	38	0–15	0·74	11.8	Welbank *et al.*
		15–30	0·04		(1974)
		30–60	0		
Spring wheat†	94	0–15	3·3	111	Welbank *et al.*
		15–30	1·1		(1974)
		30–60	1·1		
		60–100	0·3		
OTHER SPECIES					
Styloxanthes	140	0–10	32	670	Torssell *et al.*
humilis		10–20	12		(1968)
(pure plot)		20–50	4		
		50–100	2		
Clover	—	0–30	10	310	Newman (1966)
Tea	11 years	0–2·5	4	140	Barua & Dutta
(red plus		45–47·5	1		(1961)
white root)		68–70	0·5		
Pinus radiata	—	0–8	2	80	Barley (1970)
		25–45	0·8		
		91–106	0·4		
Lettuce	67	0–10	2·4	47	Rowse (1974)
(not irrigated)		10–20	1·0		
		20–40	0·5		
		40–60	0·13		

* Only results obtained by root extraction and direct measurement techniques on normal crops have been included. However, root length measurements are so scanty that some data based on root dry weight have been included, using a conversion factor of 150 m g^{-1}, and

giving values at the upper end of the range, and trees at the lower end. For most mature field crops it would seem reasonable to assume a dry weight of 1–10 mg of root per cm² of land surface. It should be borne in mind that these data have been obtained by different methods, and the recovery of all roots may often not have been achieved. Most of the values are for mature crops or vegetation stands, and root densities may be much lower for younger plants (see Fig. 7.5).

8.3   Water transfer and uptake by crops

The transfer of solutes in the profile depends heavily upon water content and movement, and it is therefore almost essential to consider this before dealing with any model of the uptake of solute. There are essentially three main problems:

(a)   The transpiration of the crop must be known. If water in the soil is not limiting and if crop cover is complete, then it may be found by the Penman formula or one of its variants (Monteith 1973), from purely meteorological data. If soil water is limiting (see Table 2.4) it is difficult to predict evapotranspiration with any accuracy (Monteith 1965). Models of the soil-plant-atmosphere continuum may be constructed, in which water stress causes stomata to close, and hence cuts water use, but these are critically dependent upon the assumptions made about stomatal response.

(b)   Water uptake is distributed throughout the rooting volume, and it has to be related in some way to the distribution of roots.

(c)   The movement of soil water from one level in the profile to another has to be calculated from a knowledge of water supply, hydraulic conductivity and the way in which water is extracted by roots. Movement of water in the absence of a crop can be dealt with readily by the normal treatment of infiltration, redistribution and drainage (Chapter 2).

Various attempts have been made to deal with these factors. A number of measurements of soil water under a transpiring crop have been made (for example, Rose & Stern, 1967). Gardner (1964) attempted to relate root distribution and water uptake by starting from the general resistance model (Fig. 2.6), so that

(Footnote continued from Table 8.2.)
these are marked with an asterisk. For earlier compilations and references, see Barley (1970), Newman (1966), Milthorpe & Moorby (1974) and Gerwitz & Page (1974).
†   Winter wheat, oats and barley root lengths were measured in the same experiment. Winter wheat root length was greatest at the early sampling dates, but by 94 days all root lengths were surprisingly similar, and there was little difference among the spring-sown cereals at any date.

$$w = \frac{\psi_{\rm s} - \psi_{\rm r}}{R_{\rm p} + R_{\rm s}} \tag{8.2}$$

where $\psi_{\rm r}$ was the potential in the root, which was assumed constant (but see Chapter 2.3), $\psi_{\rm s}$ the soil matric potential, $R_{\rm p}$ and $R_{\rm s}$ the plant and soil resistances and w the water uptake rate per unit soil volume. It was assumed that

$$R_{\rm s} = \frac{1}{BK L_{\rm V}},$$

where B is a constant (see p. 24). Hence for the i th layer of soil,

$$-h \left(\frac{d\theta}{dt}\right)_i = hw_i = (\psi_{{\rm s}i} - \psi_{\rm r} + z_i) K_i L_{{\rm V}i} Bh \tag{8.3}$$

where $\psi_{{\rm s}i}$ is the matric potential in the ith layer at depth z_i, and h its thickness. If q is the uptake per unit surface area

$$q = \sum_{i=1}^{i=n} hw_i = Bh \sum_{i=1}^{i=n} (\psi_{{\rm s}i} - \psi_{\rm r} + z_i) K_i L_{{\rm V}i} \tag{8.4}$$

If q is known, this set of equations can be solved to give θ as a function of t and z. It is possible to consider q in terms of an 'integrated conductance',

$$\bar{K} = \sum_{i=1}^{i=n} K_i L_{{\rm V}i}$$

and an 'integrated soil water potential',

$$\bar{\psi}_{\rm s} = \frac{\sum_{i=1}^{i=n} K_i L_{vi} (\psi_{{\rm s}i} + z_i)}{\sum_{i=1}^{i=n} K_i L_{{\rm V}i}} \tag{8.5}$$

so that
$$q = Bh (\bar{\psi}_{\rm s} - \psi_{\rm r}) \bar{K}. \tag{8.6}$$

The effective soil water potential and conductivity is thus weighted for the root distribution. The equation was used to predict root distribution of sorghum plants growing in cylinders, from water uptake data, but the agreement was not good, there being much more root at depth than predicted.

The relatively poor agreement is probably due to the assumption of a constant $\psi_{\rm r}$, and this aspect was dealt with by Nimah & Hanks (1973a). Their model aimed to represent the removal of water from a soil profile carrying a growing crop. It started from the general continuity equation for flow in a profile

$$\left(\frac{d\theta}{dt}\right)_z = \frac{d}{dz} \left(K \frac{d\psi}{dz}\right) + A(z, t) \tag{8.7}$$

when ψ in head units is soil matric potential, and $A(z, t)$ is a 'plant root extraction term' (negative for uptake) at depth z and time t. The latter is the essential part of the model, and it is defined as follows:

$$A (z, t) = \frac{[\psi_{\text{root}} + (1 + R_\text{c}) z - \psi(z, t) - \pi(z, t)] [RDF (z) . K]}{\Delta x \; \Delta z} \quad (8.8)$$

where ψ_{root} is the water potential in the root at soil surface level, R_c represents the frictional resistance to water movement in the root from level z to the surface (see Chapter 2.4), π the soil osmotic potential, $RDF (z)$ the fraction of the total root in the depth interval Δz at z, and Δx is the mean distance between roots. ψ_{root} is held between the limits zero and wilting point, between which it is adjusted to give a total root uptake equal to the known potential evapotranspiration. Soil surface flux (infiltration and evaporation) is represented by a simple

$$\frac{(\psi_\text{o} - \psi_1)}{\Delta z} K_{\frac{1}{2}}$$

type equation, where $K_{\frac{1}{2}}$ is the mean conductivity between the surface $(z = \text{o})$ and level 1 $(z = z_1)$. The potential evaporation is taken as an approximate percentage of potential evapotranspiration for incomplete plant cover and the lower boundary condition is set to constant water potential, or to zero flux. The model predicts the total absolute change in water content with depth during water extraction, and has achieved fair success in field tests (Nimah & Hanks 1973b). It does not deal with root systems which change properties, size or distribution with time.

Van Keulen (1975) used computer simulation methods to predict movement of water in a profile from which it was being extracted by a growing wheat crop. His model assumed uniform water potential within the root, and calculated a resistance to water uptake within each soil layer. With θ at the beginning of the season 0·23, K $0·8 \times 10^{-2}$ cm d^{-1}, and the rooting depth 6 cm, simulation indicated that water flowed easily up the profile and replaced up to 45% of that lost from the rooting zone, depending upon precise assumptions about root distribution. With θ of 0·15 and K of 2×10^{-4} cm d^{-1} later in the season, the predicted water movement through the soil into the rooted depth was quite negligible, and the plants would be wholly dependent upon soil water held near the roots.

It has been suggested that the water in the crop itself may be sufficient to supply a considerable fraction of the daily demand (see Jarvis 1975), which would invalidate the general assumption that the immediate uptake rate by the roots equals transpiration. However, with 1 cm of transpiration equalling 100 tons ha^{-1} of water, it seems that this supply must be rather short-lived. Water stress would normally be expected to show as lowered water potential

and content in the plant, but Jackson (1974) reported that comparisons between grass in lysimeters which was growing in dry soil or was irrigated
showed a difference of only 2 bars in leaf potential, less than the difference in
the topsoil, even when irrigation produced considerably larger yield. The
calculation of water uptake at different points in the drying profile is still very
difficult, and it is uncertain whether satisfactory results can be achieved. All
other parts of the soil–water balance can be quantified relatively easily
(Makkink & van Heemst 1975), but the final model must remain imperfect
until water uptake by root systems is better understood.

8.4   Transfer of salts in a profile

In any soil profile, especially if it carries vegetation, there is a constant movement of salts. This results from addition of solutes and water at the surface,
losses at the lower boundary of the profile, the generation of salts by decomposition of minerals or mineralization of organic matter, or absorption
of salts and water by plant roots. Here we consider transport in a vertical
direction only, and it is convenient to distinguish three different situations:
movement of a single band of salts, desalination of a profile by leaching with
pure water, and the balance set up in a profile supplied with water containing
dissolved salts. These cover leaching of fertilizers or other added solute,
leaching of nitrate and other salts initially present, and the use of saline
irrigation water.

The general principles of solute transport are in Chapter 4. They have
been tested on soils in the laboratory, and found to be accurate, or at least
adequate, in this situation. In the field, however, it is the exception rather
than the rule that a given system can be accurately explained or predicted,
because of the usual variations in the parameters with time and space that
are difficult to quantify. The general continuity equation for diffusion with
convection in one dimension is (Chapter 1.41):

$$\frac{\partial C}{\partial t} = \frac{\partial}{\partial z}\left(D^* \frac{dC}{dz}\right) - \frac{\partial}{\partial z}(vC_1) + R \tag{8.9}$$

where R can be a source or sink term, i.e. positive or negative. D^* is the
longitudinal dispersion coefficient, which includes dispersion due to thermal
diffusion and also hydrodynamic dispersion caused by non-uniform velocity
of flow and other effects (see p. 86 *et seq.*). In the field it is most unusual for v
to remain constant for long periods, and D^* is dependent upon both v and the
water content. There is often a substantial term R in the equation, due to
removal of salts by roots, demineralization of nitrate or precipitation of
sulphate, or production of nitrate and sulphate by mineralization. The removal of water locally by roots causes serious problems (Whisler *et al.* 1968)
since it varies with time, and it is difficult to determine where it occurs. Under

natural vegetation the dominant anion is often bicarbonate, whose concentration alters considerably with changes in temperature and carbon dioxide concentration in the soil air (see p. 57). It is therefore not surprising that most treatments of these problems aim at only rough agreement with field practice. It also accounts for the great use which has been made of numerical and computer simulation methods in this subject, since they allow great flexibility in solving equation 8.9 with boundary conditions specified by the problem.

There are no special difficulties in theory for non-adsorbed ions, such as chloride and nitrate, unless anion exclusion is serious. The latter causes 'salt sieving; in which the soil—usually clay with only very fine pores—acts as a leaky semipermeable membrane, allowing water to pass more easily than salts. Differences in salt concentration (osmotic potential) may then cause bulk movement of water (Letey *et al.* 1969; Bolt & Groenevelt 1972). If the exclusion of anions is only partial, they will be concentrated in the channels where water flow is most rapid, and will move faster than the average water flow by a factor, which was determined experimentally to be 11% in an undisturbed sandy soil and 37% in a clay soil (Frissel *et al.* 1973).

A clear account of the special problems of herbicide movement in the soil has been given by Hartley (1964).

8.4.1 Band movement

The analogy with movement in a chromatographic column at once suggests itself, and theory developed for this (Glueckauf 1955) has been freely used (Frissel *et al.* 1970) (see Nye 1974 for discussion).

(a)   Non-adsorbed solutes. It is simplest to consider movement of a single band of an applied non-absorbed solute down a profile as composed of two processes: a bulk transfer of salt in the moving water at a mean rate of v/θ, where v is water flux, in ml cm^{-2} s^{-1}, and a spreading out of the band due to diffusion and to hydrodynamic dispersion (Fig. 8.6). The situation has been dealt with simply by Gardner (1965), who gives the equation:

$$C_1 = C_{1o}\, z_o\, (4\pi D_1^*\, t)^{-\frac{1}{2}} \exp - \left[\left(z - \frac{vt}{\theta}\right)^2 \frac{1}{4D_1^*\, t}\right] \tag{8.10}$$

where C_{1o} is the initial concentration of solute in the uniform band of depth z_o, and the other symbols have their usual meaning. Fair agreement was found with field data, which is encouraging, indicating that theory can be verified despite the natural variations in rainfall and other parameters.

(b)   Adsorbed solutes. If the adsorbed ion has a linear isotherm the problems are identical to those above, with ion mobility reduced as usual by a factor equal to the buffer power, so that the mean velocity of the band is v/b. However, sulphate, phosphate and organic compounds typically have curved

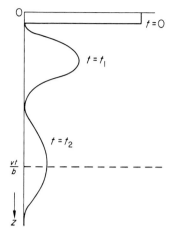

FIG. 8.6. Movement of a band of solute down a soil column, with dispersion (after Nye 1974).

isotherms. If D rises with C (see p. 82) the mobility of the solute is greatest at the centre of the band, and a skew distribution must result, with a sharp leading boundary and a diffuse trailing one.

(c) Exchangeable cations. The movement of exchangeable cations is less simple, since they must be displaced by other species of cations. If the ion in question is a very minor fraction of all exchangeable ions, and if the concentration of the competing ions in the percolating solutions remains constant, the situation reverts to that of simple adsorption, since $C_1 \propto C$. If the solution concentration varies, the situation is more complex. Thus, after a dressing of fertilizer, for example potassium chloride, is applied to the soil surface and worked into the top layer, the chloride will be washed down the profile as described for non-adsorbed species, but the nature of the cations associated with the chloride will change progressively towards that of the major cations present, and the added cations will move more slowly, and at a rate which will depend upon the subsequent anion concentration. It has been pointed out earlier (see page 21) that if solid salt is supplied to the soil, the initial process of movement is very much more complex than simply leaching and diffusion.

The leaching of sodium salt applications has received particular attention, because of the possible damage to structure caused by this ion. Bower *et al.* (1957) investigated the rate of leaching in the laboratory, and showed that, if calcium is the major cation in the leaching solutions the sodium passes down the soil in a compact band with a sharp upper boundary. If the cation in the band is more strongly held on the exchanger than the major cation in the leaching solution, then the upper boundary is diffuse, and the band will develop an elongated 'tail' (Fig. 8.7). Theories of cation chromatography (Helfferich 1962, p. 421) usually deal with a constant concentration of total salts, and consequently are not applicable directly to a salt applied to a soil surface. This situation is best dealt with by numerical solution of equation 8.9 (Frissel *et al.* 1970; de Wit & Van Keulen 1972). For very rough predictions, Tinker (1968) found it sufficient to assume chloride to be moved down the profile at the same rate as rain water, and sodium at the same rate divided by an estimated buffer power.

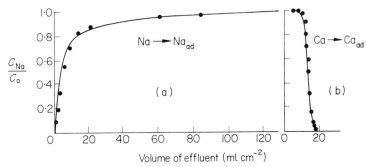

Fig. 8.7. Leaching from a soil column with (a) calcium, (b) sodium with water containing (a) sodium and (b) calcium ions (After Bower *et al.* 1957.)

8.4.2 Transfer of salt within the profile

In the previous section the uniform movement of a single band of salt was considered, but such a simple situation is rare. In practice, salts are usually distributed within the whole profile, more or less uniformly, and water additions, drainage and evaporation can cause complex sequences of movement. The most interesting cases are for nitrate in arable soils and salt (usually synonymous with the chloride ion) in naturally saline soils.

Leached nutrients are always a waste, and may cause environmental problems, nitrate being a particular risk (Gardner 1965). Its movement will follow that of water, and may be considered to be composed of three processes: downward movement, upward movement during dry spells and uptake by roots (Beek & Frissel 1973). However, water and ions are normally being absorbed by roots at different rates at various depths in the profile, and nitrate is produced by nitrification. It is difficult to describe this situation by an equation unless it is near a steady state, or if one process dominates. During prolonged dry weather upward movement may dominate in this way, and very high levels of salts have been found in the soil surface during the dry season in Africa (Wetselaar 1962). Gardner (1965) has given a simple steady-state treatment of this system, which predicted a logarithmic distribution of salt with depth, and which fitted experimental data remarkably well. Complicating processes could be expected, for example soil–water movement may be altered by the intense osmotic gradient, and a very marked temperature cycle is to be expected in such conditions, but these perturbations may be confined to a shallow layer near the surface, and are therefore, not detected in field studies.

The respective merits of ammonium and nitrate nitrogen with respect to leaching have been argued repeatedly (Bartholomew & Clark 1965). Because of its adsorption, it may be expected that ammonium ions will move much more slowly than a nitrate band, perhaps by a factor of 10, but with normal temperate agriculture it is unusual for nitrate to be applied so early, or rainfall to be so intense, that serious leaching occurs during the summer, though it may occur if spring application is followed by heavy rain. Much of the leaching loss of applied mineral nitrogen occurs in winter, by which time most ammonium will have been coverted to nitrate. In tropical climates the risk of nitrate leaching seems greater, though the literature is by no means consistent on this point. In practice the microbiological oxidation of ammonium to nitrite and then to nitrate occurs simultaneously with leaching and transfer. MacLaren (1970) has considered this situation in detail, starting from urea. His theoretical model predicts a steadily decreasing concentration of urea, and increasing concentration of nitrate, with depth, whereas the intermediates, ammonium and nitrite, have maxima at different points in the profile. Starr *et al.* (1974) included denitrification in their treatment of soil nitrogen transformations during leaching, and obtained fair agreement with laboratory experimental results.

The retention of nitrate and other solutes within large peds or aggregates in coarsely structural soils during leaching of the profile (Cunningham & Cooke 1957; Hartley 1964; Kissel *et al.* 1973; Wild 1972; Thomas & Swobada 1970) can be regarded as an extreme case of hydrodynamic dispersion, with very widely different flow speeds of water in fine and coarse pores. This argument would lead one to expect a larger dispersion coefficient with a larger flow velocity and hence less efficient leaching. A few intense storms should, therefore, cause less leaching than a steady fall. This principle has been confirmed in reclamation of saline soils, where a slow, steady application of water is recommended rather than isolated heavy irrigations (see Rhoades 1974; Bresler 1972).

If a hydrodynamic dispersion coefficient (equation 8.9) is used to describe the distribution of a non-adsorbed solute, it implies that the spread is symmetrical, as in Fig. 8.6, whereas drainage often results in bands with sharp fronts and long tails (G.W. Thomas priv. comm.); hence further theoretical and experimental work is required.

In reclamation of saline soils the whole profile initially contains salt, and this has to be removed by leaching. Pure water will move the nonadsorbed ions down at a rate equal to v/θ, but with the upper boundary made unsharp by diffusion and dispersion. This situation has been discussed in some detail by Bresler (1973b). Sodium held exchangeably on the soil must be displaced by another cation, and the course of its removal depends upon what is supplied. If it is only pure water, structural breakdown may prevent further leaching, and for this reason gypsum is often added, so that leaching is with a saturated solution of calcium sulphate.

In practice, nearly all theoretical treatments deal with the chloride ion rather than sodium chloride. The most thorough studies include both upward and downward movement of salt by convection and diffusion (Bresler & Hanks 1969; Bresler 1973b). The general equation is basically the same as 8.9,

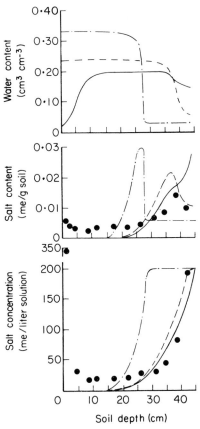

Fig. 8.8. Calculated movement of existing salts in a profile during infiltration, redistribution and evaporation of water. —.—.—, at end of infiltration (computed); ------, at end of redistribution (computed); ———, at end of evaporation (computed); •, at end of evaporation (experimental) (after Bresler & Hanks 1969).

and numerical solutions for this, with the usual equations for water movement, gave results in fair agreement with laboratory experiment. Figure 8.8 shows distributions of salt and water during a cycle of wetting—redistribution —drying. Omitting the diffusion—dispersion term, i.e. assuming that convective flow was dominant at the rather large velocity of $\sim 3 \times 10^{-3}$ cm s^{-1}, simplified the problem greatly, but gave very similar results.

A simple but effective treatment (Terkeltoub & Babcock 1971) assumes a soil column to be sub-divided into homogenous layers, and that the water containing dissolved salt is transferred from a layer to the next lower when the water content in the first exceeds some set value—for example field capacity or saturation. Good agreement with measured data was found in an irrigated field with an initially known salt distribution, when the transfer condition was selected to be a water content intermediate between field capacity and saturation. The results did not depend greatly upon the number of layers assumed. A rather similar model for predicting movements of salts in a profile under fallow—i.e. with no losses within the profile—has been published by Burns (1974). This again regards the profile as sub-divided into a series of layers, and calculates the transfer of water and salts from one to the other by standard theory.

These models have not been applied directly to profiles carrying a stand of vegetation. Water, and possibly some salt, will then be removed at different depths within the profile. If this removal is known, there is no difficulty in including this in the models discussed above, but prediction of water uptake

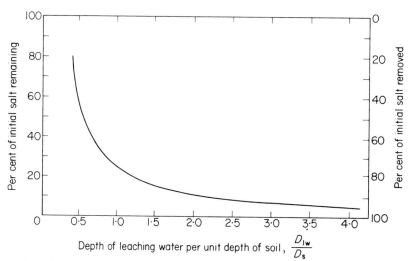

FIG. 8.9. Measured salt removal in relation to amount of leaching water from a soil profile (after Rhoades 1974).

by roots at different levels has only partial success (see p. 255). However, practical control of soil salinity appears to depend more upon empirical methods than upon models at present (see Rhoades 1974). The observed rate of removal of chloride from a soil profile by leaching (Fig. 8.9) might be explained by a considerable degree of hydrodynamic dispersion.

de Wit & Van Keulen (1972) have used simulation methods to study the special problem of desalinating a soil column by diffusion of salt into fresh water overlying the soil, as may happen when fresh water is ponded over salty land which is impermeable.

8.4.3   Use of saline water for irrigation

A variant of the above problem of leaching of soil salts is that of a profile receiving saline water: a long-term steady-state distribution will eventually be set up, with the soluble salts emerging at the bottom of the profile in a more concentrated solution, depending upon losses of water by transpiration. Over a long period, this merges into the problem of the salt balance of the catchment (Kovda *et al.* 1973).

When crops are growing, the position at which roots remove water and salt must be defined if a precise model for salt and water transfer is required. Accurate information on this is difficult to get in the field, since it will be a complex function of root distribution and of water content. Bresler (1967) has therefore put forward a simple mass-balance type of equation, where diffusion and upward movement are neglected, which is similar in principle to the 'layer' models discussed above. The equation used compared the increased salt in all layers, from o to z with the difference between the input of salt in irrigation water and the salt flowing out at z, allowing for the loss of water by transpiration and the change in concentration with time at z. The time step is conveniently taken as that between successive irrigations.

8.4.4   Dispersion of placed fertilizer

In principle there is little difficulty in predicting how a placed band or granule of fertilizer will disperse itself in soil of defined properties, since all the contributing processes have been discussed earlier. We assume that the fertilizers have a fixed solubility, which will be correct for pure salts and urea, but not for materials such as superphosphate where there is non-stoichiometric dissolution, or where there are strong common ion effects. For such a constant concentration source, in a defined line, cylinder or sphere shape, there are well-known solutions of the diffusion equation (Chapter 1) (Mokady & Zaslawsky 1967).

The factors which complicate the situation are:
(a)   The very large changes in total salt concentration that occur, leading to

large differences in dC_1/dC and hence D for cations. The very first stage of diffusion from a fertilizer granule can probably be regarded as salt or co-diffusion without adsorption, but exchange of cations will rapidly become important.

(b) Very sharp differences in pH may arise, particularly with phosphate fertilizers, or any containing or producing ammonia. The pH of a saturated solution in contact with superphosphate is 1·0 (Huffman 1962).

(c) The osmotic potential of a saturated solution of a compound fertilizer is extremely large. As an illustration, liquid fertilizer mixtures may contain 30% by weight of nitrogen, which corresponds to about 20 moles per litre. This corresponds in theory to some -400 bars water potential. The movement of water with osmotic potential as the driving force has been discussed in Chapter 2, and with such enormous gradients it could well be important for both vapour and liquid phase transfer.

(d) Some components of a fertilizer mixture may react with and precipitate ions in the soil, phosphate being the most important (Huffman 1962).

With this complexity, it is hardly surprising that there has been no accurate theory of diffusion from placed fertilizer, but a number of direct measurements have been published. For example, Burns & Dean (1964) showed how nitrate moved out from a thin band of sodium nitrate in glass beads or soil (Fig. 8.10). The movement downwards was always greater than in other directions, and this was ascribed to the higher density of the solution produced from the fertilizer compound than the soil solution, after movement of water to the fertilizer under the osmotic gradient.

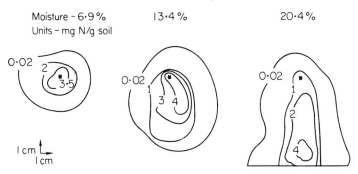

* Fertilizer band

FIG. 8.10. Movement of nitrate ion out from placed bands of fertilizer in a sand; increased moisture content leads to more rapid downward flow of dense solution (after Burns & Dean 1964).

It appears that water moves to the salt in the vapour phase, and then out again as a liquid solution (Parlange 1973). A similar effect was found by Kemper *et al.* (1975), who investigated the placement of fertilizer in ridges, with irrigation water supplied in the intervening furrows. A distance of 5 cm in a sandy, and 10 cm in a clay soil between water level and fertilizer band height prevented serious leaching, but less than this allowed water to move to the fertilizer and form a concentrated and dense solution, which then moved down the ridge and became dispersed in the irrigation water. Several workers (Lehr *et al.* 1959; Lindsay & Stevenson 1959; Gunary 1963) have investigated the movement of phosphate from superphosphate pellets; and the initial rate was always much more rapid than expected from the normal diffusion coefficient of phosphate in soil presumably due to convective flow of water towards, and solution away from the pellets.

8.4.5 Mechanical movement

The mechanical processes that may incorporate adsorbed solutes into the soil have been mentioned in Chapter 4. They are mainly responsible for re-distributing very strongly adsorbed solutes such as trace quantities of cadmium and mercury. Unfortunately, they are difficult to quantify in any simple manner. Poelstra *et al.* (1974) have tested the use of a simple 'mixing function', which assumes that a given fraction of the mercury in the surface layer will be redistributed annually over all layers below. The actual distribution of mercury to a depth of 80 cm in 15 West European pastures was erratic, and could only very roughly be reproduced by a simulation using the 'mixing function'.

Run-off and erosion are further very important methods of solute transfer, though they are difficult to predict. The movement of a strongly-adsorbed solute like phosphate will in practice be a combination of all these processes, whose outcome has been reviewed by Wadleigh (1968).

Management of the soil also affects the destiny of pollutants and pesticides. Edwards (1974), for example, states that insecticides on the soil surface usually disappear 5–10 times faster than those cultivated into the soil. Exposure to wind is important, and several workers have shown that persistent volatile insecticides disappear more slowly when shaded by crops than in open soil. Many more examples of the importance of mechanical movement on the action of pesticides are given in the two encyclopaedic volumes edited by Goring & Hamaker (1972) on organic chemicals in the soil environment.

8.4.6 Volatilization

Leistra *et al.* (1974) have given a model for the disappearance of the volatile herbicide propyzamide, which allows for diffusion out of the soil in the gas phase, for varying rates of adsorption and desorption from soil solid, and for a first order decomposition rate. The latter was determined from the experiment, not independently. Nevertheless, the form of the computed concentration profile agreed well with experiment. It is a virtue of such models that the importance of individual processes is easily tested. Figure 8.11 compares experimental data with that expected assuming different loss processes.

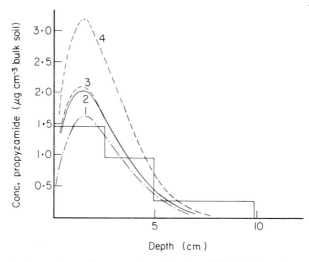

Fig. 8.11. Comparison of measured and simulated distribution of propyzamide after 40 days using various assumptions about its behaviour. (1) With decomposition and free gaseous diffusion. (2) With decomposition, no gaseous diffusion from surface. (3) With decomposition, no gaseous diffusion at all. (4) Without decomposition, full gaseous diffusion. Full line is the measured distribution (after Leistra *et al.* 1974).

8.5 The uptake of nutrients by field crops

This section examines how far the detailed information which has been outlined in earlier parts of the book can be connected to produce a general treatment of uptake by field crops. The methods dealt with here are of varying levels of complexity, but all can be regarded as 'dynamic', in that they predict or deal with rates of uptake at different times during the development of a crop. This is a considerable development from earlier methods, which tended to concentrate upon total nutrient uptake by the crop, the permanent loss of nutrients from the soil, and the relationship of these to soil composition.

The more complex methods can suitably be called 'models'. As for all such, the tests to be applied must be whether they are mechanistic and have real predictive value, or whether they are merely elaborate descriptions of the phenomena they consider. At this stage, it is best if models are relatively simple, and the number of parameters limited, since the accuracy of determination of so many of the latter is poor. All published models depend to a greater or lesser extent on data gathered from the crop itself, hence their predictive capabilities are small.

All the work discussed in this section has treated a crop as a uniform monoculture, without regard to competition between individual plants. Such effects are discussed later.

8.5.1 Nutrient flow analysis

The most basic information about crop nutrition is the rate at which nutrients are being absorbed. It is therefore pertinent to consider nutrient uptake rates first. Typical data are in Fig. 8.12 (Knowles & Watkin 1931). Nutrient contents tend to reach a maximum well before harvest, and the amounts in harvested crops (Russell 1973 p. 24) are certainly an under-estimate of the total amounts absorbed during the growing season, quite apart from continuing losses of absorbed nutrients by leaf leaching and loss of senescent leaves. We are not aware of any comprehensive data for the last two processes for arable crops.

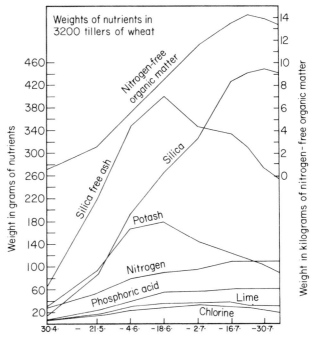

FIG. 8.12. Nutrient uptake by a wheat crop at different times during the growing season (after Knowles & Watkin 1931).

After a complete canopy with a reasonably large leaf area index has become established, the crop growth rate does not vary greatly in a constant environment, and most agricultural crops give similar values. Crop nutrient uptake is therefore reasonably constant also, and a rough estimate of uptake rate of cereals in Britain could be made by assuming that the maximum nutrient content had been absorbed over a period of about 70 days.

Viets (1965) has compiled a very useful table of maximum uptake rates of nitrogen by various crops. In agreement with the argument above, the values do not differ greatly, but lie (with one exception) between 2·1 and 4·4 kg ha^{-1} day^{-1}. However, Viets noted that the crop yields in most cases were well below the known current maxima, and uptake rates considerably larger should therefore be expected. If we assume an uptake rate of 4 kg ha^{-1} d^{-1} or 3 × 10^{-7} mol m^{-2} s^{-1}, with a surface root density L_A of 100 cm cm^{-2}, $\bar{I}$ would be 3 × 10^{-13} mol cm^{-1} s^{-1}, which is a rather low value compared to the inflows often found in laboratory experiments (see Table 5.4). From the value of $\overline{\alpha a}$ for nitrate in Table 5.4, this suggests that a mean concentration of 0·5 × 10^{-3} M in the soil solution would be sufficient, if it could be maintained indefinitely, to supply even a rather sparsely-rooted crop.

If the nutrient uptake rates per unit surface area, and root length, are known for a crop, mean inflows can easily be calculated; and Kafkafi & Bar-Yosef published a study of this type for phosphorus uptake by maize. The original feature of this work was the derivation of effective absorbing root length from the calcium uptake rate, it being assumed that there was a constant and known I value for calcium. It has been shown (Russell & Sanderson 1967) that the lengths of barley root absorbing strontium are much less than those absorbing phosphorus, and this may invalidate the method. Mengel and Barber (1974b) have also stated that I_{Ca} is not a constant during the development of the plant. Bar-Yosef *et al.* (1972b) used these I values for calcium in a calculation of the flow of phosphorus to roots by diffusion, and compared these with the predictions of a root model (Bar-Yosef *et al.* 1972a) basically similar to that of Nye (1966b). The measured and predicted fluxes were similar in magnitude, the measured values being larger, presumably because root hairs and mycorrhiza were not included in the model, and because the effective root area may have been underestimated for reasons given above.

Very complete data on root development in maize were given by Mengel & Barber (1974a), the most conspicuous feature being the sharp maximum for root density at all depths between plant ages of 75 and 90 days. Maximum topsoil density (L_V) in the 0–15 cm layer was 4 cm cm^{-3}, and the maximum L_A about 155 cm cm^{-2} to a depth of 60 cm. Root length was averaged over the crop unit cell. These data, were used (Mengel & Barber, 1974b) to calculate inflows for 10 elements (Table 8.3), which were extremely large at about 20 days age of the crop, when frequent harvests were taken. It is always difficult to measure inflows accurately at first harvest, but it is clear that the early values were in any event very large; and this agreed with results in Chapter 5. It may also reflect the high fertility level of the soil used. This gave a percentage phosphorus in the shoot of 0·6% at 20 days, which declined steadily throughout the growth period to 0·2% at 125 days. The inflows also rapidly and steadily declined (Table 8.3) and negative values were found

Table 8.3. Inflows at different times for 10 elements into field-grown maize (After Mengel & Barber 1974b.)

Plant age	Calculated nutrient uptake for									
	N	P	K	Ca	Mg	B	Cu	Mn	Zn	Fe
days	μ moles/m per day					μ moles/m per day $\times 10^2$				
20	226·9	11·3	52·9	14·4	13·8	98·1	27·0	89·0	109·8	571
30	32·4	0·9	12·4	5·2	1·61	9·00	1·85	11·09	5·78	64·8
40	18·5	0·86	8·00	0·56	0·90	5·29	1·22	7·35	3·47	46·1
50	11·2	0·66	4·75	0·37	0·78	2·55	1·10	4·61	1·97	21·4
60	5·7	0·37	1·63	0·20	0·56	0·64	0·73	2·07	0·76	13·0
70	1·2	0·17	0·15	0·047	0·28	−0·20	0·41	0·06	−0·04	11·5
80	0·46	0·08	0·06	0·060	0·19	−0·53	0·32	−0·85	−0·24	−29·3
90	2·0	0·10	0·37	0·063	0·17	−0·52	0·52	−0·39	0·50	−1·7
100	4·2	0·23	0·16	0·075	0·29	−0·41	1·03	0·51	2·19	6·7

$$1 \; \mu \text{ mol m}^{-1} \text{ d}^{-1} \sim 1 \times 10^{-13} \text{ mol cm}^{-1} \text{ s}^{-1}$$

for several trace elements after 70 days. If only roots less than 5 days old were considered to absorb nutrient, the inflow at 20 days was little changed, ($14 \cdot 6 \times 10^{-13}$ mol cm^{-1} s^{-1} for P), but then declined to a moderate level ($\sim 2 \times 10^{-13}$ mol cm^{-1} s^{-1} for P) at 35 days and remained constant for the rest of the growth period. These results emphasize the importance of an early establishment of a well-distributed and efficient root system and high initial nutrient levels in the soil solution to satisfy the plants great ability to accumulate nutrients in its early vegetative phase.

The flow of nutrients up into trees and down again to the soil is not easy to measure. Data on total contents have been determined quite frequently (see Rodin & Basilevitch 1965) but it is often unknown over how long this accumulation has taken place. Ovington (1965) measured the nutrient cycling in a stand of Scots pine, and the annual amounts of each nutrient taken up by the vegetation (Table 8.4). Nye (1961) gave full data for the cycling of nutrients in a forest ecosystem in West Africa, showing the relatively rapid turn-over of nutrients, especially potassium, each year (Table 8.4). Similar work for tropical forest by Bernhard–Reversat (1973) produced circulation rates of 93–255 kg ha^{-1} y^{-1} for potassium for three sites. Our immediate interest is with the way in which the recycled nutrients reach the roots of the vegetation, and are re-absorbed. To determine this, detailed knowledge of the root distribution and length is needed, which is extremely difficult to obtain for strongly heterogeneous and irregularly distributed vegetation.

A simpler system which has been dealt with in greater detail is the uptake of potassium into the oil palm in plantations (Tinker 1973). This crop has a simple morphology and growth pattern. Growth and leaf loss proceeds throughout the whole year—though not at a uniform rate—and the structure of the plant makes it easy to calculate the increment of stem weight and the

Table 8.4. Cycling of nutrients between soil and forest. (a) Annual nutrient balance for Scots pine (after Ovington 1965). (b) Annual return of nutrients from mature tropical forest to soil (After Nye 1961.)

(a)	Nitrogen	Potassium	Calcium	Phosphorus
		kg ha^{-1} yr^{-1}		
Nutrients contained in primary production of trees and under-storey plants	125	51	55	11
Change in nutrient capital of the plant mass	+11	+3	+7	+1
Removed in harvested tree trunks	3	2	4	0·3
Released by decomposition	111	46	44	10

(b)	Additional to soil surface from forest	Wt. of material (oven dry)	N	P	K	Ca	Mg
					Nutrient elements		
					kg ha^{-1} yr^{-1}		
Litter-fall		10,500	200	7·3	68	206	45
Timber-fall		ca. 11,200	36	2·9	6	82	8
Rain-wash			12	3·7	220	29	18
Total			248	13·9	294	317	71

number of fresh leaves each year. Previous studies allowed a calculation of the mean annual total uptake rate of nutrients by adult palms (Table 8.5), which are surprisingly similar to those for tropical forest.

Table 8.5. Annual uptake of nutrients by productive oil palms (After Tinker 1973.)

	N	P	K	Mg
		kg ha^{-1} yr^{-1}		
Bunch removal (Yield: 24 tons/ha)	71	11·6	91	22
Immobilization in palms	39	3·2	68	10
Nutrient turnover in fronds, male inflorescences, etc.	77	11·0	100	28
Mean total nutrient uptake in Malaysian conditions	187	25·8	259	60

Detailed studies of roots made by previous workers on similar soils gave the mean total weight of fine absorbing root in the top 30 cm as about 5 kg wet weight per palm. As with most tree crops and field conditions, the soil nutrient content declined very rapidly from the top 5–10 cm down, and any roots below 30 cm could probably be neglected. From other data it was clear that this weight represented a length of about 2×10^6 cm in all. This yields a value of $\bar{I}$ for potassium of 3×10^{-13} mol cm^{-1} s^{-1}, which is low in relation to typical values (see p. 118).

At present it is difficult to improve upon such very crude treatments in systems as complex as those of stands of whole trees. It is obvious that the inflow must vary with time of year—though much less in palms than in temperate tree species—and very much more detailed and laborious work will be needed to distinguish the effectiveness of different parts of the root system. The main value of such estimates of $\bar{I}$ is in comparisons between species and sites, which must await further results for tree crops. Tinker (1973) used equation 7.5 to calculate potassium depletion around palm roots and concluded that $C_{1i}-C_{1a}$ was about 10^{-4} M, which may be significant in infertile tropical soils.

8.5.2 Root system uptake models for uniform crops

Nutrient flow analysis methods, as described above, can be regarded as the simplest form of root system uptake models for crops. In effect, they regard the whole root volume as a single uniform system, without regard for horizontal or vertical changes in root or soil properties, whcih is clearly a gross simplification. It may however often be permissible to ignore changes in the horizontal direction. Generally, there will be a progressively greater degree of interpenetration of roots and shoots in a field crop. When this happens it is simplest to ignore the individual, and regard the whole population as a single unit. In some cases, as for a grass crop, it is in any event difficult to define or distinguish individuals. We then take a given area of crop as one unit, in a manner analogous to that of micrometeorologists considering the transfers of water, energy or CO_2 in a crop, and consider only changes in the vertical direction. If the root system distribution with depth is known, and the soil properties in different horizons, the profile can in principle be subdivided into layers which are sufficiently uniform. In the simplest approach, there are two compartments, topsoil and subsoil, each with a single value for L_V, $\overline{\alpha a}$, b, Db and a pattern parameter for roots. These can be dealt with exactly as the compartments in the experiment considered in Chapter 7 (p. 235). The decision on the lower boundary depth may be difficult in field work, since a small number of roots may penetrate to a considerable depth, but their contribution to crop nutrition may be very uncertain. Further, water and dissolved salts may move up to the rooting zone from below in considerable quantities—for example Stone *et al.* (1973) found a water flux of up to $0{\cdot}2$ cm d^{-1} into the root zone of sorghum at 140 cm depth in a sandy loam.

The principles thus seem to be reasonably clear, but the practical application of such models is difficult. A wide range of parameters must be measured with fair accuracy, and some, such as the solute uptake within a specified part of the rooting volume, may be extremely difficult to measure by any current method. No full field study by the principles outlined here has yet been published, though many of them have been introduced in the following model.

8.5.3 Nitrogen uptake model

The mobility of nitrate ion, and the changeable concentrations found in soil, make it difficult to construct models of the uptake of this element. The weather also affects nitrogen in the soil considerably, and must be allowed for, but on the other hand the lack of local depletion zones allows the detailed root system structure to be ignored. The prediction of nitrogen uptake is thus an ambitious project, but some success has been achieved. Greenwood *et al.* (1974) have published an ingenious whole-crop model, based upon the ideas that nitrogen in the soil affected lettuce growth both via supply of the nutrient, and by increasing the osmotic suction of the soil solution.

The model (see Fig. 8.13) can be divided into sequential steps:
(a) The nitrate-in-water content in any layer of soil was computed from Burns (1974) leaching model, the rainfall, soil nitrification rate and plant uptake.
(b) The nitrogen uptake was calculated from a mean nitrate concentration in the profile weighted for root quantity in each layer, and the uptake of an optimum crop in that locality. The estimation of nitrogen uptake as a fraction of a pre-determined 'optimum' rate, rather than as a direct-soil process, is an interesting feature of the model.
(c) Plant growth was derived from an empirical equation describing the growth of an optimum crop, and allowing for the effects of sub-optimal nitrogen supply and osmotic suction. The model could be integrated with crop environment models by modifying this equation. Growth increments were determined by nitrogen per cent in the tops, and the effect of osmotic suction from an expression which implied that growth ceased at 15 bars.
(d) The calculated nitrate uptake was allocated to the different soil layers in proportion to the weight of root and the initial nitrate concentration in the soil solution at each layer.
(e) The root weight was a set fraction of total weight, and decreased exponentially down the profile (see p. 252); the uptake properties were assumed to be of the Michaelis–Menten type (p. 106) and to be constant throughout the root system, and a 'transport coefficient' for the movement of nitrate was calculated from soil water content.

With these relationships, weather and soil data, and the predetermined optimum growth equations, the model could be re-computed at daily intervals. The growth of lettuces was modelled successfully (Fig. 8.13),

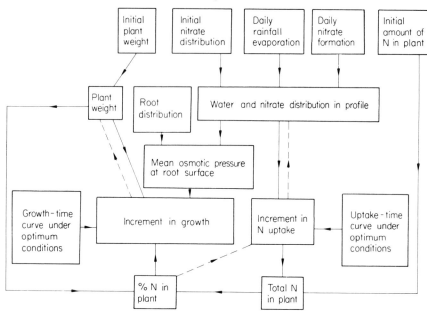

Fɪɢ. 8.13. Whole-crop model of growth and nitrogen uptake by a lettuce crop in the field—flow diagram (after Greenwood *et al.* 1974).

though the absolute values for yield were sometimes in error. The detrimental effects of the osmotic suction with high rates of fertilizer caused a sharply-peaked yield response curve, and the model predicted the best rates of nitrogen well (Fig. 8.14). When applied to a range of other experiments at different sites, the model also predicted the best nitrogen rate with fair success. In this last case it was found that the rate depended almost solely on the rate of nitrate production in the soil, and, rather surprisingly, weather variations had little effect.

8.6 Competition between plants

The methods described above treat a crop as a single plant. This is sufficient for dense uniform stands of a single species, but it does not allow a detailed treatment of the early stages of a crop when plants are isolated, grouped placing of individual plants, or stands containing more than one species. Little progress has been made in the practical development of this subject, but a fair amount of theoretical work is available. The essential point in all such situations is the competition between plants, whether they are of the same or different species. We therefore discuss this topic here.

The concept of competition in plant growth has been dealt with frequently (see Donald 1963; Milthorpe 1961). It implies that some resource is present in a quantity insufficient to allow all competitors to reach their maximum development. A variety of methods have been used in such studies, from

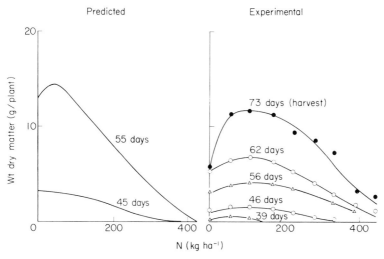

Fig. 8.14. Growth of a field lettuce crop with different rates of nitrogen, as measured and as predicted from the model of Fig. 8.13 (after Greenwood *et al.* 1974).

simple descriptions to mathematical treatments of some complexity. Usually this work has related to the competition between two species rather than two individuals, though the methods of de Wit (1960) allow interspecies competition and spacing effects in a single crop to be treated in a formally similar way. The single-species crop situation is clearly the simplest, in that all competitors may often be regarded as similar and the only issue is the efficiency of the whole monoculture system. Even here, variable germination and seedling vigour introduce irregularity. The multi-species situation introduces the additional problem of the morphological and physiological differences between the species, and growth will be determined, not only by the properties of the individuals in each species, but also by the relative distribution of the individuals. Relatively weed-free monoculture is now normal in advanced agriculture, but the competition between crops and weeds is still important. Mixed cropping is frequently used in tropical agricultural systems, and plantations and orchards usually contain at least two species. In addition, natural ecosystems are formed and controlled by the competion between their members.

Competition in the wider sense of the gradual displacement of one species by another is often considered over a period of years, in which case differences in reproductive strategies (Harper 1961) or interactions with farming operations and their timing, are often decisive. Here we consider only the short-term competition of plants growing in a single season without outside interference, in relation to the resources provided to plants by the soil. Some treatments or 'models' of competition are essentially descriptive rather than mechanistic, but a proper analysis demands consideration of the detailed methods by which the competitors interact, and not only a description of the result in terms of plant populations. We consider the possibility of such a mechanistic model later.

Competition is for light, nutrients and water, of which we are concerned with the last two. However, it is not strictly possible to separate these competitive effects, since a competitive advantage in nutrient uptake may lead to increased height, which may give a competitive advantage in intercepting light later, and vice versa. These considerations make it extremely difficult to interpret experiments in which competition operated simultaneously above and below ground. However, a number of studies have been made in which only below-ground competition has been effective (see Donald 1963), and we discuss the mechanisms by which this could operate.

The problem of deciding whether competition is above or below ground is much greater in the field situation, and there is little clear evidence on this point. For crop–weed competition there is an indication (Hewson & Roberts 1973) that the mechanism of competition in the early stages of growth— during the 'critical period' when the final crop yield is susceptible to reduction by a weed population—may be through the nitrogen demand, before competitive shading becomes important. However, there appears to be no published attempt to analyse this very complicated situation accurately. In both this case and that of natural vegetation there are usually many species growing together, developing at different times and rates, and with individuals spaced out in irregular patterns and germinating at different times, and a full understanding of such systems is still some way in the future.

8.6.1 Interactions between root systems

Competition for nutrients and water could arise in several ways. The simplest mechanism would be that the root-system of one plant absorbed water or nutrients, and hence left the roots of its neighbour in drier or poorer soil. From Chapter 7, it is clear that such soil changes might also lead to slower root growth of the second plant, hence uptake rate would be less both because the roots were fewer and the soil was poorer. Finally, one root system may excrete compounds which inhibit the extension, or change the properties of another, leading to a less efficient or less well-developed root system, and the competitive advantage of the first plant. As Kershaw (1973) pointed out, a mechanism which inhibits root extension may well appear in the plant as a deficiency of nutrients and water.

The possibility of a mutual influence between developing roots in soil has been referred to in Chapter 7, in connection with the development of a single root system. Similar considerations arise where two extending root systems begin to inter-penetrate, and effects on morphology have been noted frequently. Rogers & Head (1969) discussed the effects in fruit orchards, and concluded there was no exclusion of grass roots by tree roots. However, the grass sward roots inhabited the most fertile part of the soil, which may have compensated for any competitive effect from the tree roots. Interesting effects have been reported for the mutual antagonism of two tree root systems: pear roots from neighbouring trees intermingle freely, whereas apples show a degree of exclusion (Fig. 8.15), and peach trees almost completely exclude each others root (Israel *et al.* 1973). Similar effects have been reported for annual plants: for example soyabean root systems extended freely to the centre-line between two rows, but they then excluded each other strongly (Raper & Barber 1970).

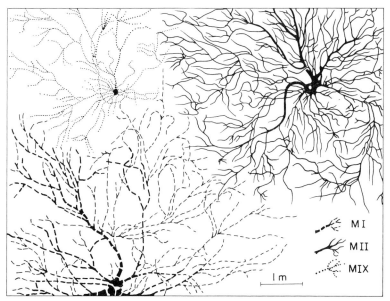

FIG. 8.15. Projection of the root systems of three neighbouring apple trees on a horizontal plane, showing mutual exclusion (after Rogers & Head 1969).

This process has been followed more exactly in the laboratory by Litav & Harper (1967), who labelled one set of plants with ^{14}C, and then calculated the degree of interpenetration of the labelled and unlabelled roots. Baldwin & Tinker (1972) carried out double-labelling experiments with onion and other plants, using ^{35}S, ^{32}P and ^{33}P in an autoradiographic technique to determine root position and exclusion. The root density distribution with one or two rows of onions is shown in Fig. 8.16. The distribution of roots from the first row show clear evidence of an exclusion mechanism. If this was caused

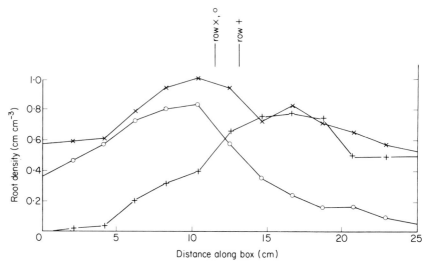

Fig. 8.16. Displacement of root system of a row of onion plants by a neighbour-ing row. x, row growing alone; o and +, rows growing together (after Baldwin & Tinker 1972).

by root-exuded toxins, a tendency towards a regular (i.e. non-random, see p. 226) pattern of roots might be expected, since the elongation of any root would be inhibited when it approached a root of the competing plant. No such regularity could be detected in the radioautographs, but it is possible that a toxin could be so mobile that local concentration differences around roots were very small, though a larger difference from within the rooting volume to outside it could persist. However, it appears equally possible that changes in nutrient concentration can cause differences in branching of root systems, which leads to the appearance of exclusion. From considerations of ion mobility, it seems that the only ion which is likely to act in this way would be nitrate, but this indicates an unexpected sensitivity of roots to small nitrate concentration differences, and a considerable difference between species in this respect.

The possibility of toxic exudates is linked with the problem of allelopathy (Rice 1974), and 'soil sickness' or replant disease (Rogers & Head 1969; Börner 1960). These are often attributed to specific compounds, such as juglones produced by walnut (Grummer 1961), and hydrogen cyanide and benzaldehyde by peach roots (see Israel *et al.* 1973).

There seems little doubt that such compounds cause root exclusion in specific cases, but it is rather surprising if other cases of root exclusion, which are found so widely, are all caused by compounds which have defied identi-fication up to the present. Woods (1960) and Kershaw (1973) concluded that toxic substances were undoubtedly produced by roots, but that they might be destroyed or adsorbed quite rapidly in soil. They are also formed by crop residues in the soil, and not only by living roots (Patrick *et al.* 1964). At present the evidence on this subject is inconclusive.

8.6.2 Competition experiments

Several authors have published the results of experiments in which plants have been grown together with below-ground interaction only, above-ground shading being prevented by a permanent screen. These experiments have always shown that one species suffers more from competition (Welbank 1964). Aspinall (1960) concluded that the competition between barley and white persicaria in sand culture was mainly via root competition, and resulted in a relative suppression of both species.

The experiments of Litav & Isti (1974) were essentially similar to the above and especially interesting because they used two very similar plants, namely two strains of *Spinacia oleracea*. When grown in pure culture, the yield differed by only some 30%, with FS strain the best. The ratio of their dry weights when grown together was largest when the supply of nutrients was large, and lessened as the nutrient supply (N and P) was diminished (Table 8.6). The authors could not explain the results of these experiments by simple competition for nutrients, since adding more of the deficient nutrient (N) simply increased the size of the dominant species, whilst the other remained constant (Table 8.6). They therefore regarded the experiments as proof of

Table 8.6. Dry weight of 2 strains of *Spinacia oleracea* grown separately and in mixed culture in soil, with different level of nitrogen and phosphorus (mg per plant) (after Litav & Isti 1974).

	Separate culture			Mixed culture		
	N + P	2N + 2P	3N + 3P	N + P	2N + 2P	3N + 3P
F_1 Selina (FS)	360	600	630	370	740	760
Early Hybrid (EH)	320	640	700	225	390	490

allelopathic effects between the root systems. However, the EH competitor was still responding to the largest rate of nitrogen, and other explanations therefore seem possible. If the relationship of R_w to nutrient composition varied between the strains, then an increase in nutrient supply, from a position where both R_w values were equal, would increase one R_w more than the other. This plant would put out relatively more and more root, compared to the poor competitor, as the nutrient level was increased, and hence take a steadily larger share of the total nutrient available in the rooting volume. In treatments at higher nutrient levels, eventually no further yield response would occur, and the competitor suppressed at lower levels should then perform equally. Thus, on theoretical grounds, it is possible that one competitor could steadily increase its advantage as nutrient supplies are increased, but lose it at even higher nutrient levels. In tests of the mechanism of root competitive effects, it is therefore necessary to make quite sure that excess quantities of nutrient are given in some treatments. Such experiments also need very detailed measurements, of the type discussed in the next section, to be fully interpretable.

8.7   Theoretical aspects of uptake of nutrients by groups of plants

If the individual plants in a crop are widely spaced out, they will, at least during the seedling stage, in effect be isolated. Their uptake and growth can be treated as explained in Chapter 7 during this stage, but presently the roots systems will interact.

8.7.1   Competition between simple root systems for nutrients

The aspects of plant 'competition' which we deal with here are limited to the mutual effects of two plants growing as neighbours. These effects on nutrient uptake can arise in essentially two different ways: the amount or absorbing power of active root may be changed, or the effective concentration of nutrient in the soil exploited by those roots can be varied. The two mechanisms are not, of course, separate, in that differences in nutrient concentrations may cause difference in root length, and vice versa, but it is convenient to analyse the situation in this way.

We assume, in the first instance that uptake rates of parts of two root systems (P and Q) in a given unit volume of soil, are

$$2\pi \, (\overline{\alpha a})_{\mathrm{P}} \, L_{\mathrm{P}} \, C_{1\mathrm{P}} \; \& \; 2\pi \, (\overline{\alpha a})_{\mathrm{Q}} \, L_{\mathrm{Q}} \, C_{1\mathrm{Q}}$$

respectively; this is likely to be true for mobile ions where C_1 may be assumed constant throughout the root volume. Length and absorbing power would then be the only points of interest. Andrews & Newman (1970) carried out an interesting experiment on the effect of root length on competitive uptake. They root-pruned wheat plants, and showed that the total uptake of nitrogen by pruned plants was slightly larger than that of unpruned plants (Table 8.7). $\overline{\alpha a}$ thus had changed to compensate for loss of root length and weight, though there are not sufficient data to calculate this change. Despite this, when pruned and unpruned plants were grown together, the unpruned plants absorbed 60% more than pruned. This implies that the pruned root system was unable to re-adjust its $\overline{\alpha a}$ sufficiently rapidly to make it an equal competitor, or that its root distribution was inefficient relative to that of the unpruned plants. In contrast, the pruning reduced phosphate uptake by roughly the same ratio whether plants were growing alone or in competition, presumably because αa is not increased by pruning or because changes in αa have little effect for this ion except at high phosphate levels (see p. 135 *et seq*). Given sufficient time, a limited root system can absorb as much of a mobile solute like nitrogen as a larger one, but if the two grow together, the larger system may absorb nitrogen more rapidly than the smaller one and the latter then suffers a permanent disadvantage. The efficiency of a root system, in

Table 8.7. Effect of root-pruning on the uptake of nitrogen and phosphorus by wheat plants in pots growing alone or in competition (after Andrews & Newman 1970).

	Alone	Mixed
Nitrogen	mg/plant	
Shoot		
Unpruned	4·10	7·75
Pruned	4·61	3·07
Whole plant		
Unpruned	8·43	13·47
Pruned	10·11	8·36
Ratio unpruned : pruned	0·83	1·61
Phosphorus		
Shoot		
Unpruned	1·74	1·98
Pruned	1·29	1·31
Whole plant		
Unpruned	2·83	3·12
Pruned	2·50	2·46
Ratio unpruned: pruned	1·13	1·27

terms of its ability to fulfil an uptake function, therefore may depend greatly upon whether it is in a competitive environment or not. Using a similar argument, Lee (1960) explained the relative success of Atlas barley in competition with Vaughan barley by the more dense root system produced by the former at an early stage. However, Vose (1962) tested a number of rye-grass strains for uptake of nitrogen, phosphorus and potassium and found large differences in plot trials where individual plants were spaced out, and rapid root extension was of considerable advantage, but there was little difference in growth in mixed sward conditions, where root extension did not provide any new untapped soil. It is therefore difficult to make any statement about the 'value' of a particular type of root system without specifying the exact conditions under which it is to be tested (see p. 279).

No really detailed experimental analysis of competition between plants has yet been published, and it is clearly a considerable task. Here we consider a theoretical basis for such an analysis.

It is useful first to identify the various competitive situations which can arise. These are

(a) Isolated pairs of plants, of similar (i) or different (ii) type. This situation has been used frequently in experimental work, but it is difficult to see any practical field situation in which it could occur.

(b) Regularly-spaced vegetation, which can be (i) monoculture, as in most row crops or (ii) mixed, as in orchards with ground cover, plantations with shade trees or some forms of mixed cropping. Such situations require the identification of a unit repeating cell which is completely representative of the whole field. With appropriately placed boundaries, it may be assumed there is no significant net material transfer across them, though roots will of course pass in both directions.

(c) Irregularly-spaced vegetation which again can be (i) monoculture, as in a pine forest or a pure grass sward of one species. More frequently, it will be (ii) mixed vegetation, as in most natural associations. This will also include spaced crops with a heavy weed population, rough grazing, mixed leys and most planted forests. In these cases the decision on the appropriate unit cell may be very difficult, and in natural high forest, or grassland with infrequent trees of variable size and species, the unit cell might be impractically large. However, the standard quantitative techniques of describing mixed vegetation (see Greig-Smith 1964) should allow a rational decision on the size of unit cell needed. If any significant soil variability occurs, it may well be impossible to define any area as truly representative in the sense used here.

8.7.2 Competitive uptake of a mobile solute

A sophisticated whole plant model has been developed by Baldwin (1976) with the express purpose of testing competitive effects between two plants of different species. The conceptual framework is in Fig. 8.17, and illustrates the underlying idea of 'pools' of carbohydrate and nutrient (nitrogen in this case) which combine to form the structural materials of the plant. If carbohydrate is in excess, root extension is enhanced, leading to more rapid uptake; if nitrogen is in excess, root extension is slowed and more of the photosynthate is diverted into the leaves. By this means the overall balance of the plant is preserved, and the known reactions to increased nitrogen supply or photosynthesis are modelled (see p. 236). The partition between root and shoot is controlled by an allometric relationship, modified by the N : C ratio. The soil part of the system is a cylinder divided into layers, and vertical root

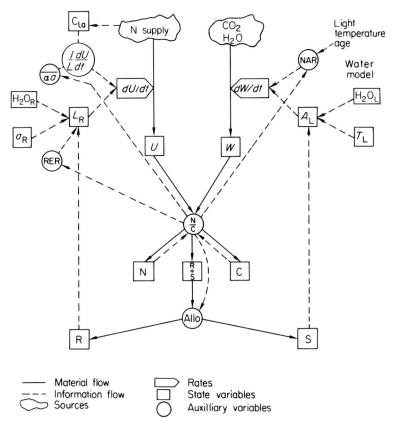

FIG. 8.17. Structure of whole-plant model for growth and uptake of nitrogen (after Baldwin 1976). C_{la}, solution concentration at root surface; $(1/L)(dU/dt)$, inflow; $\overline{\alpha a}$, root demand coefficient; H_2O_R, root water content; a_R, root radius; L_R, root length; RER, root extension rate; R, root weight; U, nutrient uptake; Allo, allometric relation; N, free nitrogen in tissue; C, unbound carbohydrate in tissue; NAR, net assimilation rate; A_L, leaf area; H_2O_L, shoot water content; T_L, leaf thickness; W, plant weight; dU/dt, nutrient uptake rate; dW/dt, plant growth rate; N/C, ratio of free nitrogen to soluble carbohydrate; S, shoot weight.

extension determines how many layers are exploited. The model contains a water flow sub-model based on that of Nimah & Hanks (see p. 255), but this is not discussed here. The methods of Baldwin *et al.* (1973) are used to model nutrient uptake. The model in this form is only suitable for application to mobile nutrients such as nitrate. Within each soil layer the roots from both plants are randomly mixed, and the rate of uptake is proportional to the lengths of root and αa for each. The extension rates of the main root axes are defined, and this determines the times at which each soil layer is first exploited by each plant. The relative advantages of producing root systems of different density and depth, but the same total dry matter content, can therefore be explored in detail.

The model was used first to determine the importance of a range of plant and soil parameters on the production of dry matter of single plants. The results were expressed as sensitivity coefficients (Table 8.8). Next, a plant with variable characteristics was simulated in competition with one of constant characteristics, and the outcome shown in the form of replacement diagrams (Fig. 8.18).

In summary, the important quantity was maximum root extension rate, whereas the number of seminal roots had little effect, and a larger number could be detrimental due to their lower extension rate. Larger seed weight or net assimilation rates had considerable effects, as might be expected. An important point stressed in this work was that competitive effects could

Table 8.8. Sensitivity coefficients for some plant and soil parameters in the whole-plant model. They indicate that fractional difference in growth of a single plant caused by a standard fractional change in each parameter (after Baldwin 1976).

Parameters		Sensitivity coefficient $\left(\dfrac{dY}{Y}\right)\bigg/\left(\dfrac{dp}{p}\right)$ (Y is yield, p is value of parameter)
Standard plant		
Seed wt	0·01 g	0·42
N A R	0·01 g cm^{-2} d^{-1}	1·54
Root radius	0·01 cm	0·32
Root extension rate (maximum)	3 cm d^{-1}	0·53
Soil		
Cylinder diameter	10 cm	1·7
Nitrate conc. in soil soln. initially	5×10^{-3} M	0·62

depend quite strongly upon environmental variables, such as nitrate concentration and that situations could readily be found in which the most rapidly growing species in pure culture did less well than its competitor when they were grown together. In addition, some of the systems modelled gave a larger total dry matter yield when two plants of different species were grown together than when either was grown in pure stand (Fig. 8.18). This possibility has frequently been discussed in relation to mixed stands of vegetation, and it is very useful to show that such situations can be predicted from theory.

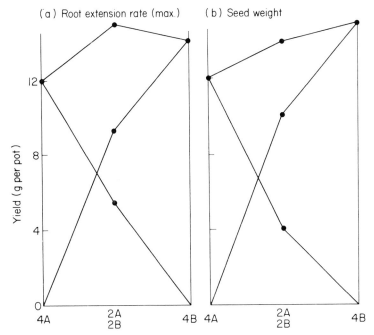

FIG. 8.18. Dry matter yield per pot of four standard plants (4A), four plants with one differing parameter (4B), or two plants (A) in competition with two plants (B). (a) root extension rate (maximum): A 3 cm d^{-1}, B 4 cm d^{-1}; (b) seed weight: A 0.010 g, B 0.015 g.

The competitive situation with two plants could be modelled in more detail by assigning different degrees of overlap of root systems, so that if plants A and B are assigned discrete but adjacent cells, roots from B invade A's cell, and form a new subdivision of it, which is occupied by both A and B roots—as discussed in the next section.

8.7.3 Competitive uptake of non-mobile solutes

The situation is complex for less mobile ions, since a distinct depletion zone around each root will develop, with an effective radius dependent, as will be shown, upon $\overline{\alpha a}/Db$. For simplicity, we assume that $\overline{\alpha a}$ is similar at all points on the roots of plant. The relative position of the individual roots belonging to the two plants is important in this case. If the root systems interpenetrate randomly, the proportions of neighbouring roots belonging to the other root system will be in proportion to the relative lengths of root. If any of the mechanisms causing root exclusion operate, which have been discussed earlier (p. 276 *et seq.*), then this ratio will be diminished. Mechanisms tending to the opposite (i.e. causing neighbours to be from the other root systems more often than expected from pure chance) are not easy to suggest, but could possibly arise if one root system excreted nutrients, e.g. legumes excreting nitrogen compounds. It seems most likely that the relationship of the roots in the two systems, within any defined volume, would be random or contagious, i.e. exclusion of the neighbouring system.

In extreme cases of root exclusion, we have complete separation of the root systems, which can be treated as individuals in separate soil volumes by the methods given earlier (Chapter 7), with the exception that diffusion or mass flow may move mobile solutes from one root volume to the other. At the other extreme, the roots interpenetrate randomly. If $\overline{\alpha a}$ values of the roots in the two systems are similar, little error should be caused by assuming the zero transfer boundary between roots to be equidistant from either, and in this case the steady-state treatment of Chapter 7 can be used. Thus L_P and $\overline{\alpha a}_P$ are used for the calculation of uptake by plant P, but the interroot distance x is calculated from L_P plus L_Q. It is possible (Wild *et al.* 1974) that species growing in the same environment may have fairly similar $\overline{\alpha a}$ values, but it cannot be assumed that this will always be the case. If it is not, an approximate method of correcting for this is to allocate different volumes of soil to the roots of different $\overline{\alpha a}$, in the steady-state treatment. The approximate position of the zero transfer boundary can be found from the point at which the steady-state concentration profiles of the individual roots cross. If the distance between roots of type P and Q (based on

$$1/\sqrt{\pi(L_P + L_Q)}$$

is x, let the zero transfer boundary be distance y from root P. We then have

$$C_{1i} = C_{1P}\left(1 + \frac{(\overline{\alpha a})_P}{Db}\ln\frac{y}{a}\right) \tag{8.11}$$

and for root Q

$$C_{1i} = C_{1Q}\left(1 + \frac{(\overline{\alpha a})_Q}{Db}\ln\frac{x-y}{a}\right) \tag{8.12}$$

if we assume the same root radius for both.

Hence

$$\frac{C_{1P}}{C_{1Q}} = \frac{Db + (\overline{\alpha a})_Q\ln\dfrac{x-y}{a}}{Db + (\overline{\alpha a})_P\ln\dfrac{y}{a}}. \tag{8.13}$$

The mean rate of decrease in concentration must be the same in all compartments to maintain a constant zero concentration boundary; hence

$$\frac{y^2}{I_P} = \frac{(x-y)^2}{I_Q} \quad \text{and} \quad \frac{C_{1P}}{C_{1Q}} = \frac{(\overline{\alpha a})_Q \, y^2}{(\overline{\alpha a})_P \, (x-y)^2}. \tag{8.14}$$

Combination of the two equations 8.13 and 8.14 allows y and C_{1P}/C_{1Q} to be found, though the final equation cannot be made explicit in y (Fig. 8.19).

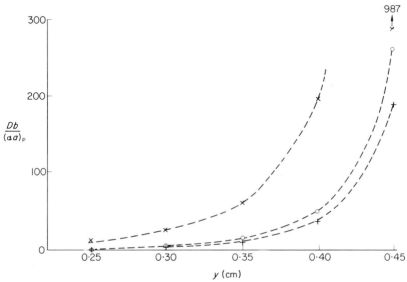

FIG. 8.19. Calculated values for position of zero-transfer boundary, between two roots of different αa values, for steady state treatment.

$$\frac{Db}{(\alpha a)_Q} = +, 0{\cdot}1; 0, 1{\cdot}0; \times 10; \, x = 0{\cdot}5 \text{ cm (see text)}$$

If the population is not uniformly mixed, so that there are volumes of soil in which P or Q roots preponderate, they must be considered as separate soil volumes, as discussed for soil volumes with varying nutrient concentration in Chapter 7. This method allows any two different types of roots to be dealt with in a similar model, and it could equally well be used for two different types of root from the same plant, such as seminal and nodal roots, if these are known to have different absorbing power.

With this, the treatment of two competing plants is simple in principle: their total soil volume is sub-divided into compartments containing different measured lengths of roots belonging to each plant. Each compartment is dealt with as described above, and transfer between compartments is carried

out as discussed earlier (see p. 233). The measurement of all required parameters is in practice a considerable task, but it cannot be baulked. Experiments on the mechanism of root competition are unlikely to yield clear
answers unless the plant and soil variables that control the issue have first
been identified and steps taken to measure them.

8.8   Conclusion

We are conscious of the incomplete state of the subjects dealt with in this
book. This does not mean that they are new—on the contrary, there is hardly
a topic that has not been studied for a century. The changes which give
freshness to the subject are, firstly, that plant nutrition is now seen as a
dynamic process, rather than as a static comparison of plant demand and
soil supply. Secondly, there has been more readiness to overlap the boundaries of traditional subjects, and to draw new ideas from wherever they can
be found. Modelling techniques are simply the tools with which these two
trends can be best expressed. The word 'integrate' has been much as used in
recent times, but the concepts which are developing can properly be said to
integrate plant physiology and soil science. It remains to be seen whether
they can deal in detail with the truly formidable physical and biological complexity of natural vegetation and soils. The mathematical techniques are now
amply good enough for predicting any process we wish. The major factors
impeding further progress are the inaccuracy of soil physical measurements;
the natural heterogeneity of soils; and the difficulty of predicting root development and properties, when these are so closely linked with the growth of a
whole plant or community in a fluctuating environment.

REFERENCES

ADAMS F. & PEARSON R.W. (1970) Differential response of cotton and peanuts to subsoil acidity. *J. Agron.* **62**, 9–12.

ALBERSHEIM P. (1965) The substructure and function of the cell wall. In *Plant Biochemistry*, eds. Bonner J.F. & Varner J.E., pp. 151–186. London: Academic Press.

ALBRECHT W.A., GRAHAM E.R. & SHEPPARD H.R. (1942) Surface relationships of roots and colloidal clay in plant nutrition. *Amer. J. Bot.* **29**, (3). 210–213.

ALLISON F.E. (1965) Evaluation of incoming and outgoing processes that affect soil nitrogen. In *Soil Nitrogen* eds. Bartholomew V.W. & Clark F.E., pp. 573–606. Agronomy Monograph 10.

AMER F., BOULDIN D.R., BLACK C.A. & DUKE F.R. (1955) Characterisation of soil phosphorus by anion exchange resin adsorption and ^{32}P equilibration. *Plant Soil* **6**, 391–408.

ANDERSON W.P. (1972) Ion transport in higher plant cells. *Ann. Rev. Pl. Physiol.* **23**, 51–72.

ANDERSON W.P. (1975) Long distance transport in roots. In *Ion Transport in Plant Cells and Tissues*, eds. Baker D.A. & Hall J.L., pp. 231–266. Amsterdam: North Holland Publishing Co.

ANDERSSEN R.S., HALE R.P. & RADOK R.R.M. (1969) The simulation with mathematical models of ion uptake by growing roots. *Plant Soil*, **30**, 271–289.

ANDREWS R.E. & NEWMAN E.I. (1970) Root density and competition for nutrients. *Oecologia Plantarum*. **5**, 319–334.

ARAMBARRI P. & TALIBUDEEN O. (1959) Factors influencing the isotopically exchangeable phosphate in soils. Pt. I. The effect of low concentrations of organic anions. *Plant Soil*, **11**, 343–354.

ARMSTRONG W. (1972) A re-examination of the functional significance of aerenchyma. *Physiol. Plant.* **27**, 173–177.

ARNOLD P.W. (1970) The behaviour of potassium in soils. *Fertil. Soc. Proc.* **115**, 3–30.

ASHER C.J., OZANNE P.G. & LONERAGAN J.F. (1965) A method for controlling the ionic environment of plant roots. *Soil Sci.* **100**, 149–156.

ASLYNG H.C. (1954) The lime and phosphate potential of soils; the solubility and availability of phosphates. *R. Vet. Agric. Coll. Copenhagen Yearbook*, 1–50.

ASPINALL D. (1960) An analysis of competition between barley and white Persicaria II. Factors determining the course of competition. *Ann. Appl. Biol.* **48**, 637–654.

ATKINSON D. (1973) Root studies. *Ann. Rep. East Malling Res. St.* 1972 pp. 56–60.

ATKINSON D. (1974) Some observations on the distribution of root activity in apple trees. *Plant Soil*, **40**, 333–342.

AUDUS L.J. (1969) Geotropism. In *Physiology of Plant Growth and Development*, ed. M.J. Wilkins, pp. 205–244. New York: McGraw-Hill.

AUNG L.H. (1974) Root–shoot relationships. In *The plant root and its environment*, ed. E.W. Carson, pp. 29–62. University Press of Virginia.

AVNIMELECH Y. & SCHERZER S. (1971) The effect on yield of P uptake by young plants. In *Recent advances in plant nutrition*, ed. Samish R.M., pp. 365–384.

BABCOCK K.L. (1963) Theory of the chemical properties of soil colloid systems at equilibrium. *Hilgardia*, **34**, 417–542.

BACHE B.W. (1963) Aluminium and iron phosphate studies relating to soils. I Solution and hydrolysis of variscite and strengite. *J. Soil Sci.* **14**, 113–123.

BACHE B.W. (1970) Determination of pH, lime potential and aluminium hydroxide potential of acid soils. *J. Soil Sci.* **21**, 28–37.

BACHE B.W. & WILLIAMS E.G. (1971) A phosphate sorption index for soils. *J. Soil Sci.* **22**, 289–301.

BAGSHAW R., VAIDYANATHAN L.V. & NYE P.H. (1969) An evaluation of the properties of soil potassium influencing its supply by diffusion to plant roots in soil. *J. Agric. Sci.*, (*Camb*), **73**, 1–14.

BAGSHAW R., VAIDYANATHAN L.V. & NYE P.H. (1972) The supply of nutrient ions by diffusion to plant roots in soil. VI. Effects of onion plant roots on pH and phosphate desorption characteristics in a sandy soil. *Plant Soil*, **37**, 627–639.

BAILEY G.W., WHITE J.L. & ROTHBERG T. (1968) Adsorption of organic herbicides by montmorillonite: Role of pH and chemical character of adsorbate. *Soil Sci. Soc. Amer. Proc.* **32**, 222–234.

BAKER D.A. & HALL J.L. (eds.) (1975) *Ion Transport in Plant Cells and Tissues*. Amsterdam: North Holland Publishing Co.

BALDWIN J.P. (1972) *Nutrient uptake by competing roots in soil*. D.Phil. Thesis, Oxford.

BALDWIN J.P. (1976) Competition for plant nutrients in soil; a theoretical approach. *J. Agric. Sci.* **87**, 341–356.

BALDWIN J.P. & NYE P.H. (1974) A model to calculate the uptake by a developing root system or root hair system of solutes with concentration variable diffusion coefficients. *Plant Soil*, **40**, 703–706.

BALDWIN J.P. & TINKER P.B. (1972) A method for estimating the lengths and spatial pattern of two interpenetrating root systems. *Plant Soil*, **37**, 209–213.

BALDWIN J.P., NYE P.H. & TINKER P.B. (1973) Uptake of solutes by multiple root systems from soil. III. A model for calculating the solute uptake by a randomly dispersed root system developing in a finite volume of soil. *Plant Soil*, **38**, 621–635.

BALDWIN J.P., TINKER P.B. & MARRIOTT F.H.C. (1971) The measurement of length and distribution of onion roots in the field and the laboratory. *J. appl. Ecol.*, **8**, 543–554.

BALDWIN J.P., TINKER P.B. & NYE P.H. (1972) Uptake of solutes by multiple root systems from soil. II. The theoretical effects of rooting density and pattern on uptake of nutrients from soil. *Plant Soil*, **36**, 693–708.

BARBER D.A. (1969) The influence of the microflora on the accumulation of ions by plants. In *Ecological aspects of the mineral nutrition of plants* ed. Rorison I.H., Oxford: Blackwell Scientific Publications Ltd.

BARBER D.A. (1972) 'Dual isotherms' for the absorption of ions by plant tissue. *New Phytol.* **71**, 255–262.

BARBER D.A. & GUNN K.B. (1974) The effect of mechanical forces on the exudation of organic substances by the roots of cereal plants grown under sterile conditions. *New Phytol.* **73**, 39–45.

BARBER D.A. & LOUGHMAN B.C. (1967) The effect of microorganisms on the absorption of inorganic nutrients by intact plants: II. Uptake and utilisation of phosphate by barley plants grown under sterile and non-sterile conditions. *J. exp. Bot.* **18**, 170–176.

BARBER D.A. & MARTIN J.K. (1976) The release of organic substances by cereal roots into soil. *New Phytol.* **76**, 69–80.

BARBER D.A. & ROVIRA A.D. (1975) Rhizosphere microorganisms and the absorption of phosphate by plants. *Ann. Rep. ARC Letcombe Laboratory*, 1974, 27–28.

BARBER S.A. (1962) A diffusion and mass-flow concept of soil nutrient availability. *Soil Sci.* **93**, 39–49.

BARBER S.A. & OZANNE P.G. (1970) Autoradiographic evidence for the differential effect of four plant species in altering the Ca content of the rhizosphere soil. *Soil Sci. Soc. Amer. Proc.* **34**, 635–637.

BARBER S.A., WALKER J.M. & VASEY E.H. (1962) Principles of ion movement through the soil to the plant root. In *Transactions International Society Soil Science, Commissions IV and V*. International Soil Conference: Soil Bureau, P.B. Lower Hutt, New Zealand. **1963**. 121–124.

BARBER S.A., WALKER J.M. & VASEY E.H. (1963) Mechanisms for the movement of plant nutrients from the soil and fertiliser to the plant root. *J. Agric. Food Chem.* **11**, 204–207.

BARLEY K.P. (1970) The configuration of the root system in relation to nutrient uptake. *Adv. Agron.* **22**, 159–201.

BARLEY K.P. & GREACEN E.L. (1967) Mechanical resistance as a soil factor influencing the growth of roots and underground shoots. *Adv. Agron.* **19**, 1–43.

BARLEY K.P. & ROVIRA A.D. (1970) The influence of root hairs on the uptake of phosphate. *Comm. Soil Sci. Plant Anal.* **1**, 287–292.

BARRACLOUGH D. (1976) *The diffusion of macromolecules in the soil pore space.* D.Phil. Thesis, Oxford.

BARRER R.M. (1969) Diffusion and permeation in heterogeneous media. In *Diffusion in Polymers*, eds. Crank J. & Park G. S., Ch. 6, New York: Acad. Press.

BARUA D.N. & DUTTA K.N. (1961) Root growth of cultivated tea in the presence of shade trees and nitrogenous manure. *Emp. J. exp. Agric.* **29**, 287–298.

BAR-YOSEF B., KAFKAFI U. & BRESLER E. (1972a) Uptake of phosphorus by plants growing under field conditions. I. Theoretical model and experimental determination of its parameters. *Soil Sci. Soc. Amer. Proc.* **36**, 783–788.

BAR-YOSEF B., BRESLER E. & KAFKAFI U. (1972b) Uptake of phosphorus by plants growing under field conditions. II. Computed and experimental results for corn plants. *Soil Sci. Soc. Amer. Proc.* **36**, 789–794.

BARTHOLOMEW W.V. & CLARK F.E. (1965) Soil Nitrogen. *Am. Soc. Agron. Monograph* **10**.

BARTHOLOMEW W.V. & KIRKHAM D. (1960) Mathematical descriptions and interpretation of culture induced soil nitrogen changes. *Trans. 7th Int. Cong. Soil Sci.*, **2**, 471–477.

BASSETT D.M., STOCKTON J.R. & DICKENS W.L. (1970) Root growth of cotton as measured by P^{32} uptake. *J. Agron.* **62**, 200–203.

BAVER L.D., GARDNER W.H. & GARDNER W.R. (1972) *Soil Physics.* New York: Wiley & Sons.

BAYLIS G.T.S. (1970) Root hairs and phycomycetous mycorrhizas in phosphorus deficient soil. *Plant Soil*, **33**, 713–716.

BAYLIS G.T.S. (1975) The magnolioid mycorrhiza and mycotrophy in root systems derived from it. In *Endomycorrhizas* eds. by Sanders F.E., Mosse B. & Tinker P.B., pp. 373–389. London: Academic Press.

BECKETT P.H.T. (1964) Potassium–calcium exchange equilibria in soils: specific adsorption sites for potassium. *Soil Sci.*, **97**, 376–383.

BECKETT P.H.T. (1964) Potassium–calcium exchange equilibria in soils: specific adsorption 150–154.

BECKETT P.H.T. (1969) Residual potassium and magnesium: a review. Tech. Bull. No. 20, Min. of Agric. Fish. & Food, H.M.S.O. (London) 183–196.

BECKETT P.H.T. (1971) Potassium potentials. A review. *Potash Reviews. Subject* 5, 1–41.

BECKETT P.H.T. & CRAIG J.B. (1964) The determination of potassium potentials. *Trans. 8th Int. Cong. Soil Sci.*, **2**, 249–255.

BECKETT P.H.T. & NAFADY M.H.M. (1967a) Studies on soil potassium. VI. The effect of K fixation and release on the form of the K: (Ca + Mg) exchange isotherm. *J. Soil Sci.* **18**, 244–262.

BECKETT P.H.T. & NAFADY M.H.M. (1967b) Potassium—calcium exchange equilibria in soils: the location of non-specific (Gapon) and specific exchange sites. *J. Soil Sci.* **18**, 263–281.

BECKETT P.H.T. & WHITE R.E. (1964) Studies on the phosphate potentials of soils. III. The pool of labile inorganic phosphate. *Plant Soil.* **21**, 253–282.

BEEK J. & FRISSEL M.J. (1973) Simulation of nitrogen behaviour in soils. PUDOC, Wageningen, Netherlands.

BEEVER J.E. & WOOLHOUSE H.W. (1975) Changes in the growth of roots and shoots when *Perilla frutescens* Britt. is induced to flower. *J. exp. Bot.* **26**, 451–464.

BENIANS G.J. & BARBER D.A. (1974) The uptake of phosphorus by barley plants from soil under aseptic and non-sterile conditions. *Soil Biol. Biochem.* **6**, 195–200.

BERGMANN W. (1958) Uber die Beeinflussung der Wurzelbehaarung von Roggenkeim pflanzen durch verschiedener aussenfaktoren. *Zeit Pfl. Dung.* **80**, 218–224.

BERNHARD-REVERSAT F. (1973) Le cycle du potassium en forêt tropicale humide. In *Potassium in Tropical Crops and Soils.* pp. 321–327. Proc. of the 10th Coll. of the International Potash Institute. Abidjan.

BERNSTEIN L. & HAYWARD H.E. (1958) Physiology of salt tolerance. *Ann. Rev. Pl. Physiol.* **9**, 25–46.

BESSEL D., VAN'T WOUDT & HAGAN R.M. (1957) Crop responses at excessively high soil moisture levels. In *Drainage of Agricultural lands,* ed. by Luthin J.N., pp. 514–611. American Soc. of Agronomy. Monograph No. 7.

BHAT K.K.S. & NYE P.H. (1973) Diffusion of phosphate to plant roots in soil. I. Quantitative autoradiography of the depletion zone. *Plant Soil,* **38**, 161–175.

BHAT K.K.S. & NYE P.H. (1974a) Diffusion of phosphate to plant roots in soil. II. Uptake along the roots at different times and the effect of different levels of phosphorus. *Plant Soil,* **41**, 365–382.

BHAT K.K.S. & NYE P.H. (1974b) Diffusion of phosphate to plant roots in soil. III. Depletion around onion roots without root hairs. *Plant Soil,* **41**, 383–394.

BHAT K.K.S., NYE P.H. & BALDWIN J.P. (1976) Diffusion of phosphate to plant roots in soil. IV. The concentration distance profile in the rhizosphere of roots with root hairs in a low-P soil. *Plant Soil,* **44**, 63–72.

BIELESKI R.L. (1973) Phosphate pools, phosphate transport and phosphate availability. *Ann. Rev. Pl. Physiol.* **24**, 225–252.

BIGGAR J.W. & NIELSEN D.R. (1963) Miscible displacement: V. Exchange processes. *Soil Sci. Soc. Amer. Proc.* **27**, 623–627.

BIRD R.B., STEWART W.E. & LIGHTFOOT E.N. (1960) *Transport Phenomena.* New York: Wiley.

BLACK C.A. (1968) *Soil–Plant Relationships.* London: Wiley.

BLAKEMORE M. (1966) Seasonal changes in the amount of phosphorus and potassium dissolved from soils by dilute calcium chloride solutions. *J. Agric. Sci.* **66**, 139–146.

BLOODWORTH M.E., BURLESON C.A. & COWLEY W.R. (1958) Root distribution of some irrigated crops using undisrupted soil cores. *J. Agron.* **50**, 317–320.

BODMAN G.B. & COLMAN E.A. (1943) Moisture and energy conditions during downward entry of water into soils. *Soil Sci. Soc. Amer. Proc.* **8**, 116–122.

BOGGIE R., HUNTER R.F. & KNIGHT A.H. (1958) Studies of root development of plants in the field using radioactive tracers. *J. Ecol.* **46**, 621–639.

BOGGIE R. & KNIGHT A.H. (1962) An improved method for the placement of radioactive isotopes in the study of root systems of plants growing in deep peat. *J. Ecol.* **50**, 461–462.

BOKEN E. (1969) Nutrient concentration curves for oats and barley at different times of the period of growth. *Plant Soil,* **31**, 311–320.

BOLE J.B. (1973) Influence of root hairs in supplying soil phosphorus to wheat. *Can. J. Soil Sci.* **53**, 169–175.

BOLT G.H. (1955) Ion adsorption by clays. *Soil Sci.* **79**, 267–276.

BOLT G.H. (1967) Cation exchange equations used in soil science—a review. *Nether. J. agric. Sci.* **15**, 81–103.

BOLT G.H. & FRISSEL M.J. (1960) Thermodynamics of soil moisture. *Nether. J. agric. Sci.* **8**, 57–78.

BOLT G.H. & GROENEVELT P.H. (1972) Coupling between transport processes in porous media. *Proc. 2nd IAHR–ISSS Symp. on Transport Phenomena in Porous Media. Guelph,* 630–652.

BOLTON J. (1971) Quantity–Intensity relationships for labile sodium in field soils. *J. Soil Sci.* **22**, 417–429.

BÖRNER H. (1960) Liberation of organic substances from higher plants and their role in the soil sickness problem. *Bot. Rev.* **26**, 393–424.

BOULDIN D.R. (1961) Mathematical description of diffusion processes in the soil–plant system. *Soil Sci. Soc. Amer. Proc.* **25**, 476–480.

BOULTER D., JEREMY J.J. & WILDING M. (1966) Amino acids liberated into the culture medium by pea seedling roots. *Plant Soil*, **24**, 121–127.

BOWEN G.D. (1973) Mineral nutrition of Ectomycorrhizas. In *Ectomycorrhizae*, eds. Marks G.C. & Kozlowski T.T., pp. 151–206. New York: Academic Press.

BOWEN G.D. & ROVIRA A.D. (1961) The effects of microorganisms on plant growth. I. Development of roots and root hairs in sand and agar. *Plant Soil*, **15**, 166–188.

BOWEN G.D. & ROVIRA A.D. (1966) Microbial factor in short-term phosphate uptake studies with plant roots. *Nature Lond.* **211**, 665–666.

BOWEN G.D. & ROVIRA A.D. (1973) Are modelling approaches useful in rhizosphere biology? In *Modern methods in the study of microbial soil ecology*, ed. Rosswall E., p. 443. Bull 17, Ecological Research Committee, Stockholm.

BOWER C.A., GARDNER W.R. & GOERTZEN J.O. (1957) Dynamics of cation exchange in soil columns. *Soil Sci. Soc. Amer. Proc.* **21**, 20–24.

BOWER C.A. & GOERTZEN J.O. (1955) Negative adsorption of salts by soils. *Soil Sci. Soc. Amer. Proc.* **19**, 147–151.

BOYLE J.R., VOIGT G.K. & SAWHNEY B.L. (1967) Biotite flakes; alteration by chemical and biological treatment. *Science*, **155**, 193–195.

BRAMS E. (1969) The mucilaginous layer of citrus roots—its delineation in the rhizosphere and removal from roots. *Plant Soil*, **30**, 105–108.

BRAY R.H. (1954) A nutrient mobility concept of soil–plant relationships. *Soil Sci.* **78**, 9–22.

BRESLER E. (1967) A model for tracing salt distribution in the soil profile and for estimating the efficient combination of water quantity under varying field conditions. *Soil Sci.* **104**, 227–233.

BRESLER E. (1972) Control of soil salinity. In *Optimizing the soil physical environment towards greater crop yields*, ed. by Hillel D., pp. 101–132. New York: Academic Press.

BRESLER E. (1973) Anion exclusion and coupling effects in non-steady transport through unsaturated soils. I. Theory. *Soil Sci. Soc. Amer. Proc.* **37**, 663–669.

BRESLER E. & HANKS R.J. (1969) Numerical method for estimating simultaneous flow of water and salt in unsaturated soils. *Soil Sci. Soc. Amer. Proc.* **33**, 827–840.

BREWSTER J.L. (1971) *Some factors affecting the uptake of plant nutrients from the soil.* D.Phil. Thesis, Oxford.

BREWSTER J.L., BHAT K.K.S. & NYE P.H. (1975a) The possibility of predicting solute uptake and plant growth response from independently measured soil and plant characteristics. II. The growth and uptake of onions in solutions of constant phosphate concentration. *Plant Soil*, **42**, 171–195.

BREWSTER J.L., BHAT K.K.S. & NYE P.H. (1975b) The possibility of predicting solute uptake and plant growth response from independently measured soil and plant characteristics. III. The growth and uptake of onions in a soil fertilized to different initial levels of phosphate and a comparison of the results with model predictions. *Plant Soil*, **42**, 197–226.

BREWSTER J.L., BHAT K.K.S. & NYE P.H. (1976) The possibility of predicting solute uptake and plant growth response from independently measured soil and plant characteristics. V The growth and phosphorus uptake of rape in soil at a range of phosphorus concentrations and a comparison of results with the predictions of a simulation model. *Plant Soil*, **44**, 295–328.

BREWSTER J.L. & TINKER P.B. (1970) Nutrient cation flows in soil around plant roots. *Soil Sci. Soc. Amer. Proc.* **34**, 421–426.

BREWSTER J.L. & TINKER P.B. (1972) Nutrient flow rates into roots. *Soil Ferts.* **35**, 355–359.

BRIGGS G.F., HOPE A.B. & ROBERTSON R.N. (1961) *Electrolytes and Plant Cells*, Oxford: Blackwell Scientific Publications Ltd.

BROECKER W.S. & OLSON E.A. (1960) Radiocarbon from nuclear tests II. *Science* **132**, 712–721.

BROESHART H. & NETHSINGHE D.A. (1972) Studies on the pattern of root activity of tree crops using isotope techniques. In *Isotopes and Radiation in Soil–Plant Relationships including Forestry*. Int. At. Energy Agency, Vienna. 453–461.

BROUWER R. (1965) Water movement across the root. In *The state and movement of water in living organisms*. Society of Experimental Biology Symposium. No. XIX, pp. 121–150.

BROUWER R. & DE WIT C.T. (1969) A simulation model of plant growth with special attention to root growth and its consequences. In *Root growth*, ed. by Whittington W.J., pp. 224–244. 15th Easter School, Nottingham.

BROWN D.A., FULTON B.E. & PHILLIPS R.E. (1964) Ion diffusion. I. A quick-freeze method for the measurement of ion diffusion in soil and clay systems. *Soil Sci. Soc. Amer. Proc.* **28**, 628–631.

BROWN M.E. (1972) Plant growth substances produced by microorganisms of soil and rhizosphere. *J. appl. Bact.* **35**, 443–451.

BROWN M.E. (1974) Seed and root bacterization. *Ann. Rev. Phytopath.* **12**, 181–197.

BRUGGEMAN D.A.G. (1935) The calculation of various physical constants of heterogeneous substances. I. The dielectric constants and conductivities of mixtures composed of isotopic substances. *Ann. Physik.* **24**, 636–664.

BUCHOLTZ K.P. (1971) The influence of allelopathy on mineral nutrition. In *Biochemical interactions between plants*, pp. 86–89. *Nat. Acad. Sci.*, Washington, D.C.

BUCKINGHAM E. (1904) Contributions to our knowledge of the aeration of soils. *U.S. Bur. Soils Bull.* **25**.

BURD J.S. & MARTIN J.C. (1923) Water displacement of soils and soil solution. *J. agric. Sci.* **13**, 265–295.

BURNS G.R. & DEAN L.A. (1964) The movement of water and nitrate around bands of sodium nitrate in soils and glass beads. *Soil Sci. Soc. Amer. Proc.* **28**, 470–474.

BURNS I.G. (1974) A model for predicting the redistribution of salts applied to fallow soils after excess rainfall or evaporation. *J. Soil Sci.* **25**, 165–178.

BYRNE J.M. (1974) Root Morphology. In *The Plant Root and its Environment*, ed. by Carson E.W., pp. 3–27. University Virginia Press, Charlottesville.

CAILLOUX M. (1972) Metabolism and absorption of water by root hairs. *Can. J. Bot.* **50**, 557–575.

CAMPBELL G.S. (1968) *Soil water distribution near absorbing root hairs, as affected by unsaturated conductivity and transpiration*, Ph.D. Thesis. Washington State University.

CANNELL R.Q. & FINNEY J.R. (1973) Effects of direct drilling and reduced cultivation on soil conditions for root growth. *Outlook on Agric.* **7**, 184–189.

CANNON W.A. (1949) A tentative classification of root systems. *Ecology*, **30**, 542–548.

CARLEY H.E. & WATSON R.D. (1966) A new gravimetric method for estimating root surface areas. *Soil. Sci.* **102**, 289–291.

CARRODUS B.B. (1967) Absorption of nitrogen by mycorrhizal roots of beech. II. Ammonia and nitrate as sources of nitrogen. *New Phytol.* **66**, 1–4.

CARSLAW M.S. & JAEGER J.C. (1959) *Conduction of Heat in Solids*, 2nd Edition, Oxford: Clarendon Press.

CARTWRIGHT B. (1972) The effect of P deficiency on the kinetics of phosphate absorption by sterile excised barley roots and some factors affecting the ion uptake efficiency of roots. *Comm. Soil Sci. Plant Analysis.* **3**, (4) 313–322.

CARY J.W. & MAYLAND H.F. (1972) Salt and water movement in unsaturated frozen soil. *Soil Sci. Soc. Amer. Proc.* **36**, 549–555.

CASSEL D.K., NELSON D.R. & BIGGAR J.W. (1969) Soil water movement in response to imposed temperature gradients. *Soil Sci. Soc. Amer. Proc.* **33**, 493–500.

CHAMBERLAIN G.T. (1962) Trace elements in some East African soils and plants. Part II. Manganese. *E. Afr. Agric. For. J.* **29**, 114–119.

CHAMBERS E.F. & HOLM L.L. (1965) Phosphorus uptake as affected by associated plants. *Weeds.* **13**, 312–314.

CHAMPION R.A. & BARLEY K.P. (1969) Penetration of clay by root hairs. *Soil Sci.* **108**, 402–407.

CHAPMAN S. & COWLING T.G. (1951) *Mathematical Theory of Non-Uniform Gases*, 2nd Edition, Cambridge: Cambridge University Press.

CHILDS E.C. (1969) *An introduction to the physical basis of soil water phenomena.* London: Wiley, Interscience.

CHILDS E.C. & COLLIS-GEORGE N. (1950) The permeability of porous materials. *Proc. R. Soc. (A)* **201**, 392–405.

CHRISTIE E.K. & MOORBY J. (1975) Physiological responses of arid grasses. I. The influence of phosphorus supply on growth and phosphorus absorption. *Aust. J. Agric. Res.* **26**, 423–436.

CLAPHAM A.R. (1969) Introduction. In *Ecological aspects of the mineral nutrition of plants*, ed. Rorison I. Oxford: Blackwell Scientific Publications Ltd.

CLARK F.E. (1949) Soil microorganisms and plant roots. *Adv. Agron.* **1**, 242–282.

CLARK F.E. & PAUL E.A. (1970) The microflora of grassland. *Adv. Agron.* **22**, 375–436.

CLARKE A.L. & BARLEY K.P. (1968) The uptake of nitrogen from soils in relation to solute diffusion. *Aust. J. Soil Res.* **6**, 75–92.

CLARKSON D.T. (1974) *Ion transport and cell structure in plants.* London: McGraw-Hill.

CLARKSON D.T., DREW M.C., FERGUSON I.B. & SANDERSON J. (1975) The effect of the Take-all fungus, *Gaeumannomyces graminis*, on the transport of ions by wheat plants. *Physiol. Plant Path.* **6**, 75–84.

CLARKSON D.T., GRAHAM J. & SANDERSON J. (1974) Water uptake by the roots of marrow and barley plants. *Letcombe Laboratory Ann. Rep.* 1973. pp. 9–11.

CLARKSON D.T., SANDERSON J. & RUSSELL R.S. (1968) Ion uptake and root age. *Nature*, **220**, 805–806.

CLEMENT C.R., HOPPER M.J., CANAWAY R.J. & JONES L.H.P. (1974) A system for measuring the uptake of ions by plants from flowing solutions of controlled composition. *J. exp. Bot.* **25**, 81–99.

CLOWES F.A.L. (1969) Anatomical aspects of structure and development. In *Root Growth.* ed. by Whittington W.J., pp. 3–18. 15th Easter School, London: Butterworth.

COCKCROFT B., BARLEY K.P. & GREACEN E.L. (1969) The penetration of clays by fine probes and root tips. *Aust. J. Soil Res.* **7**, 333–348.

COLE D.W., GESSEL S.P. & DICE S.F. (1967) Distribution and cycling of nitrogen, phosphorus, potassium and calcium in a second growth Douglas Fir ecosystem. *Proc. Ann. Meet. Amer. Ass. Advancement Sci.* 197–231.

COLE P.I. & ALSTON A.M. (1974) Effect of transient dehydration of absorption of chloride by wheat roots. *Plant Soil.* **40**, 243–247.

COLLINS J.C. & HOUSE C.R. (1969) The exchange of sodium ions in the root of *Zea Mays. J. exp. Bot.* **20**, 497–506.

COLLIS-GEORGE N. & BOZEMAN J.M. (1970) A double layer theory for mixed ion systems as applied to the moisture content of clays under restraint. *Aust. J. Soil Res.* **8**, 239–258.

COOKE G.W. (1954) Recent advances in fertilizer placement. *J. Sci. Fd. Agric.* **5**, 429–440.

COOKE G.W. (1967) *The control of soil fertility*, Ch. 15. London: Crosby Lockwood.

COOKE G.W. (1969) Plant nutrient cycles. *Proc. VIIth Colloquium Int. Potash Inst. Berne·* 75–95.

COOPER A.J. (1973) Root temperature and plant growth. Research Review No. 4, *Commonwealth Bur. Hort. and Plantation Crops.* Farnham Royal. *Comm. Agric. Bur.*

CORMACK R.G.H. (1962) Development of root hairs in Angiosperms II. *Bot. Rev.* **28**, 446–464.

CORMACK R.G.H., LEMOY P. & MACLACHLAN G.A. (1963) Calcium in root hair wall. *J. exp. Bot.* **14**, 311–315.

CORNFORTH I.S. (1968) The effect of the size of soil aggregates on nutrient supply. *J. agric. Sci.* **70**, 83–85.

COTTENIE A. & KIEKENS L. (1972) Exchange of Zn, Mn, Cu and Fe in relation to saturation of the soil complex. *Proc. 9th Colloquium Int. Potash Int. Berne.* 113–123.

COWAN I.R. (1965) Transport of water in the soil–plant atmosphere system. *J. appl. Ecol.* **2**, 221–239.

COWAN I.R. & MILTHORPE F.L. (1968) Plant factors influencing the water status of plant tissues, In *Water deficits and plant growth*, ed. Kozlowski T.T., pp. 137–193.

CRANK J. (1975) *Mathematics of Diffusion.* 2nd Edition, Oxford: Clarendon Press.

CROSSETT R.N., CAMPBELL D.J. & STEWART H.E. (1975) Compensatory growth in cereal root systems. *Plant Soil*, **42**, 673–683.

CROWDY S.H. (1972) Translocation. In *Systemic Fungicides*, ed. by Marsh R.W., pp. 92–115.

CUNNINGHAM R.K. (1964) Cation–anion relationships in crop nutrition: III Relationships between the ratios of sum of the cations: sum of the anions and nitrogen concentrations in several plant species. *J. agric. Sci.* **63**, 109–111.

CUNNINGHAM R.K. & COOKE G.W. (1957) Inorganic nitrogen in soils. *Rep. Rothamsted exptl. Sta.*, **1956**, 53–54.

CURRIE J.A. (1960) Gaseous diffusion in porous media. Pt. II. Dry granular materials. *Br. J. appl. Phys.* **11**, 318–324.

CURRIE J.A. (1961) Gaseous diffusion in porous media. Pt. III. Wet granular materials. *Br. J. appl. Phys.* **12**, 275–281.

CURRIE J.A. (1970) Movement of gases in soil respiration. *Soc. Chem. Ind. Monograph, No.* **37**, 152–169.

CURTISS C.F. & HIRSCHFELDER J. (1949) Transport properties of multicomponent gas mixtures. *J. chem. Phys.* **17**, 550–555.

CUTTER E.G. (1971) *Plant Anatomy: Experiment and Interpretation. Vol. II.* London: Edward Arnold.

CZAPEK F. (1896) Zur lehre von den Wurzelausscheidungen. *Jahrb. wiss. Bot.* **29**, 321–390.

DAFT M.J. & NICOLSON T.H. (1966) Effect of *Endogone* mycorrhiza on plants growth. *New Phytol.* **65**, 343–350.

DAINTY J. (1969a) The Ionic Relations of Plants. In *Physiology of plant growth and development*, ed. by Wilkins M.B., pp. 455–488. New York: McGraw-Hill.

DAINTY J. (1969b) The water relations of plants. In *Physiology of plant growth and development*, ed. by Wilkins M.B., pp. 421–454. New York: McGraw-Hill.

DAKSHINAMURTI C. (1959) Effective pore space measurements in porous media by ionic diffusion. *Soil Sci.* **88**, 209–212.

DALTON F.N., RAATS P.A.C. & GARDNER W.R. (1975) Simultaneous uptake of water and solutes by plant roots. *Agron. J.* **67**, 334–339.

DANIELSON R.E. (1967) Root systems in relation to irrigation. In *Irrigation of Agricultural Lands*, eds. Hogan R.M., Haise H.R. and Edminster T.W., pp. 390–424. Agronomy Monograph No. 11, Am. Soc. Agron. Madison, USA.

DANIELSON R.E. & RUSSELL M.B. (1957) Ion absorption by corn roots as influenced by moisture and aeration. *Soil Sci. Amer. Proc.* **21**, 3–6.

DART P.J. & MERCER F.V. (1964) The legume rhizosphere. *Arch. Mikrobiol.* **47**, 344–378.

DAUBENY C.G.B. (1846) On the distinction between the dormant and active ingredients of the soil. *J. R. Agric. Soc. Eng.* **7**, 237–245.

DE HAAN F.A.M., BOLT G.H. & PIETERS B.G.M. (1965) Diffusion of ^{40}K into an illite during prolonged shaking. *Soil Sci. Soc. Amer. Proc.* **29**, 528–530.

DE ROO H.C. (1969) Tillage v root growth. In *Root Growth.* ed. Whittington W.J. pp. 339–356. Proc. 15th Easter School, Nottingham. London: Butterworths.

DE WIT C.T. (1960) On Competition. *Versl. landbouwk. Onder* 2. **66**, 1–88.

DE WIT C.T., BROUWER R. & PENNING DE VRIES F.W.T. (1969) The simulation of photosynthetic systems. In *Prediction and measurement of photosynthetic productivity*, ed. by Setlik I., pp. 47–50. Proc. IBP/PP technical meeting, Trebon, 1969.

DE WIT C.T. & VAN KEULEN H. (1972) *Simulation of transport processes in soils.* Centre for Agricultural Publishing and Documentation, Wageningen.

DEBYE P. & HUCKEL E. (1923) *Phys. Z.* **24**, 185.

DEIST J. & TALIBUDEEN O. (1967a) Ion exchange in soils from the ion pairs K–Ca, K–Rb, and K–Na. *J. Soil Sci.* **18**, 125–137.

DEIST J. & TALIBUDEEN O. (1967b) Thermodynamics of K–Ca ion exchange in soils. *J. Soil Sci.* **18**, 138–148.

DEIST J. & TALIBUDEEN O. (1967c) ^{86}Rb as a tracer for exchangeable K in soils. *Soil Sci.* **104**, 119–122.

DEVAUX H. (1916) Action rapide des solutions salines sur les plantes vivantes: déplacement reversible d'une partie des substances basiques contenues dans la plante. *C.R. Acad. Sci. Paris.* **162**, 561–563.

DITTMER H.J. (1938) A comparative study of the subterranean members of three field grasses. *Science*, **88**, 482.

DITTMER H.J. (1940) A quantitative study of the subterranean members of soyabean. *Soil Cons.* **6**, 33–34.

DITTMER H.J. (1949) Root hair variation in plant species. *Amer. J. Bot.* **36**, 152–155.

DOBEREINER J. & DAY J.M. (1975) Nitrogen fixation in the rhizosphere of tropical grasses. In *Nitrogen fixation by free-living micro-organisms*. Ed. Stewart W.D.P. Cambridge: Cambridge University Press.

DONALD C.M. (1963) Competition among crop and pasture plants. *Adv. Agron.* **15**, 1–118.

DRAKE M. (1964) Soil chemistry and plant nutrition. In *Chemistry of the soil*, ed. Bear F.E., pp. 395–444. New York: Reinhold.

DRAYCOTT A.P., DURRANT M.J. & WEBB D.J. (1974) Effects of plant density, irrigation and potassium and sodium fertilizers on sugar beet. I. Yields and nutrient composition. *J. Agric. Sci.* **82**, 251–259.

DREW M.C. (1966) *Uptake of plant nutrients by roots growing in the soil.* D.Phil. Thesis, Oxford.

DREW M.C. & GOSS M.J. (1973) Effect of soil physical factors on root growth. *Chem. Ind.* 679–684.

DREW M.C. & NYE P.H. (1969) The supply of nutrient ions by diffusion to plant roots in soil. II. The effect of root hairs on the uptake of potassium by roots of rye-grass (*Lolium multiflorum*). *Plant Soil*, **31**, 407–424.

DREW M.C. & NYE P.H. (1970) The supply of nutrient ions by diffusion to plant roots in soil. III. Uptake of phosphate by roots of onion, leek and rye-grass. *Plant Soil*, **33**, 545–563.

DREW M.C., NYE P.H. & VAIDYANATHAN L.V. (1969) The supply of nutrient ions by diffusion to plant roots in soil. I. Absorption of potassium by cylindrical roots of onion and leek. *Plant Soil*, **30**, 252–270.

DREW M.C. & SAKER L.R. (1975a) Nutrient supply and the growth of the seminal root system in barley. II. Localized compensatory increases in lateral root growth and rates of nitrate uptake. *J. exp. Bot.* **26**, 79–90.

DREW M.C. & SAKER L.R. (1975b) Further studies on the modification to root growth and ion uptake caused by localized enrichment of phosphate in the rooting zone of barley. *Ann. Report. ARC, Letcombe Lab.* 1974. 8–10.

DUNCAN W.G. & OHLROGGE A.J. (1958) Principles of nutrient uptake from fertilizer bands II. *Agron. J.* **50**, 605–608.

DUNHAM R.J. & NYE P.H. The influence of soil water content on the uptake of ions by roots. I. Soil water content gradients near a plane of onion roots. *J. appl. Ecol.* **10**, 585–598 (1973). II. Chloride uptake and concentration gradients in soil. *J. appl. Ecol.* **11**, 581–596 (1974). III. Phosphate, potassium, calcium and magnesium uptake and concentration gradients in soil. *J. appl. Ecol.* **13**, 967–984.

DYER B. (1894) On the analytical determination of probably available 'mineral' plant food in soil. *J. Chem. Soc.* **65**, 115–167.

DYER B. (1901) A chemical study of the phosphoric acid and potash contents of wheat soils of Broadbalk Field, Rothamsted. *Phil. Trans. R. Soc. B.* **194**, 235–290.

EATON F.M., HARDING R.B. & GANGE T.J. (1960) Soil solution extractions at 1/10 bar moisture percentages. *Soil Sci.* **90**, 253–258.

EAVIS B.W. (1972) Soil physical conditions affecting seedling growth I. Mechanical impedance, aeration and moisture availability as influenced by bulk density and moisture levels in a sandy loam soil. *Plant Soil*, **36**, 613–622.

EAVIS B.W. & PAYNE D. (1969) Soil physical conditions and root growth. In *Root Growth*, ed. Whittington W.J., pp. 315–338. 15th Easter School, Nottingham. London: Butterworths.

EAVIS B.W., TAYLOR H.M. & HUCK M.G. (1971) Radicle elongation of pea seedlings as affected by oxygen concentration and gradients between shoot and root. *Agron. J.* **63**, 770–772.

EDWARDS C.A. (1974) Factors affecting the persistence of pesticides in the soil. *Chem. Ind.* **5**, 190–193.

EKDAHL I. (1957) The growth of root hairs and roots in nutrient media and by distilled water and the effects of oxalate. *Kongl. Lantbrukshogskol. Anm.* **23**, 497–518.

ELGAWHARY S.M., MALZER G.L. & BARBER S.A. (1972) Calcium and strontium transport to plant roots. *Soil Sci. Soc. Amer. Proc.* **36**, 794–799.

ELLIS F.B. & BARNES B.T. (1973) Estimation of the distribution of living roots of plants under field conditions. *Plant Soil*, **39**, 81–91.

ELRICK D.E. & FRENCH L.K. (1966) Miscible displacement patterns on disturbed and undisturbed soil cores. *Soil Sci. Soc. Amer. Proc.* **30**, 153–156.

EPSTEIN E. (1972) *Mineral Nutrition of Plants: Principles and Perspectives.* New York: Wiley.

EPSTEIN E., Rains D.W. & Elzam O.E. (1963) Resolution of dual mechanisms of potassium absorption by barley roots. *Proc. Nat. Acad. Sci.* **49**, 684–692.

EPSTEIN E. & RAINS D.W. (1965) Carrier-mediated transport in barley roots—kinetic evidence for a spectrum of active sites. *Proc. Nat. Acad. Sci.* **53**, 1320–1324.

ERIKSSON E. (1952) Cation exchange equilibria on clay minerals. *Soil Sci.* **74**, 103–113.

ESAU K. (1960) *Anatomy of seed plants.* New York: Wiley.

ESHEL A. & WAISEL Y. (1973) The variations in uptake of Na and Rb along barley roots. *Physiol Plant.* **28**, 557–600.

EVANS G.C. (1972) *The quantitative analysis of plant growth.* Oxford: Blackwell Scientific Publications Ltd.

EVERETT D.H. & WHITTON W.I. (1952) A general approach to hysteresis. I. *Trans. Faraday Soc.* **48**, 749–757.

FAHN A. (1967) *Plant Anatomy*. Oxford: Pergamon Press Ltd.

FARR E., VAIDYANATHAN L.V. & NYE P.H. (1969) Measurement of ionic concentration gradients in soil near roots. *Soil Sci.* **107**, 385–391.

FARR E., VAIDYANATHAN L.V. & NYE P.H. (1970) The measurement and mechanism of ion diffusion in soils. V. Diffusion of hydrogen ion in soils. *J. Soil Sci.* **21**, 1–14.

FARR E. & VAIDYANATHAN L.V. (1972) The supply of nutrient ions by diffusion to plant roots in soil. IV.—Direct measurement of changes in labile phosphate content in soil near absorbing roots. *Plant Soil.* **37**, 609–616.

FARRELL D.A., GREACEN E.L. & GURR C.G. (1966) Vapour transfer in soil due to air turbulence. *Soil Sci.* **102**, 305–313.

FAXEN H. (1922) Der Widerstand gegen die Bewegung einen starren Kugel in einer zahen Flussigkeit, die zwischen zwei parallelen ebenen Wanden eingeschlossen ist. *Ann. Physik*. **68**, 89–119.

FEHRENBACHER J.B. & ALEXANDER J.D. (1955) A method for studying corn root distribution using a soil core sampling machine and a shaker-type washer. *Agron. J.* **47**, 468–472.

FINNEY J.R. (1973) Root growth in relation to tillage. *Chem. Ind.* 676–678.

FINNEY J.R. & KNIGHT B.A.G. (1973) The effect of soil physical conditions produced by various cultivation systems on the root development of winter wheat. *J. Agric. Sci.* **80**, 435–442.

FISHER R.A. (1925) *Statistical Methods for Research Workers*, 1st Edition, Edinburgh: Oliver & Boyd.

FITTER A.H. (1974) A relationship between phosphorus requirement, the mobilization of added phosphate, and the phosphate buffering capacity of colliery shales. *J. Soil Sci.* **25**, 41–50.

FLOCKER W.J., TIMMS H. & VOMOCIL J.A. (1960) Effect of soil compaction on tomato and potato yields. *Agron. J.* **52**, 345–348.

FORRESTER S.D. & GILES C.H. (1971) From manure heaps to monolayers: the earliest development of solute–solid adsorption studies. *Chem. Ind.* 1314–1321.

FOTH H.D. (1962) Root and top growth of corn. *Agron. J.* **54**, 49–52.

FOX J.E. (1969) The Cytokinins, In *Physiology of plant growth and development* ed. by Wilkins M.B., pp. 85–126. New York: McGraw-Hill.

FOY C.D. (1974) Effects of aluminium on plant growth. In *The plant root and its environment*. ed. Carson E.W., pp. 601–642. University Press, Virginia, Charlottesville, USA.

FOY C.D., BURNS G.R., BROWN J.C. & FLEMING A.L. (1965) Differential aluminium tolerance of two wheat varieties associated with plant induced pH changes around their roots. *Soil Sci. Soc. Amer. Proc.* **29**, 64–67.

FRANKLIN R.E. (1969) Effect of adsorbed cations on phosphorus uptake by excised roots. *Plant Physiol.* **44**, 697–710.

FRIED J.J. & UNGEMACH P.O. (1971) Determination in situ de coefficient de dispersion longitudinale d'un milieu poreux natural. *C-R. Acad. Sci. de Paris*. **172**, A. 1327.

FRIED M. & BROESHART H. (1967) *The soil–plant system*. New York: Academic Press.

FRIED M., ZSOLDOS F., VOSE P.B. & SHATOKHIN I.L. (1965) Characterizing the NO_3 and NH_4 uptake process of rice roots by use of ^{15}N labelled NH_4NO_3. *Plant Physiol.* **18**, 313–320.

FRISSEL M.J. & POELSTRA P. (1967a) Chromatographic transport through soils I. Theoretical evaluations. *Plant Soil*, **26**, 285–302.

FRISSEL M.J. & POELSTRA P. (1967b) Chromatographic transport through soils II. Column experiments with Sr and Ca isotopes. *Plant Soil*. **27**, 20–32.

FRISSEL M.J., POELSTRA P. & REINIGER P. (1970a) Chromatographic transport through soils. III. A simulation model for the evaluation of the apparent diffusion coefficient in undisturbed soils with tritiated water. *Plant Soil*, **33**, 161–176.

FRISSEL M.J., POELSTRA P. & REINIGER P. (1970b) Sorption and transport in soils. In *Sorption and transport processes in soils.* pp. 135–151. S.C.I. Monograph No. 37.

FRISSEL M.J. *et al.* (1973) Tracing soil moisture migration with ^{36}Cl, ^{60}Co and tritium. Proc. Symp. on isotopes and radiation techniques in studies of soil physics, irrigation and drainage in relation to crop production. *I.A.E.A., Vienna,* 145–151.

FURLAN V. & FORTIN J.A. (1973) Formation of endomycorrhizae by *Endogone calospora* on *Allium cepa* under three temperature regimes. *Naturaliste Canadien,* **100,** 467–477.

GAINES G.L. & THOMAS H.C. (1953) Adsorption studies on clay minerals. II. A formulation of the thermodynamics of exchange adsorption. *J. Chem. Phys.* **21,** 714–718.

GAMS W. (1967) *Mikroorganismen in der Wurzelregion in Weizen. Mitt. aus der Biologischer Bundesanstalt fur Land und Forst-wirtschaft.* Berlin-Dahlem. Heft 123.

GAPON E.N. (1933) Theory of exchange adsorption in soils. *J. Gen. Chim. Moscow,* **3,** 144–163.

GARBOUCHEV I.P. (1966) Changes occurring during a year in the soluble phosphorus and potassium in soil under crops in rotation experiments at Rothamsted, Woburn and Saxmundham. *J. Agric. Sci.* **66,** 399–412.

GARDNER W.R. (1960) Dynamic aspects of water availability to plants. *Soil Sci.* **89,** 63–73.

GARDNER W.R. (1964) Relation of root distribution to water uptake and availability. *J. Agron.* **56,** 41–45.

GARDNER W.R. (1965) Movement of nitrogen in soil. In *Soil Nitrogen,* eds. Bartholomew W.V. & Clark F.E., pp. 550–572. Am. Soc. Agron. Monograph 10.

GARDNER W.R., JURY W.A. & KNIGHT J. (1975) Water uptake by vegetation. In *Heat and mass transfer in the biosphere,* eds. Vries D.A. & Afgan N.A., pp. 443–456. New York: John Wiley & Son.

GARWOOD E.A. (1967) Seasonal variation in appearance and growth of grass roots. *J. Br. Grassland Soc.* **22,** 121–130.

GARWOOD E.A. & WILLIAMS T.E. (1967) Soil water use and growth of a grass sward. *J. Agric. Sci.* **68,** 281–292.

GASSER J.K.R. (1961) Transformation, leaching and uptake of fertilizer nitrogen applied in autumn and spring to winter wheat on heavy soil. *J. Sci. Fd Agric.* **12,** 375–380.

GASSER J.K.R. (1962) Transformation, leaching and uptake of fertilizer—N applied to winter and to spring wheat grown on a light soil. *J. Sci. Fd Agric.* **13,** 367–375.

GASSER J.K.R. (1969) Available soil nitrogen—its measurement, and some factors affecting its correlation with crop performance. *Welsh Soils Discussion Group Rep. No.* 10, 76–92.

GASSER J.K.R. & THORBURN M.A.P. (1972) The growth composition and nutrient uptake of spring wheat. *J. Agric. Sci.* **78,** 393–409.

GAUCH H.G. (1972) *Inorganic Plant Nutrition,* Pennsylvania: Dowden, Hutchinson & Ross.

GAZZERI G. (1823) *A textbook of manuring,* quoted by Orth, A., *Landwirtsch. Vers-Sta.* 1873, **16,** 56.

GEERING H.R. (1967) M.Sc. Thesis, Cornell Univ. (quoted in Olsen S.R. & Kemper W.D., *Adv. Agron.* 1968, **20,** 91–149).

GERDEMANN J.W. (1968) Vesicular-arbuscular mycorrhiza in plant growth. *Ann. Rev. Phytopath.* **6,** 397–418.

GERDEMANN J.W. & TRAPPE J.M. (1975) Taxonomy of the *Endogonaceae.* In *Endomycorrhizas,* eds. Sanders F.E., Mosse B. & Tinker P.B., pp. 35–52. London: Academic Press.

GERRETSEN F.C. (1948) The influence of microorganisms on the phosphate intake by the plant. *Plant Soil.* 1948. **1,** 51–81.

GERWITZ A. & PAGE E.R. (1973) Estimation of root distribution in soil, by labelling with ^{86}Rb and counting with commercially available equipment. *Lab. Pract.* **22**, 35–36.

GERWITZ A. & PAGE E.R. (1974) An empirical mathematical model to describe plant root systems. *J. appl. Ecol.* **11**, 773–781.

GLASSTONE S. (1940) *Textbook of Physical Chemistry.* London: MacMillan.

GLASSTONE S., LAIDLER K.J. & EYRING H. (1941) *Theory of Rate Processes.* New York: McGraw-Hill.

GLUEKAUF E. (1955) Theory of chromatography. Pt. 10. Formulae for diffusion into spheres and their application to chromatography. *Trans. Faraday Soc.* **51**, 1540–1551.

GORE A.J.P. & OLSON J.S. (1967) Preliminary models for accumulation of organic matter in an *Eriophorum/Calluna* ecosystem. *Aquilo, Ser. Bot. Soc. Amic. Nat. Oulensis.* **6**, 297–313.

GORING C.A.I. & HAMAKER J.W. (eds.) (1972) *Organic Chemicals in the Soil Environment.* 2 Vols. New York: Marcel Dekker.

GORING C.A.I. (1968) The size, shape and origin of lignin macromolecules. In *Solution Properties of Natural Polymers. Chem. Soc. London. Spec. Pub. No.* **27**, 115–134.

GORING R.L. & CHURCHILL S.W. (1961) Thermal conductivity of heterogeneous materials. *Chem. Eng. Prog.* **57**, 53–59.

GRABLE A.R. (1966) Soil aeration and plant growth. *Advan. Agron.* 58–106.

GRAF H., REICHENBACK V. & RICH C.I. (1968) Preparation of dioctahedral vermiculite from muscovite and subsequent exchange properties. *Trans 9th Int. Cong. Soil Sci. Adelaide.* **1**, 709–719.

GRAHAM-BRYCE I.J. (1965) Diffusion of cations in soils. *Tech. Rep. Ser. No.* 48, 42–56. Int. Atom. Ener. Agency. Vienna.

GRAHAM-BRYCE I.J. (1969) Diffusion of organo-phosphorus insecticides in soil. *J. Sci. Fd Agric.* **20**, 489–492.

GRAHAM-BRYCE I.J. & BRIGGS G.G. (1970) Pollution of soils. *R. Inst. Chem.* **3**, 87–104.

GREACEN E.L., FARRELL D.A. & COCKCROFT B. (1968) Soil resistance to metal probes and plant roots. *Trans. Int. Soc. Soil Sci. (Adelaide)* **1**, 769–778.

GREACEN E.L. & OH J.S. (1972) Physics of root growth. *Nature, N.B. (Lond.)* **235**, 24–25.

GREAVES M.C. & WEBLEY D.M. (1969) The hydrolysis of *myo* inositol hexaphosphate by soil microorganisms. *Soil Biol. Biochem.* **1**, 37–43.

GREAVES M.P. & DARBYSHIRE J.F. (1972) The ultrastructure of the mucilaginous layer on plant roots. *Soil Biol. Biochem.* **4**, 443–449.

GREEN J.S. (1976) *An investigation of the self diffusion of chloride ions within soil aggregates.* Thesis, Chemistry Pt 2, Oxford.

GREENLAND D.J. (1958) Nitrate fluctuations in tropical soils. *J. agric. Sci.* **50**, 82–89.

GREENLAND D.J. (1965a) Interaction between clays and organic compounds in soils. Pt. I. Mechanisms of interaction between clays and defined organic compounds. *Soils Fert.* **28**, 415–425.

GREENLAND D.J. (1965b) Interaction between clays and organic compounds in soils. Pt. II. Adsorption of solid organic compounds and its effects on soil properties. *Soils Fert.* **28**, 521–532.

GREENLAND D.J. (1970) Sorption of organic compounds by clays and soils. *Soc. Chem. Ind. Monograph No.* **37**, 79–88.

GREENLAND D.J. (1971) Changes in the nitrogen status and physical condition of soil under pastures, with special reference to the maintenance of the fertility of Australian soils used for growing wheat. *Soils Fert.* **34**, 237–251.

GREENLAND D.J. & NYE P.H. (1959) Increases in the carbon and nitrogen contents of tropical soils under natural fallows. *J. Soil Sci.* **9**, 284–299.

GREENWAY H., HUGHES P.G. & KLEPPER B. (1969) The effect of water deficits on phosphorus nutrition of tomato plants. *Physiol. Plant.* **22**, 199–207.

GREENWOOD D.J. (1969) Effect of oxygen distribution in the soil on plant growth. In *Root Growth*. ed. Whittington W.J., pp. 202–220. 15th Easter School, Nottingham, London: Butterworths.

GREENWOOD D.J., WOOD J.T. & CLEAVER T.J. (1974) A dynamic model for the effects of soil and weather conditions on nitrogen response. *J. agric. Sci.*, **82**, 455–467.

GREIG-SMITH P. (1964) *Quantitative Plant Ecology*. Ch. 3. London: Butterworths.

GRIFFIN D.M. & QUAIL G. (1968) Movement of bacteria in moist particulate systems. *Aust. J. Biol. Sci.*, 1968. **21**, 5–79.

GRÜMMER G. (1961) The role of toxic substances in the interrelationships between higher plants. In *Mechanisms in Biological competition. Symp. Soc. exp. Biol.* **15**, 219–228.

GUGGENHEIM E.A. (1933) *Modern Thermodynamics by the methods of Willard Gibbs*. London: Methuen.

GUGGENHEIM E.A. (1935) *Phil. Mag.* **19**, 588.

GUGGENHEIM E.A. & STOKES R.H. (1958) Activity coefficients of 2 : 1 and 1 : 2 electolytes in aqueous solution from isopiestic data. *Trans. Faraday Soc.* **54**, 1646–1649.

GUNARY D. (1963) Behaviour of carrier free phosphorus –32 in natural soils in relation to the measurement of labile soil phosphorus. *J. Sci. Fd Agric.* **14**, 319–324.

GUNARY D. (1966) Pyrophosphate in soil—some physico-chemical aspects. *Nature, Lond.*, **210**, 1297–1298.

GUNARY D. (1970) A new adsorption isotherm for phosphate in soil. *J. Soil Sci.* **21**, 72–77.

GUR A. & SHULMAN Y. (1971) Influence of high root temperatures on the potassium nutrition and on certain organic constituents of apple plants. In *Recent advances in plant nutrition, Volume II* ed. Samish R.M., pp. 643–656. New York: Gordon & Breach.

GURR C.G., MARSHALL T.J. & HUTTON J.T. (1952) Movement of water in soil due to a temperature gradient. *Soil Sci.*, **74**, 335–342.

HACKETT C. (1968) A study of the root system of barley. I. Effects of nutrition on two varieties. *New Phytol.* **67**, 287–299.

HACKETT C. (1969) A study of the root system of barley. II.Relationships between root dimensions and nutrient uptake. *New Phytol.* **68**, 1023–1030.

HACKETT C. (1972) A method of applying nutrients locally to roots under controlled conditions and some morphological effects of locally applied nitrate on the branching of wheat roots. *Austr. J. Biol. Sci.* **25**, 1169–1180.

HACKETT C. & ROSE D.A. (1972a) A model of the extension and branching of a terminal root of barley, and its use in studying relations between root dimensions. I. The model. *Aust. J. Biol. Sci.* **25**, 669–679.

HACKETT C. & ROSE D.A. (1972b) II. Results and inferences from manipulation of the model. *Aust. J. Biol. Sci.*, **25**, 681–690.

HALE M.G., FOY C.L. & SHAY F.J. (1971) Factors affecting root exudation. *Adv. Agron.* **23**, 89–109.

HALE M.G., LINDSAY D.L. & HARMEED K.M. (1973) Gnotobiotic culture of plants and related research. *Bot. Rev.* **39**, 261–273.

HALL N.S., CHANDLER W.F., van BAVEL C.H.M., REID P.H. & ANDERSON J.H. (1953) A tracer technique to measure growth and activity of plant root systems. *North Carolina Agr. Exp. Sta. Tech. Bull.* **101**, 3–40.

HALLSWORTH E.G. & CRAWFORD D.V. eds. (1965) *Experimental Pedology*. London: Butterworths.

HALSTEAD E.H., BARBER S.A., WARNCKE D.D. & BOLE J.B. (1968) Supply of Ca, Sr, Mn and Zn to plant roots growing in soil. *Soil Sci. Soc. Amer. Proc.* **32**, 69–72.

HALSTEAD R.L. & MCKERCHER R.B. (1975) Biochemistry and cycling of phosphorus. In *Soil Biochemistry*. Vol. 4. eds. Paul E.A. & MacLaren A.D., pp. 31–63.

HAMAKER J.W. (1972) Diffusion and volatilization. In Organic chemicals in the Soil environment, eds. Goring C.A.I. & Hamaker J.W. pp. 341–397. New York: Marcel Dekker.

HAMAKER J.W., GORING C.A.I. & YOUNGSON C.R. (1966) Sorption and leaching of 4-amino-3,5,6-trichloropicolinic acid in soils. In *Organic Pesticides in the Environment*. Adv. Chem. No. 60, 23–37.

HAMAKER J.W. & THOMPSON J.M. (1972) Adsorption. In *Organic Chemicals in the Soil Environment*, eds. Goring C.A.I. & Hamaker J.W., Ch. 2 Vol. 1. New York: Marcel Dekker.

HANCE R.J. (1965) The adsorption of urea and some of its derivations by a variety of soils. *Weed Res.* **5**, 98–107.

HANCE R.J. (1970) Influence of sorption on the decomposition of pesticides. *Soc. Chem. Ind. Monograph No.* **37**, 92–104.

HANNAPEL R.J., FULLER W.H. & FOX R.H. (1964a) Phosphorus movement in a calcareous soil. I. Predominance of organic forms of P in P movement. *Soil Sci.* **97**, 350–357.

HANNAPEL R.J., FULLER W.H. & FOX R.H. (1964b) Phosphorus movement in a calcareous soil. II. Soil microbial activity and organic P movement. *Soil Sci.* **97**, 421–427.

HARLEY J.L. (1969) *The Biology of Mycorrhiza.* London: Leonard Hill.

HARLEY J.L. & MCCREADY C.C. (1952) Uptake of phosphate by excised mycorrhizal roots of beech. II. *New Phytol.* **51**, 56–64.

HARMSEN G.W. & JAEGAR G. (1962) Determination of the quantity of carbon and nitrogen in the rhizosphere of young plants. *Nature (Lond.)* **195**, 1119–1120.

HARPER J. (1961) Approaches to the study of plant competition. *Symp. Soc. exp. Biol.* **15**, 1–39. London: Cambridge University Press.

HARRISON-MURRAY R.S. & CLARKSON D.T. (1973) Relationship between structural development and absorption of ions by the root system of *Cucurbita pepo. Planta (Berl.)* **114**, 1–16.

HARTLEY G.S. (1964) Herbicide behaviour in the soil. In *Physiology and Biochemistry of Herbicides*, ed. Audus L.J., pp. 111–161. London: Academic Press.

HAYMAN D.S. (1974) Plant growth responses to vesicular-arbuscular mycorrhiza. VI. Effect of light and temperature. *New Phytol.* **73**, 71–80.

HAYMAN D.S. (1975) The occurrence of mycorrhiza in crops as affected by soil fertility. In *Endomycorrhizas* eds. Sanders F.E., Mosse B. & Tinker P.B., pp. 495–510. London: Academic Press.

HAYMAN D.S. & MOSSE B. (1972) Plant growth responses to vesicular-arbuscular mycorrhiza. III. Increased uptake of labile P from soil. *New Phytol.* **71**, 41–47.

HAYWARD D.O. & TRAPNELL B.M.W. (1964) *Chemisorption (2nd Edition)* London: Butterworth.

HEAD G.C. (1968) Seasonal changes in the diameter of secondarily thickened roots of fruit trees in relation to growth of other parts of the tree. *J. Hort. Sci.* **43**, 275–282.

HELFFERICH F. (1962) *Ion exchange.* New York: McGraw-Hill.

HEWSON R.T. & ROBERTS H.A. (1973) Some effects of weed competition on the growth of onions. *J. Hort. Sci.* **48**, 51–57.

HIATT A.J. (1968) Electostatic association and Donnan phenomena as mechanisms of ion accummulation. *Plant Physiol.* **43**, 893–901.

HIGINBOTHAM N. (1973) Mineral absorption process in plants. *Bot. Rev.* **39**, 16–70.

HILLEL D. (1971) *Soil and Water. Principles and Processes.* New York: Academic Press.

HILL-COTTINGHAM D.G. & LLOYD JONES H.P. (1965) The behaviour of ion chelating agents with plants. *J. exp. Bot.* **16**, 233–242.

HINGSTON F.J., ATKINSON R.J., POSNER A.M. & QUIRK J.P. (1968) Specific adsorption of anions on goethite. *Proc. 9th Int. Cong. Soil Sci. (Adelaide)*, **1**, 669–678.

HOAGLAND D.R. (1944) *Lectures on the Inorganic Nutrition of Plants.* Waltham, Mass: Chronica Botanica.

HODGES T.K. (1973) Ion absorption by plant roots. *Adv. Agron.* **25**, 163–207.

HODGSON J.F. (1963) Chemistry of the micronutrient elements in soils. *Adv. Agron.* **15**, 119–159.

HODGSON J.F. (1968) Theoretical approach for the contribution of chelates to the movement of iron to roots. *Int. Soc. Soil Sci. Trans. 9th Cong. (Adelaide)*, **2**, 229–241.

HODGSON J.F. (1969) Contribution of metal-organic complexing agents to the transport of metals to roots. *Soil Sci. Soc. Amer. Proc.* **33**, 68–75.

HODGSON J.F., LINDSAY W.L. & TRIERWEILER J.F. (1966) Micronutrient cation complexing in soil solution. II. Complexing of zinc and copper in displaced solution from calcareous soils. *Soil Sci. Soc. Amer. Proc.* **30**, 723–726.

HOOYMANS J.J.M. (1969) The influence of the transpiration rate on uptake and transport of potassium ions in barley plants. *Planta, (Berl.)* **88**, 369–371.

HOPKINS H.T. (1956) Absorption of ionic species of orthophosphate by barley roots: effects of 2–4 dinitrophenol and oxygen tension. *Plant Physiol.* **31**, 155–161.

HOUSE P.R. & FINDLAY N. (1966) Analysis of transient changes in fluid exudation from isolated maize roots. *J. exp. Bot.* **17**, 627–640.

HSIAO T.C. (1973) Plant response to water stress. *Ann. Rev. Plant Physiol.* **24**, 519–570.

HUCK M.G. (1970) Variation in tap root elongation rate as influenced by composition of soil air. *Agron. J.* **62**, 815–818.

HUDSON J.P. (1965) Evaporation from lucerne under advective conditions in the Sudan. *Exptal. Agric.* **1**, 23–32.

HUNT R. (1973) A method of estimating root efficiency. *J. appl. Ecol.* **10**, 157–164.

HUFFMAN E.O. (1968) The reactions of fertilizer phosphate with soils. *Outlook Agric.* **5**, 202–207.

HUFFMAN E.O. (1962) Reactions of phosphate in soils: Recent research by TVA. *Fertil. Soc. Proc.* **71**, 1–48.

INTERNATIONAL SOCIETY OF SOIL SCIENCE (1974) Soil physics terminology. *I.S.S.S. Bull.* **44**, 10–17.

ISCHTSCHERIKOW W. (1907) Die Gewinnung der Bodenlösung in unveränderten Zustande. *Russ. J. exp. Agric.* **8**, 147–166.

ISENSEE A.R., BERGER K.C. & STRUCKMEYER B.E. (1966) Anatomical and growth responses of primary corn roots to several fertilizers. *J. Agron.* **58**, 94–97.

ISRAEL D.W., GIDDENS J.E. & POWELL W.W. (1973) The toxicity of peach tree roots. *Plant Soil*, **39**, 103–112.

JACKSON D.K. (1974) The course and magnitude of water stress in *Lolium perenne* and *Dactylis glomerata J. Agric. Sci.* **82**, 19–27.

JAEGER J.C. & CLARK M.A. (1942) A short table of I (01 ; x). *Proc. R. Soc. Edinb.* **61A**, 229–230.

JARVIS P.G. (1975) Water transfer in plants. In *Transfer processes in the Plant Environment*, eds. de Vries D.A. & Afgan N.H., pp. 369–394. New York: Wiley and Sons.

JEFFERS J.N.R. (ed.) (1972) *Mathematical Models in Ecology.* Oxford: Blackwell Scientific Publications Ltd.

JEFFREY D.W. (1967) Phosphate nutrition of Australian heath plants. I. The importance of proteoid roots in *Banksia* (Proteacea). *Aust. J. Bot.* **15**, 403–411.

JENKINSON D.S. (1966) The turnover of organic matter in soil. In *The use of Isotopes in Soil Organic Matter Studies.* FAO. 187–197. Oxford: Pergamon Press Ltd.

JENKINSON D.S. (1968) Chemical test for potentially available nitrogen in soil. *J. Sci. Fd Agric.* **19**, 160–168.

JENNY H. (1960) *Growth in Living Systems*, ed. by Zarrow M.X., International Symposium on Growth. Basic Books Inc.

JENNY H. & GROSSENBACHER K. (1963) Root–soil boundary zones as seen in the electron microscope. *Soil Sci. Soc. Amer. Proc.* **27**, 273–277.

JENNY H. & OVERSTREET R. (1939a) Cation interchange between plant roots and soil colloids. *Soil Sci.* **47**, 257–272.

JENNY H. & OVERSTREET R. (1939b) Surface migration of ions and contact exchange. *J. Phys. Chem.* **43**, 1185–1196.

JESSUP R.W. (1969) Soil salinity in saltbush country of north-eastern South Australia. *Trans. R. Soc. S. Aust.* **93**, 69–78.

JONES L.H.P. (1957) The relative content of manganese in plants. *Plant Soil*, **8**, 328–336.

JOST W. (1960) *Diffusion in Solids, Liquids and Gases.* New York: Academic Press.

JUNGK A. & BARBER S.A. (1975) Plant age and the phosphorus uptake characteristics of trimmed and untrimmed corn root systems. *Plant Soil* **42**, 227–239.

JUNIPER B.E., GROVES S., LANDAU-SCHACHAR B. & AUDUS L.J. (1966) Root cap and the perception of gravity. *Nature, (Lond.)* **209**, 93–94.

JURGENSEN M.F. & DAVEY C.B. (1970) Non-symbiotic N fixing microorganisms in acid soils and the rhizosphere. *Soils Ferts.* **33**, 435–446.

KARPLUS W.J. (1958) *Analog simulation, solution of field problems.* New York: McGraw-Hill.

KATO T. (1963) Physiological studies on the bulking and dormancy of onion plants. I. The process of bulb formation and development. *J. Jap. Soc. Hort. Sci.* **32**, 81–89.

KATZNELSON H. (1965) Nature and importance of the rhizosphere. In *The ecology of soil-borne plant pathogens.* eds. Baker K.F. & Snyder W.C., pp. 187–209. London: Murray.

KATZNELSON H., ROUATT J.W. & PETERSON E.A. (1962) The rhizosphere effect of mycorrhizal and non-mycorrhizal roots of yellow birch seedlings. *Can. J. Bot.* **40**(3), 377–382.

KAUFFMAN M.D. & BOULDIN D.R. (1967) Relationships of exchangeable and non-exchangeable potassium in soils adjacent to cation exchange resins and plant roots. *Soil Sci.* **104**, 145–150.

KAUTSKY J., BARLEY K.P. & FIDDAMAN D.K. (1968) Ion uptake from soils by plant roots, subject to the Epstein—Hagen relation. *Aust. J. Soil Res.* **6**, 159–167.

KEARNEY P.C. & KAUFMANN D.D. (1969) (Eds.) *The degradation of herbicides*, New York: Dekker.

KEEN E.A. (1931) *The physical properties of the soil.* London: Longmans.

KEENEY D.R. & BREMNER J.M. (1966) A chemical index of soil nutrient availability. *Nature, London,* **211**, 892–893.

KEMPER W.D., MAASLAND D.E.L. & PORTER L.K. (1964) Mobility of water adjacent to mineral surfaces. *Soil Sci. Soc. Amer. Proc.* **28**, 164–167.

KEMPER W.D., OLSEN J. & HODGSON A. (1975) Fertilizer or salt leaching as affected by surface sloping and placement of fertilizer and irrigation water. *Soil Sci. Soc. Amer. Proc.* **39**, 115–119.

KEMPER W.D. & ROLLINS J.B. (1966) Osmotic efficiency coefficients across compacted clays. *Soil Sci. Soc. Amer. Proc.* **30**, 529–534.

KENDALL M.G. & MORAN P.A.P. (1963) *Geometrical Probability.* London: Griffin.

KERRIDGE P.C. & KRONSTAD W.E. (1968) Evidence of genetic resistance to aluminium toxicity in wheat (*Triticum aestivum*, Vill., Host). *J. Agron.* **60**, 710–711.

KERSHAW K.A. (1973) *Quantitative and Dynamic Plant Ecology.* London: Edward Arnold.

KIRBY E.J.M. & RACKHAM O. (1971) A note on the root growth of barley. *J. appl. Ecol.* **8**, 919–924.

KIRKBY E.A. (1969) Ion uptake and ionic balance in plants in relation to the form of nitrogen nutrition. In *Ecological aspects of the mineral nutrition of plants.* ed. by Rorison I.H., pp. 215–235. Oxford: Blackwell Scientific Publications Ltd.

KIRKBY E.A. (1974) Recycling of potassium in plants considered in relation to ion uptake and organic acid accumulation. In *Proc. 7th Int. Coll. on Plant Analysis and Fertilizer Problems* 557–568. Hanover.

KIRKBY E.A. & MENGEL K. (1967) Ionic balance in different tissues of the tomato plant in relation to nitrate, urea and ammonium nutrition. *Plant Physiol.* **42**, 6–14.

KIRKHAM D. & POWERS W.L. (1972) *Advanced Soil Physics.* New York: Wiley.

KISSEL D.F., RITCHIE J.T. & BURNETT E. (1973) Chloride movement in undisturbed swelling clay soil. *Soil Sci. Soc. Amer. Proc.* **37**, 21–24.

KNIGHT B.A.G. & TOMLINSON T.E. (1967) The interaction of paraquat (1 : 1—Dimethyl 4 : 4—Dipyridylium Dichloride) with mineral soil. *J. Soil Sci.* **18**, 233–243.

KNOWLES F. & WATKIN J.E. (1931) The assimilation and translocation of plant nutrients in wheat during growth. *J. Agric. Sci.,* **21**, 612–637.

KOVDA V.A., VAN DEN BURG C. & HAGAN R.M. (1973) *Irrigation, drainage, and salinity.* FAO/UNESCO London: Hutchinson.

KRAMER P.J. (1969) *Plant and Soil Water Relationships: A Modern Synthesis.* New York: McGraw-Hill.

KRAMER P.J. & BULLOCK H.C. (1966) Seasonal variations in the proportions of the suberised and unsuberised roots of trees in relation to the absorption of water. *Am. J. Bot.* **53**, 200–204.

KRAMER P.J. & COILE J.S. (1940) An estimation of the volume of water made available by root extension. *Plant Physiol.* **15**, 743–747.

KRASILNIKOV N.A. (1958) *Soil microorganisms and higher plants.* Moscow 1958. Israel Programme for Scientific Translations 1961.

KUTSCHERA L. (1960) *Wurzelatlas mitteleuropäischer Ackerunkräuter und Kulturpflanzen.* Frankfurt a.M.: DLG Verlag GMBH.

LA RUE C.D. (1935) Regeneration in monocotyledinous seedlings. *J. Amer. Bot.* **22**, 486–492.

LAGERWERFF J.V. (1960) The contact-exchange theory amended. *Plant Soil,* **13**, 253–264.

LAI T.M. & MORTLAND M.M. (1962) Self-diffusion of exchangeable cations in bentonite. *Clays and Clay Min.* **9**, 229–247.

LAI T.M. & MORTLAND M.M. (1968) Cationic diffusion in clay minerals. I. Homogeneous and heterogeneous systems. *Soil Sci. Soc. Amer. Proc.* **32**, 56–61.

LAMBRUSCHINI R. (1830) *Firenze Acc. Georgofili Atti.* **9**, 330 quoted by Sestini, F. *Landwirtsch. Vers—Sta.* 1873, **16**, 409.

LAMONT B. (1972) The morphology and anatomy of proteoid roots in the genus *Hakea. J. Aust. Bot.* **20**, 155–174.

LANG A.R.G. & GARDNER W.R. (1970) Limitation to water flux from soils to plants. *J. Agron.* **62**, 693–695.

LARSEN S. (1967) Soil Phosphorus. *Adv. Agron.* **19**, 151–210.

LARSEN S. (1973) Recycling of phosphorus in relation to long term soil reserves. *Phos. Agric. No.* **16**, 1–6.

LARSEN S., GUNARY D. & SUTTON C.D. (1965) The rate of immobilization of applied phosphate in relation to soil properties. *J. Soil Sci.* **16**, 141–148.

LARSEN S. & WIDDOWSON A.E. (1964) Effect of soil:solution ratio on determining the chemical potentials of phosphate ions in soil solutions. *Nature,* **203**, 942.

LARSEN S. & WIDDOWSON A.E. (1968) Chemical composition of soil solution. *J. Sci. Fd. Agric.* **19**, 693–695.

LAST F.T. *et al.* (1969) Build-up of tomato brown root rot caused by *Pyrenochaeta lycopersici* Schneider and Gerlach. *Ann. Appl. Biol.* **64**, 449–459.

LAST P.J. & TINKER P.B. (1968) Nitrate nitrogen in leaves and petioles of sugar beet in relation to yield of sugar and juice purity. *J. Agric. Sci. Camb.* **71**, 383–392.

LAUCHLI A. (1972) Translocation of inorganic solutes. *Ann. Rev. Plant Physiol.* **23**, 197–218.

LAUDELOUT H., VAN BLADEL R., BOLT G.H. & PAGE A.L. (1968) Thermodynamics of heterovalent cation exchange reactions in a montmorillonite clay. *Trans. Faraday Soc.* **64**, 1477–1488.

LAWLOR D.W. (1972) Growth and water use of *Lolium perenne* I. Water Transport. *J. appl. Ecol.* **9**, 79–98.

LEE J.A. (1960) A study of plant competition in relation to development. *Evolution,* **14**, 18–24.

LEHR J.R., BROWN W.E. & BROWN E.H. (1959) Chemical behaviour of monocalcium phosphate monohydrate in soils. *Soil Sci. Soc. Amer. Proc.* **23**, 3–7.

LEISTRA M., SMELT J.H., VERLAAT J.G. & ZANDVOORT R. (1974) Measured and computed concentration patterns of propyzamide in field soils. *Weed Research,* **14**, 87–95.

LETEY J., KEMPER W.D. & NOONAN L. (1969) The effect of osmotic pressure gradients on water movement in unsaturated soil. *Soil Sci. Soc. Amer. Proc.* **33**, 15–18.

LEWIS D.G. & QUIRK J.P. (1967) Phosphate diffusion in soil and uptake by plants. III. ^{31}P movement and uptake by plants as indicated by ^{32}Pauto radiography. *Plant Soil,* **26**, 445–453; IV. Computed uptake by model roots as a result of diffusive-flow, *ibid.,* **26**, 454–468.

LEWIS E.T. & RACZ G.J. (1969) Phosphorus movement in some calcareous and non-calcareous Manitoba soils. *J. Can. Soil Sci.* **49**, 305–312.

LIEBIG J.F. von (1840) *Chemistry in its Application to Agriculture and Physiology.* London: Taylor & Walton.

LINDSAY W.L. (1974) Role of chelation in micronutrient availability. In *The Plant Root and its Environment,* ed. by Carson E.W., pp. 507–524. Charlottesville: University Press of Virginia.

LINDSAY W.L. & STEPHENSON H.F. (1959) Nature of the reactions of monocalcium phosphate monohydrate in soils. I. The solution that reacts with the soil. *Soil Sci. Soc. Amer. Proc.* **23**, 12–17.

LIPPS R.C. & FOX R.L. (1964) Root activity of subirrigated alfalfa as related to soil moisture, temperature and oxygen supply. *Soil Sci.* **97**, 4–12.

LIPPS R.C., FOX R.L. & KOHLER F.E. (1957) Characterizing root activity of alfalfa by radioactive tracer techniques. *Soil Sci.* **84**, 195–204.

LITAV M. & ISTI D. (1974) Root competition between two strains of *Spinacia oleraceae* II. Effects of nutrient supply and of non-simultaneous emergence. *J. appl. Ecol.* **11**, 1017–1026.

LITAV M. & HARPER J.L. (1967) A method of studying spatial relationships between the root systems of two neighbouring plants. *Plant Soil* **26**, 390–391.

LONERAGAN J.E., CARROLL M.D. & SNOWBALL K. (1966) Phosphorus toxicity in cereal crops. *J. Aust. Inst. Agric. Sci.* **32**, 221–223.

LONERAGAN J.F., SNOWBALL K. & SIMMONS W.J. (1968) Response of plants to calcium concentration in solution culture. *Aust. J. Agric. Res.* **19**, 845–847.

LONERAGAN J.F. & ASHER C.J. (1967) Response of plants to phosphate concentration in solution culture: II. Rate of phosphate absorption and its relation to growth. *Soil Sci.* **103**, 311–318.

LOUGHNAN F.C. (1969) *Chemical weathering of the silicate minerals.* New York: Elsevier.

LOUW M.A. & WEBLEY D.M. (1959) The bacteriology of the root region of the oat plant grown under controlled pot culture conditions. *J. appl. Bacteriol.* **22**, 216–226.

Low A.J. (1972) Some aspects of soil structure. *Chem. Ind.* 373–378.

Lund Z.F. (1965) A technique for making thin section of soil with roots in place. *Soil Sci. Soc. Amer. Proc.* **29**, 633–635.

Lungley D.R. (1973) The growth of root systems—a numerical computer simulation model. *Plant Soil*, **38**, 145–159.

Luttge U. & Pitman M.G. (eds.) (1976) *Transport in Plants II, Part A Cells; Part B Tissues and Organs.* Encycl. Pl. Physiol. N.S. Vol. 2. Berlin: Springer-Verlag.

Luxmoore R.J., Stolzy L.H. & Letey J. (1970) Oxygen diffusion in the soil–plant system: I. A model. *J. Agron.* **62**, 317–322.

Luxmoore R.J. & Stolzy L.H. (1972) Oxygen diffusion in the soil–plant system. IV. Oxygen concentration and temperature effects on oxygen relations predicted for maize roots. *J. Agron.* **64**, 720–725.

McAuliffe C.D., Hall N.S., Dean L.A. & Hendricks S.B. (1947) Exchange reactions between phosphate and soils: hydroxylic surfaces of soil minerals. *Soil Sci. Soc. Amer. Proc.* **12**, 119–123.

McColl J.G. & Cole D.W. (1968) A mechanism of cation transport in forest soil. North-West Science **42**, 134–140.

McComb A.L. (1938) The relation between mycorrhizae and the development and nutrient absorption of pine seedlings in a prairie nursery. *J. For.* **36**, 1148–1154.

McLaren A.D. (1970) Temporal and vectorial reactions of nitrogen in soil: a review. *Can. J. Soil Sci.* **50**, 97–109.

Macklon A.E.S. & Weatherley P.E. (1965) Controlled environment studies of the nature and origins of water deficits in plants. *New Phytol.* **64**, 414–427.

Macpherson D.C. (1939) Cortical gas spaces in the roots of *Zea mays.* L. *New Phytol.* **38**, 190–202.

Magistad O.C. (1941) Ion and plant relationships in Western arid soils. *Soil Sci.*, **51**, 461–471.

Makkink G.F. & Van Heemst H.D.J. (1975) *Simulation of the water balance of arable land and pastures.* Wageningen: Pudoc.

Malcolm R.L. & Kennedy V.C. (1969) Rate of cation exchange on clay minerals as determined by specific—ion electrode techniques. *Soil Sci. Soc. Amer. Proc.* **33**, 247–253.

Marriott F.H.C. (1972) Buffon's problem for non-random distribution. *Biometrics*, **28**, 621–624.

Marriott F.H.C. & Nye P.H. (1968) The importance of mass flow in the uptake of ions by roots from soil. *Trans. 9th Int. Cong. Soil Sci. (Adelaide).* **1**, 127–134.

Marschner H. & Richter Ch. (1973) Akkumulation und translokation von K, Na und Ca bei angebot zu Wurzelzonen von Maiskeimpflanzen. *Zeitsch. Pflzen Dung. Bodenk.* **135**, 1–15.

Marschner H., Saxena M.C. & Michael G. (1965) Phosphate uptake by barley seedling. *Z. Pflanzen Dung. Bodenk. Vol.* **111**, 82–94.

Marsh B.a'B. (1971) Measurement of length in a random arrangement of lines. *J. appl. Ecol.* **8**, 265–272.

Marshall T.J. (1958) A relation between permeability and size distribution of pores. *J. Soil Sci.* **9**, 1–8.

Martin H. & Laudelout H. (1963) Thermodynamique de l'echange des cations alcalins dans les argiles. *J. Chim. Physique* **60**, 1086–1099.

Martin J.T. & Juniper B.E. (1970) *The cuticles of plants.* London: Edward Arnold.

May L.H., Chapman F.H. & Aspinall D. (1965) Quantitative studies of root development. I. The influence of nutrient concentration. *Aust. J. Biol. Sci.*, **18**, 25–35.

MELHUISH F.W. & LANG A.R.G. (1968) Quantitative studies of roots in soil. I. Length and diameters of cotton roots in a clay loam soil by analysis of surface-ground blocks of resin-impregnated soil. *Soil Sci.* **106**, 16–22.

MELHUISH F.W. & LANG A.R.G. (1971) Quantitative studies of roots in soil. Analysis of non-random populations. *Soil Sci.* **112**, 161–166.

MELIN E. & NILLSON H. (1958) Translocation of nutritive elements through mycorrhizal mycelia to pine seedlings. *Bot. Notis.* **111**, 251–256.

MENGEL D.B. & BARBER S.A. (1974a) Development and distribution of the corn root system under field conditions. *J. Agron.* **66**, 341–344.

MENGEL D.B. & BARBER S.A. (1974b) Rate of nutrient uptake per unit of corn root under field conditions. *J. Agron.* **66**, 399–402.

MEREDITH R.E. & TOBIAS C.W. (1962) Conduction in heterogeneous systems. *Adv. Electrochem. Engng.* **2**, 15–47.

METZGER W.H. (1928) The effect of growing plants on solubility of soil nitrients. *Soil Sci.* **25**, 273–280.

MEYERHOFF G. & SCHULTZ G.V. (1952) Molecular weight determination of polymethacrylate esters by means of sedimentation in an ultracentrifuge and diffusion. *Makromol. Chem.* **7**, 294–319.

MICHEL B.E. (1971) Further comparisons between Carbowax 6000 and mannitol as suppressants of cucumber hypocotyl elongation. *Plant Physiol.* **48**, 513–516.

MIKOLA P. (1973) Application of mycorrhizal symbiosis in forestry practice. In *Ectomycorrhizae*. eds. Marks G.C. & Kozlowski T.T., pp. 383–412. New York: Academic Press.

MILLER M.H. (1974) Effects of nitrogen on phosphorus absorption by plants. In *The plant root and its environment*. ed. Carson E.W., pp. 643–668. Charlottesville: University Virginia Press.

MILLER M.H., MAMARIL C.P. & BLAIR G.J. (1970) Ammonium effects on phosphorus absorption through pH changes and phosphorus precipitation at the soil–root interface. *J. Agron.* **62**, 524–527.

MILLINGTON R.J. (1959) Gas diffusion in porous media. *Science* **130**, 100–102.

MILLINGTON R.J. & QUIRK J.P. (1960) Transport in porous media. *Trans. 7th Int. Congr. Soil Sci. (Madison)*, **I**. 97–106.

MILLINGTON R.J. & QUIRK J.P. (1961) Permeability of porous solids. *Trans. Farad. Soc.* **57**, 1–8.

MILLINGTON R.J. & SHEARER R.C. (1971) Diffusion in aggregated porous media. *Soil Sci.* **111**, 372–378.

MILTHORPE F.L. (1961) The nature and analysis of competition between plants of different species. Mechanisms in biological competition. *Soc. exp. biol. Symp.* **15**, 330–348.

MILTHORPE F.L. & IVINS J.D. (1966) (eds.) *The growth of cereals and grasses*. Proc. 12th Easter School, Nottingham. London: Butterworths.

MILTHORPE F.L. & MOORBY J. (1974) *An introduction to crop physiology*. Cambridge: Cambridge University Press.

MITCHELL A.R. (1969) *Computational methods in partial differential equations*. London: Wiley.

MITCHELL R.L. & RUSSELL W.J. (1971) Root development and rooting patterns of soyabean (Glycine max (L.) Merrill) evaluated under field conditions. *J. Agron.* **63**, 313–316.

MOKADY R.S. & ZASLAVSKY D. (1967) Movement and fixation of phosphates applied to soils. *Soil Chemistry and Fertility*. Meeting Comm. II & IV, Int. Soc. Soil Sci., Aberdeen, pp. 329–334.

MONTEITH J.L. (1965) Evaporation and environment. *Symp. Soc. exp. Biol.* **19**, 205–234.

MONTEITH J.L. (1973) *Principles of environmental physics*. London: Arnold.

MOORE D.P. (1972) Mechanisms of micronutrient uptake by plants, In *Micronutrients in Agriculture*, eds. Mortvedt J.J., Giordano P.M. & Lindsay W.L. pp. 171–192. Soil Sci. Soc. Amer. Madison USA.

Moss P. (1963) Some aspects of the cation status of soil moisture. Pt. I. The ratio law and moisture content. *Plant Soil*, **18**, 99–113.

Moss P. (1969) A comparison of potassium activity ratios derived from equilibration procedures and from measurements on displaced soil solution. *J. Soil Sci.* **20**, 297–306.

Moss P. & Beckett P.H.T. (1971) Sources of error in the determination of soil potassium activity ratios by the Q/I procedure. *J. Soil Sci.* **22**, 514–536.

Mosse B. (1973) Advances in the study of vesicular-arbuscular mycorrhiza. *Phytopath.* **11**, 171–196.

Mott C.J. (1967) *Cationic mobility in oriented bentonite.* D.Phil. thesis, Oxford.

Mott C.J.B. (1970) Sorption of anions by soils. *Soc. Chem. Ind. Monograph No.* **37**, 40–52.

Mott C.J.B. & Nye P.H. (1968) Contribution of adsorbed strontium to its self-diffusion in a moisture-saturated soil. *Soil Sci.* **105**, 18–23.

Muljadi D., Posner A.M. & Quirk J.P. (1966) The mechanism of phosphate adsorption by kaolinite, gibbsite and pseudo boehmite. Pt. I. The isotherms and the effect of pH on adsorption. *J. Soil Sci.* **17**, 212–247.

Muller W.H. (1965) Volatile materials produced by *Salvia laucophylla*: effects on seedling growth and soil bacteria. *Bot. Gaz. (Chicago)* **126**, 195–200.

Muller C.H. (1946) Root development and ecological relations of guayule. U.S.D.A. Tech. Bull. 923.

Murdoch C.L., Jackobs J.A. & Gerdemann J.W. (1967) Utilization of phosphorus sources of different availability by mycorrhizal and non-mycorrhizal maize. *Plant Soil*, **27**, 329–334.

Murrmann R.P. & Peech M. (1969) Effect of pH on labile and soluble phosphate in soils. *Soil. Sci. Soc. Amer. Proc.* **33**, 205–210.

Nagarajah S. (1969) *The desorption of phosphate from kaolinite and hydrous oxides.* Thesis. University of Western Australia.

Nagarajah S., Posner A.M. & Quirk J.P. (1970) Competitive adsorption of phosphate with poly-galacturonate and organic anions on kaolinite and oxide surfaces. *Nature, Lond.* **228**, 83–85.

Nearpass D.C. (1965) Effects of soil acidity on the adsorption, penetration and persistence of simazine. *Weeds*, **13**, 341–346.

Newbould P., Ellis F.B., Barnes B.T. & House K.R. (1969) The distribution of roots under rye-grass swards. *Ann. rep. Letcombe Laboratory.* ARC, R.L. **20**, 45–47.

Newman A.C.D. (1970) In discussion *Sorption and Transport Processes in Soils.* p. 32. Monograph No. 37. Soc. Chem. Ind. London.

Newman E.I. (1966) A method of estimating the total root length in a sample. *J. appl. Ecol.*, **3**, 139–145.

Newman E.I. (1969) Resistance to water flow in soil and plant I. Soil resistance in relation to amount of root : theoretical estimates. *J. appl. Ecol.* **6**, 1–12.

Newman E.I. (1974) Root and soil–water relations. In *The Plant Root and its Environment*, ed. Carson E.W., pp. 363–440. Charlottesville: University Press of Virginia.

Newman E.I. (1976) Water movement through root systems. *Phil. trans. R. Soc. Lond.* **B273**, 463–478.

Newman E.I. & Andrews R.E. (1973) Uptake of P and K in relation to root growth and root density. *Plant Soil*, **38**, 49–69.

Nielsen C. (1972) Diffusion of potassium in relation to physical and chemical properties of soils. *Roy. Vet. Agric. Univ. Copenhagen Yearbook*, 142–159.

Nielsen D.R., Kirkham D. & van Wijk W.R. (1961) Diffusion equation calculations of field soil water infiltration profiles. *Soil Sci. Soc. Amer. Proc.* **25**, 165–168.

NIELSEN K.F. & HUMPHRIES E.C. (1966) Effects of root temperature on plant growth. *Soils Ferts.* **29**, 1–7.

NIELSEN N.E. (1972) A transport kinetic concept of ion uptake from soil by plants. I. A method for isolating soil solution from soils with or without plant cover. *Plant Soil*, **36**, 505–520.

NIMAH M.N. & HANKS R. (1973a) Model for estimating soil–water, plant and atmospheric inter-relations. I. Description and sensitivity. *Soil Soc. Sci. Amer. Proc.* **37**, 522–527.

NIMAH M.N. & HANKS R. (1973b) Model for estimating soil–water, plant and atmospheric inter-relations. II. Field test of model. *Soil Soc. Sci. Amer. Proc.* **37**, 528-532.

NISSEN P. (1971) Uptake of sulphate by roots and leaf slices of barley : mediated by single multiphasic mechanisms. *Physiol. Plant*, **24**, 315–324.

NISSEN P. (1974) Uptake mechanism: organic and inorganic. *Ann. Rev. Plant Physiol.* **25**, 38–80.

NOGGLE J.C. & FRIED M. (1960) A kinetic analysis of P absorption by excised roots of millet, barley and alfalfa. *Soil Sci. Soc. Amer. Proc.* **24**, 33–35.

NUTMAN P.S. & DART P.J. (1967) Time lapse cinematographic studies of clover root hairs by nodule bacteria. *Ann. Rep. Rothamsted exptal. Stn.* 1966, p. 80.

NYE P.H. (1961) Organic matter and nutrient cycles under moist tropical forest. *Plant Soil*, **13**, 333–346.

NYE P.H. (1966a) The measurement and mechanisn of ion diffusion in soil : I. The relation between self-diffusion and bulk diffusion. *J. Soil Sci.* **17**, 16–23.

NYE P.H. (1966b) The effect of nutrient intensity and buffering power of a soil, and the absorbing power, size and root-hairs of a root, on nutrient absorption by diffusion. *Plant Soil*, **25**, 81–105.

NYE P.H. (1966c) Changes in the concentration of nutrients in the soil near planar absorbing surfaces when simultaneous diffusion and mass flow occur. *Trans. Comm. II & IV. Int. Soc. Soil Sci.* (Aberdeen) 317–327.

NYE P.H. (1968) The use of exchange isotherms to determine diffusion coefficients in soil. *9th Int. Cong. Soil Sci. Trans.* (*Adelaide*) **1**, 117–126.

NYE P.H. (1968) Processes in the root environment. *J. Soil Sci.* **19**, 205–215.

NYE P.H. (1973) The relation between the radius of a root and its nutrient absorbing power (α). *J. exp. Bot.* **24**, 783–786.

NYE P.H. (1974) A theoretical aperçu of the movement of nutrients in the soil profile. *J. Sci. Fd Agric.* **25**, 709–716.

NYE P.H., BREWSTER J.L. & BHAT K.K.S. (1975) The possibility of predicting solute uptake and plant growth response from independently measured soil and plant characteristics. I. The theoretical basis of the experiments. *Plant Soil*, **42**, 161–170.

NYE P., CRAIG D., COLEMAN N.T. & Ragland J.L. (1961) Ion exchange equilibria involving aluminium. *Soil Sci. Soc. Amer. Proc.* **25**, 14–17.

NYE P.H. & FOSTER W.N.M. (1961) Relative uptake of phosphorus by crops and natural fallow from different parts of their root zone. *J. Agric. Sci.* **56**, 299–306.

NYE P.H. & GREENLAND D.J. (1960) *The Soil under Shifting Cultivation*, Commonwealth Agric. Bur., Farnham, England.

NYE P.H. & MARRIOTT F.H.C. (1969) A theoretical study of the distribution of substances around roots resulting from simultaneous diffusion and masss flow. *Plant Soil*, **30**, 459–472.

NYE P.H. & SPIERS J.A. (1964) Simultaneous diffusion and mass flow to plant roots. *8th Int. Congr. Soil Sci.*, (*Bucharest*), **11**, 535–542.

NYE P.H. & TINKER P.B.H. (1969) The concept of a root demand coefficient. *J. appl. Ecol.*, **6**, 293–300.

OBERLANDER H.E. & ZELLER A. (1964) Phosphorus uptake in different zones of the root system of lucerne. *Bodenkultur*, **15**, 317–328.

OBIHARA C.H. & RUSSELL E.W. (1972) Specific adsorption of silicate and phosphate by soils. *J. Soil Sci.* **23**, 105–117.

OGATA G., RICHARDS L.A. & GARDNER W.R. (1960) Transpiration of alfalfa determined from soil water content changes. *Soil Sci.*, **89**, 179–182.

O'LEARY J.W. & KRAMER P.T. (1964) Root pressure in conifers. *Science*, **145**, 284–285.

OLIVER S. & BARBER S.A. (1966a) An evaluation of the mechanisms governing the supply of Ca, Mg, K and Na to soyabean roots. *Soil Sci. Soc. Amer. Proc.* **30**, 82–86.

OLIVER S. & BARBER S.A. (1966b) Mechanisms for the movement of Mn, Fe, B, Cu, Zn, Al and Sr from one soil to the surface of soybean roots (*Glycine max*). *Soil Sci. Soc. Amer. Proc.* **30**, 468–470.

OLSEN S.R. & KEMPER W.D. (1968) Movement of nutrients to plant roots. *Adv. Agron.* **20**, 91–151.

OLSEN S.R., KEMPER W.D. & JACKSON R.D. (1962) Phosphate diffusion to plant roots. *Soil Sci. Soc. Amer. Proc.*, **26**, 222–227.

OLSEN R.A. & PEECH M. (1960) The significance of the suspension effect in the uptake of cations by plants from soil–water systems. *Soil Sci. Soc. Amer. Proc.* **24**, 257–261.

OLSEN S.R. & WATANABE F.S. (1957) A method to determine a phosphorus adsorption maximum of soils as measured by the Langmuir isotherm. *Soil Sci. Soc. Amer. Proc.* **21**, 144–149.

OLSEN S.R. & WATANABE F.S. (1970) Diffusive supply of phosphorus in relation to soil textural variations. *Soil Sci.* **110**, 318–327.

ONDERDONK J.J. & KETCHESON J.W. (1973) Effect of soil temperature on direction of corn root growth. *Plant Soil*, **39**, 177–186.

OSWALT D.L., BERTRAND A.R. & TEEL M.R. (1959) Influence of nitrogen fertilization and clipping on grass roots. *Soil Sci. Soc. Amer. Proc.* **23**, 228–230.

OVINGTON J.D. (1965) Nutrient cycling in woodlands. In *Experimental Pedology*, eds. Hallsworth E.G. & Crawford D.W., pp. 208–215. Proc. 11th Easter School, Nottingham. London: Butterworth.

PAPAVIZAS G.C. & DAVEY C.B. (1961) Extent and nature of the rhizosphere of *Lupinus*. *Plant Soil*, **14**, 215–236.

PARKER F.W. (1924) Carbon dioxide production of plant roots as a factor in the feeding power of plants. *Soil Sci.* **17**, 229–247.

PARKINSON D. (1967) Soil microorganisms and plant roots. In *Soil Biology*, eds. Burgess A. & Raw F., pp. 449–473. London: Academic Press.

PARLANGE J.Y. (1973) Movement of salt and water in relatively dry soils. *Soil Sci.* **116**, 249–255.

PASSIOURA J.B. (1963) A mathematical model for the uptake of ions from the soil solution. *Plant Soil*, **18**, 225–238.

PASSIOURA J.B. (1966) The amounts of nutrients contacted by roots. *Tech. Rep. Ser. No.* 65. p. 82–84. Int. Atom. En. Agency, Vienna.

PASSIOURA J.B. (1971) Dispersion in aggregated media. *I.* Theory. *Soil Sci.* **111**, 339–344.

PASSIOURA J.B. (1973) Sense and nonsense in crop simulation. *J. Aust. Inst. Agric. Sci.* **39**, 181–183.

PASSIOURA J.B. & COWAN I.R. (1968) On solving the non-linear diffusion equation for the radial flow of water to roots. *Agr. Meteorol.* **5**, 129–134.

PASSIOURA J.B. & FRERE M.H. (1967) Numerical analysis of the convection and diffusion of solute to roots. *Aust. J. Soil Res.* **5**, 149–159.

PASSIOURA J.B. & ROSE D.A. (1971) Hydrodynamic dispersion in aggregated media : 2. Effects of velocity and aggregate size. *Soil Sci.* **111**, 345–351.

PATRICK Z.A., Tousson, T.A. & KOCH L.W. (1964) Effect of crop residue decomposition products on plant roots. *Ann. Rev. Phytopath.* **2**, 267–292.

PAUL J.L. (1965) Influence of soil moisture on chloride uptake by wheat seedlings at low rates of transpiration. *Agrochimica* **9**, 368–379.

PEARSON R.W. (1974) Significance of rooting pattern to crop production and some problems of root research. In *The Plant Root and its Environment*, ed. Carson G.W., pp. 247–270. Charlottesville: University Virginia Press.

PEARSON R.W. & ADAMS F. (eds.) (1967) *Soil acidity and liming*. Agronomy No. 12 Madison USA.

PEARSON V. & TINKER P.B. (1975) Measurement of fluxes in the external hyphae of endomycorrhizas. In *Endomycorrhizas*, eds. Sanders F.E., Mosse B. & Tinker P.B., pp. 277–288. London: Academic Press.

PEEL A.J. (1974) *Transport of nutrients in plants*, p. 258. London: Butterworth.

PENMAN H.L. (1963) Vegetation and hydrology. *C.A.B. Tech. Comm. No. 53, Farnham Royal*.

PEREIRA H.C. (1957) Field measurements of water use for irrigation control in Kenya coffee. *J. agric. Sci.* **49**, 459–466.

PFANNKUCH H.O. (1963) Contribution à l'étude des déplacements de fluides miscibles dans un milieu poreux. *Revue de l'Institut Francais de Petrole.* 18, 215–268.

PHILIP J.R. (1957a) The theory of infiltration; 4. Sorptivity and algebraic infiltration equations. *Soil Sci.* **84**, 257–264.

PHILIP J.R. (1957b) The physical principles of soil–water movement during the irrigation cycle. *3rd Congr. Inter. Comm. on Irrig. & Drain. Quest.* 8, pp. 8.125–8.154.

PHILIP J.R. (1969) Theory of flow and transport processes in pores and porous media. *Ciba Foundation Symposium on circulatory and respiratory mass transport.* eds. Wolstenholme G.E.W. & Knight J., pp. 25–44. London: Churchill.

PHILIP J.R. (1973) On solving the unsaturated flow equation. I. The flux-concentration relation. *Soil Sci.*, **116**, 328–335.

PHILIP J.R. & DE VRIES D.A. (1957) Moisture movement in porous materials under temperature gradients. *Trans. Amer. Geophys. Union*, **38**, 222–232.

PHILLIPS R.E. & BROWN D.A. (1965) Ion diffusion. III. The effect of soil compaction on self-diffusion of ^{86}Rb and ^{89}Sr. *Soil Sci. Soc. Amer. Proc.* **29**, 657–661.

PIELOU E.C. (1969) *An introduction to mathematical ecology*. New York: Wiley, Interscience.

PITMAN M.G. (1965) Ion exchange and diffusion in roots of *Hordeum vulgare. Aust. J. Biol. Sci.* **18**, 541–546.

PITMAN M.G. (1967) Conflicting measurements of sodium and potassium uptake by barley roots. *Nature, (Lond.)* **30**, 1343–1344.

PLACE G.A. & BARBER S.A. (1964) The effect of soil moisture and rubidium concentration on diffusion and uptake of rubidium—86. *Soil Sci. Soc. Amer. Proc.* **28**, 239–243.

POELSTRA P., FRISSEL M.J., VAN DER KLUYT N. & TAP W. (1974) *Behaviour of mercury compounds in soils: accumulation and evaporation.* FAO/IAEA/WHO. Symposium on comparative aspects of food and environmental contamination, pp. 281–292. Helsinki.

PONNAMPERUMA F.N. (1972) The chemistry of submerged soils. *Adv. Agron.* **24**, 29–96.

PORTER L.K., KEMPER W.D., Jackson R.D. & STEWART B.A. (1960) Chloride diffusion in soils as influenced by moisture content. *Soil Sci. Soc. Amer. Proc.* **24**, 460–463.

POULOVASSILIS A. (1969) The effect of hysteresis of pore-water on the hydraulic conductivity. *J. Soil Sci.* **20**, 52–56.

POWER J.F., GRUNES D.L., WILLIS W.O. & REICHMAN G.A. (1963) Soil temperature and phosphorus effects upon barley growth. *Agron. J.* **55**, 389–392.

PRESTON R.D. (1974) *Physical Biology of Plant Cell Walls*, New York: Chapman Hall.

RACZ G.J., RENNIE D.A. & HUTCHISON W.L. (1964) The ^{32}P injection method for studying the root system of wheat. *Can. J. Soil Sci.* **44**, 100–108.

RAMZAN M. (1971) *Changes in soil pH as a result of cation and anion diffusion.* D.Phil. Thesis, Oxford.

RAPER C.D. & BARBER S.A. (1970a) Rooting systems of soyabeans. I. Differences in morphology among varieties. *Agron. J.* **62**, 581–584.

RAPER C.D. & BARBER S.A. (1970b) Rooting systems of soyabeans. II. Physiological effectiveness as nutrient absorption surfaces. *Agron. J.* **62**, 585–588.

REITEMEIER R.F. & RICHARDS L.A. (1944) Reliability of pressure membrane method for extraction of soil solution. *Soil Sci.* **57**, 119–136.

REICOSKY D.C., MILLINGTON R.J. & PETERS D.B. (1970) A comparison of three methods of estimating root length. *Agron. J.* **62**, 451–453.

REICOSKY D.C., MILLINGTON R.J., KLUTE A. & PETERS D.B. (1972) Patterns of water uptake and root distribution of soybeans (*Glycine max.*) in the presence of a water table. *Agron. J.* **64**, 292–297.

REID M.S. & BIELSKI R.L. (1970) Response of *Spirodela oligorrhiza* to phosphorus deficiency. *Plant Physiol.* **46**, 609–613.

RENKIN E.M. (1954) Filtration, diffusion and molecular sieving through porous cellulose membranes. *J. Gen. Physiol.* **38**, 225–243.

RENNIE P.J. (1955) The uptake of nutrients by mature forest growth. *Plant Soil*, **7**, 49–95.

REYNOLDS E.R.C. (1970) Root distribution and the cause of its spatial variability in *Pseudotsuga taxifolia* (Poir) Britt. *Plant Soil*, **32**, 501–517.

RHOADES J.D. (1974) Drainage for salinity control. In *Drainage for Agriculture*, ed. van Schilfgaarde J.H., No. 17 Agron. Series Amer. Soc. Agron. Madison, USA.

RIBBLE J.M. & DAVIS L.E. (1955) Ion exchange in soil columns, *Soil Sci.* **79**, 41–47.

RICE E.L. (1974) *Allelopathy*, New York: Academic Press.

RICHARDS L.A. (1954) Diagnosis and improvement of saline and alkali soils. *USDA Agriculture Handbook No. 60.*

RILEY D. & BARBER S.A. (1969) Bicarbonate accumulation and pH changes at soyabean (*Glycine max* (L.) Merr.) root–soil interface. *Soil Sci. Soc. Amer. Proc.* **33**, 905–908.

RILEY D. & BARBER S.A. (1970) Salt accumulation at the soyabean (*Glycine max* (L.) Merr.) root–soil interface. *Soil Sci. Soc. Amer. Proc.* **34**, 154–155.

RILEY D. & BARBER S.A. (1971) Effect of ammonium and nitrate fertilization on phosphorus uptake as related to root–induced pH changes at the root–soil interface. *Soil Sci. Soc. Amer. Proc.* **35**, 301–306.

RIVIERE J. (1959) *Contributions a l'étude de la rhizosphere du blé.* D.Sc. Thesis, Univ. Paris.

RIVIERE J. (1960) Étude de la rhizosphere du blé. *Ann. Agron.* **11**, 397–440.

ROBINSON R.A. & STOKES R.H. (1959) *Electrolyte Solutions.* London: Butterworths.

RODIN L.E. & BAZILEVICH N.I. (1965) *Production and Mineral Cycling in Terrestrial Vegetation.* London: Oliver & Boyd.

RODIN L.E. & BAZILEVICH N.I. (1965) *Dynamics of organic matter and biological turnover of ash elements and nitrogen in main types of the earth's vegetation.* Nanka, Leningrad.

ROELOFSEN P.A. & KREGER D.R. (1951) The sub-microscopic structure of pectin in collenchyma cell walls. *J. exp. Bot.* **2**, 332–343.

ROGERS W.S. (1939) Apple growth in relation to rootstock, soil, seasonal and climatic factors. *J. Pomol.* **17**, 99–130.

ROGERS W.S. & HEAD G.C. (1969) Factors affecting the distribution and growth of roots of perennial woody species. In *Root Growth.* ed. Whittington W.J., pp. 280–291. 15th Easter School, Nottingham, London: Butterworths.

ROGOWSKI A.S. (1972) Estimation of the soil moisture characteristics and hydraulic conductivity: comparison of models. *Soil Sci.* **114**, 423–429.

Romell L.G. (1922) Luftväxlingen i marken som ekologisk faktor. *Medd. Statens Skogsforsöks-anstalt* **19**, No. 2.

Rorison I. (1969) Ecological inferences from laboratory experiments on mineral nutrition. In *Ecological aspects of the mineral nutrition of plants*, ed. Rorison I., pp. 155–157. Oxford: Blackwell Scientific Publications Ltd.

Rose C.W. (1966) *Agricultural Physics*. Oxford: Pergamon Press Ltd.

Rose C.W. & Stern W.R. (1967) Determination of withdrawal of water from soil by crop roots as a function of depth and time. *Aust. J. Soil Res.* **5**, 11–19.

Rose D.A. (1963a) Water movement in porous materials: Part 1. Isothermal vapour transfer. *Br. J. appl. Physics.* **14**, 256–262.

Rose D.A. (1963b) Water movement in porous materials: Part 2. The separation of the components of water movement. *Br. J. appl. Physics.* **14**, 491–496.

Rose D.A. (1963c) Water movement in porous materials: Part 3. Evaporation of water from soil. *Br. J. appl. Physics. Ser.* **2**, **1**, 1779.

Rosene H.F. (1943) Quantative measurements of the velocity of water absorption in individual root hairs by a microtechnique. *Plant Physiol.* **18**, 588–607.

Rovira A.D. (1962) Plant root exudates in relation to the rhizosphere microflora. *Soils Ferts.* **25**, 167–172.

Rovira A.D. (1969) Plant root exudates. *Bot. Rev.* **35**, 35–57.

Rovira A.D. & Bowen G.D. (1966) Phosphate incorporation by sterile and non-sterile plant roots. *Aust. J. Biol. Sci.* **19**, 1167–1169.

Rovira A.D. & Bowen G.D. (1970) Translation and loss of phosphate along roots of wheat seedlings. *Planta (Ber.)* **93**, 18–25.

Rovira A.D. & Davey C.B. (1974) Biology of the rhizosphere, In *The Plant Root and its Environment*, ed. Carson E.W., pp. 153–240. Charlottesville: University of Virginia Press.

Rovira A.D. & McDougall B.M. (1956) Microbial and biochemical aspects of the rhizosphere. In *Soil biochemistry*. Vol. I. eds. McLaren A.D. & Peterson G.H., pp. 417–463. London: Edward Arnold.

Rovira A.D., Newman E.I., Bowen H.J. & Campbell R. (1974) Quantitative assessment of the rhizosphere microflora by direct microscopy. *Soil Biol. Biochem.* **6**, 211–216.

Rowell D.L., Martin M.W. & Nye P.H. (1967) The measurement and mechanism of ion diffusion in soils. III. The effect of moisture content and soil solution concentration on the self-diffusion of ions in soils. *J. Soil Sci.* **18**, 204–222.

Rowse H.R. (1974) The effect of irrigation on the length, weight and diameter of lettuce roots. *Plant Soil* **40**, 381–391.

Rowse H.R. & Phillips D.A. (1974) An instrument for measuring the total length of root in a sample. *J. appl. Ecol.* **11**, 309–314.

Ruer P. (1967) Morphologie et anatomie du system radiculaire du palmier à huile. *Oleagineaux*, **22**, 535–537.

Russell Sir E.J. (1937) *Soil conditions and plant growth*. (7th Edition) London: Longmans.

Russell Sir E.J. (1966) *A History of Agricultural Science in Great Britain*. London: Allen & Unwin.

Russell Sir E.J. & Russell E.W. (1950) *Soil Conditions and Plant Growth* (8th Edition) London: Longmans.

Russell E.W. (1956) The effects of very deep ploughing and of subsoiling on crop yields. *J. agric. Sci.* **48**, 129–144.

Russell E.W. (1961) *Soil conditions and plant growth*. 9th Edition p. 452 London: Longmans.

Russell E.W. (1973) *Soils Conditions and Plant Growth* (10th Edition) London: Longmans.

Russell J.S. (1975) Systems Analysis—Soil Ecosystems, In *Soil Biochemistry*, Vol. 3 Ch. 2 eds. Paul E.A. & McLaren A.D. New York: Marcel Dekker.

RUSSELL R.S. & BARBER D.A. (1960) The relationship between salt uptake and the absorption of water by intact plants. *Ann. Rev. Plant Physiol.* **11**, 127–140.

RUSSELL R.S. & ELLIS F.B. (1968) Estimation of the distribution of plant roots in soil. *Nature, Lond.* **217**, 582–583.

RUSSELL R.S. & NEWBOULD P. (1969) The pattern of nutrient uptake in root systems. In *Root Growth.* ed. Whittington W.J., pp. 148–166. Proc. 15th Easter School, Nottingham, London: Butterworths.

RUSSELL R.S. & SANDERSON J. (1967) Nutrient uptake by different parts of the intact roots of plants. *J. exp. Bot.* **18**, 491–508.

RUSSELL R.S. & SHORROCKS V.M. (1959) The relationship between transpiration and the absorption of inorganic ions by intact plants. *J. Exptl Bot.* **10**, 301–316.

SALMON R.C. (1964) Cation exchange reactions. *J. Soil Sci.* **15**, 273–283.

SAMTSEVICH S.A. (1968) Gel-like excretions of plant roots and their influence upon soil and rhizosphere microflora. In *Methods of productivity studies in root systems and rhizosphere organisms.* eds. Ghilasov M.S., Kanda V.A., Noichkova L.N., Rodin L.E. & Shveshnikova V.M. Soviet Natnl. Comm. I.B.P.

SANDERS F.E. (1971) *Effect of root and soil properties on the uptake of nutrients by competing roots.* D.Phil. Thesis, Oxford.

SANDERS F.E. (1975) The effect of foliar applied phosphate on the mycorrhizal infections of onion roots. In *Endomycorrhizas*, eds. Sanders F.E., Mosse B. & Tinker P.B., pp. 261–276, London: Academic Press.

SANDERS F.E. & TINKER P.B. (1971) Mechanism of absorption of phosphate from soil by *Endogone* mycorrhizas. *Nature, London,* **232**, 278–279.

SANDERS F.E. & TINKER P.B. (1973) Phosphate flow into mycorrhizal roots. *Pestic. Sci.* **4**, 385–395.

SANDERS F.E. & TINKER P.B. (1975) Mycorrhizas. *Spectrum.* **126**, 13–15.

SANDERS F.E., TINKER P.B. & NYE P.H. (1971) Uptake of solutes by multiple root systems from soil. I. An electrical analog of diffusion to root systems. *Plant Soil* **34**, 453–466.

SANDERS F.E., TINKER P.B. & SPARLING G. (1975) Mycorrhiza-forming fungi of the Endogenaceae and their influence on the phosphate nutrition of plants. *Proc. 10th Int. Conf. Soil Sci.* Moscow.

SAUSSURE TH. DE (1804) *Récherches Chimiques sur la Végétation.* p. 327 Paris.

SCAIFE M.A. & SMITH R. (1973) The phosphorus requirements of lettuce. II. A dynamic model of phosphorus uptake and growth. *J. Agric. Sci.* **80**, 353–361.

SCHANDER H. (1941) The displacement of optimum soil reaction during development of *Lupinus lutens. Bodenk. Pfl. Ernahr.* **20**, 129–151.

SCHEFFER F., ULRICH B. & KAUFMANN H.J. (1961) Uber die Phosphationen-aufnahme durch Pflanzen. *Plant Soil,* **14**, 264–276.

SCHLOESING M. TH. (1866) Sur l'analyse des principes soluble de la terre végétable. *Compt. Rend.* **63**, 1007–1012.

SCHNAPPINGER M.G., BANDEL V.A. & KRESGE C.B. (1969) Effect of phosphorus and potassium on Alfalfa root anatomy. *Agron. J.* **61**, 805–808.

SCHOFIELD R. K. (1955) Can a precise meaning be given to available soil phosphorus? *Soils Fert.* **18**, 373–375.

SCHOFIELD R.K. & GRAHAM-BRYCE I.J. (1960) Diffusion of ions in soils. *Nature* **188**, 1048–1049.

SCHOFIELD R.K. & TAYLOR A.W. (1955) The measurement of soil pH. *Soil Sci. Soc. Amer. Proc.* **19**, 164–167.

SCHUURMANN J.J. & GOEDEWAAGEN M.A.J. (1971) *Methods for the examination of root systems and roots.* 2nd Edition. Centre for Agricultural Publishing and Documentation, Wageningen.

Scott F.M. (1963) Root hair zone of soil grown roots. *Nature* **199**, 1009–1010.

Scotter D.R. & Raats P.A.C. (1970) Movement of salt and water near crystalline salt in relatively dry soil. *Soil Sci.* **109**, 170–178.

Shamoot S., McDonald L. & Bartholomew W.V. (1968) Rhizo-deposition of organic debris in soil. *Soil Sci. Soc. Amer. Proc.* **32**, 817–820.

Shay F.J. & Hale M.G. (1973) Effect of low levels of calcium on exudation of sugars and sugar derivations from intact peanut roots under axenic conditions. *Plant Physiol.* **51**, 1061–1063.

Shone M.G.T. & Wood A.V. (1972) Factors affecting absorption and translocation of simazine by barley. *J. exp. Bot.* **23**, 141–151.

Shone M.G.T. (1966) The initial uptake of ions by barley roots. II. Applications of measurements of adsorption of ions to elucidate the structure of free space. *J. exp. Bot.* **17**, 89–95.

Slatyer R.O. (1967) *Plant–water relationships.* London: Academic Press.

Slatyer R.O. & Taylor S.A. (1960) Terminology in plant and soil–water relations. *Nature* **187**, 922–924.

Smith J.H., Allison F.E. & Saulides D.A. (1962) Phosphobacterin as a soil inoculant. *USDA Tech. Bull.* **1263**.

Smith K.A. & Robertson P.D. (1971) Effect of ethylene on root extension of cereals. *Nature, London.* **234**, 148–149.

Smith W.H. (1969) Release of organic materials from the roots of tree seedlings. *Forest Sci.* **15**, 138–143.

Smith W.H. (1970) Root exudates of seedlings and mature sugar maple. *Phytopathology*, **60**, 701–703.

Soileau J.M. (1973) Activity of barley seedling roots as measured by strontium uptake. *Agron. J.* **65**, 625–628.

Soper K. (1958) Effect of flowering on the root system and summer survival of rye-grass. *N. Zealand J. Agric. Res.* **1**, 329–340.

Sparling G.P. (1976) Effects of vesicular-arbuscular mycorrhizas on Pennine grassland vegetation. *Ph.D. Thesis.* University of Leeds, Leeds.

Stanford G. & Pierre W.H. (1947) The relation of potassium fixation to ammonium fixation. *Soil Sci. Soc. Amer. Proc.* **11**, 155–160.

Staple W.J. (1965) Moisture tension, diffusivity and conductivity of a loam soil during wetting and drying. *Can. J. Soil Sci.* **45**, 78–86.

Starr J.L., Broadbent F.E. & Nielsen D.R. (1974) Nitrogen transformations during continuous leaching. *Soil Sci. Soc. Amer. Proc.* **38**, 283–289.

Stevenson F.J. (1965) Origin and distribution of nitrogen in soil. In *Soil Nitrogen.* eds. Bartholomew F.W. & Clark F.E., Agronomy Monograph 10, 1–42.

Stone L.R., Horten M.L. & Olsen T.C. (1973) Water loss from an irrigated sorghum field. I. Water flux within and below the root zone. *Agron. J.* **65**, 492–479; II. Evapo-transporation and root extraction. Loc. Cit. 495–497.

Struik G.J. & Bray J.R. (1970) Root–shoot ratios of native forest herbs and *Zea mays* at different soil moisture levels. *Ecology*, **52**, 892–893.

Stuckey I.H. (1941) Seasonal growth of grass roots. *Amer. J. Bot.* **28**, 486–491.

Sutcliffe J.F. (1962) *Mineral Salts Absorption in Plants.* Oxford: Pergamon Press Ltd.

Sutton P.D. (1969) The effect of low soil temperature on P nutrition of plants—a review. *J. Sci. Fd Agric.* **20**, 1–3.

Swaby R.J. (1962) Effect of microorganisms on nutrient availability. *Trans. Int. Soc. Soil Sci.* Comm. IV and V, *New Zealand* 159–172.

Swaby R.J. & Shurber J. (1958) Phosphate dissolving microorganisms in the rhizosphere of legumes. In *Nutrition of legumes*, ed. Hallsworth E.G., 5th Easter School, Nottingham. London: Butterworths.

TACKETT J.L. & PEARSON R.W. (1964) Oxygen requirements of cotton seedling roots for penetration of compacted soil cores. *Soil Sci. Soc. Amer. Proc.* **28**, 600–605.

TALIBUDEEN O. (1957) Isotopically exchangeable phosphorus in soils. II. Factors influencing the estimation of labile phosphorus. *J. Soil Sci.* **8**, 89–96.

TANFORD C. (1961) *Physical Chemistry of Macromolecules.* Ch. 6. Transport Processes. New York: Wiley.

TAYLOR H.M. (1974) Root behaviour as affected by soil structure and strength. In *The Plant Root and its Environment.* ed. Carsen E.W., pp. 271–292.

TAYLOR H.M., HUCK M.G., KLEPPER B. & LUND Z.F. (1970) Measurement of soil-grown roots in a rhizotron. *J. Agron.* **62**, 807–809.

TAYLOR H.M. & KLEPPER B. (1975) Water uptake by cotton root systems: an examination of assumptions in the single root model. *Soil Sci.* **120**, 57–67.

TAYLOR, H.M. & RATLIFF L.F. (1969) Root elongation rates of cotton and peanuts as a function of soil strength and soil water content. *Soil Sci.* **108**, 113–119.

TAYLOR S.A. & CARY J.W. (1964) Linear equations for the simultaneous flow of matter and energy in a continuous soil system. *Soil Sci. Soc. Amer. Proc.* **28**, 167–172.

TEPE W. & LEIDENFROST E. (1958) Ein Vergleich zwischen pflanzenphysiologischen, kinetischen und statischen Bodenuntersuchung-werten. I. *Mitt. Landw. Forsch.* **11**, 217–229.

TERKELTOUB R.W. & BABCOCK K.L. (1971) A simple method for predicting salt movement through soil. *Soil Sci.* **111**, 182–187.

THENG B.K.G., GREENLAND D.J. & QUIRK J.P. (1967) Adsorption of alkylammonium cations by montmorillonite. *Clay Miner.* **7**, 1–17.

THOMPSON E.J. & BLACK C.A. (1970) Changes in organic phosphorus in soil in the presence and absence of plants. II. Soil in a simulated rhizosphere. *Plant Soil.* **32**, 161–168.

THOMAS G.W. & SWOBODA A.R. (1970) Anion exclusion effects on chloride movement in soil. *Soil Sci.* **110**, 163–166.

THORNLEY J.H.M. (1972) A balanced quantitative model for root: shoot ratios in vegetative plants. *Ann. Bot.* **36**, 431–441.

THORNLEY J.H.M. (1976) *Mathematical models in plant physiology.* London: Academic Press.

THORUP R.M. (1969) Root development and P uptake by tomato plants under controlled soil moisture conditions. *J. Agron.* **61**, 808–811.

TIESSEN H. & CAROLUS R.L. (1963) Effect of different analyses and concentrations of fertilizer solutions on initial root growth of tomato and tobacco plants. *Proc. Amer. Soc. Hort. Sci.* **83**, 680–683.

TIMONIN M.T. (1946) Microflora of the rhizosphere in relation to the manganese deficiency disease of oats. *Soil Sci. Soc. Amer. Proc.* **11**, 284–292.

TINKER P.B. (1968) Changes in soil sodium following its addition in fertilizers. In: *Residual value of applied nutrients. Proc. Open Conf. Soil Chemistry, MAFF Tech. Bull.* **20**.

TINKER P.B. (1969a) A steady state method for determining diffusion coefficients in soil. *J. Soil Sci.* **20**, 336–345.

TINKER P.B. (1969b) The transport of ions in the soil around plant roots. In *Ecological aspects of the mineral nutrition of plants,* ed. Rorison I., pp. 135–147. Oxford: Blackwell Scientific Publications Ltd.

TINKER P.B. (1973) Potassium uptake rates in tropical crops. *Proc. 10th Coll. Int. Potash Inst. Abidjan,* 169–175.

TINKER P.B. (1975a) Effects of vesicular-arbuscular mycorrhizas on higher plants. *Symp. Soc. exp. Biol.* **29**, 325–350.

TINKER P.B. (1975b) The soil chemistry of phosphorus and mycorrhizal effects on plant growth. In *Endomycorrhizas.* eds. Sanders F.E., Mosse B. & Tinker P.B., pp. 353–371. London, Academic Press.

TINKER P.B. (1976) Transport of water to plant roots in soil. *Phil. Trans. R. Soc. Lond.* **B273**, 445–461.

TINKER P.B. & BOLTON J. (1966) Exchange equilibria of sodium on some British soils. *Nature, London* **212**, 548.

TINKER P.B. & SANDERS F.E. (1975) Rhizosphere microorganisms and plant nutrition. *Soil Sci.* **119**, 363–368.

TORII K. & LATIES G.G. (1966) Dual mechanism of ion uptake in relation to vacuolation of corn roots. *Plant Physiol.* **41**, 865–870.

TORREY J.G. & CLARKSON D.T. (eds.) (1975) *The Development and Function of Roots.* New York: Academic Press.

TORSSELL B.W.R., BEGG J.E., ROSE C.W. & BYRNE G.F. (1968) Stand morphology of Townsville lucerne (*Stylosanthes humilis*). Seasonal growth and root development. *Austr. J. exp. Agric. Anim. Husb.* **8**, 533–543.

TROLLDEMIER G. (1967) Vergleich zwischen fluoreszmikroskopischer Direktzählung, Plattengussverfahren und Membranfiltermethode bei Rhizosphärenuntersuchungen. In *Progress in Soil Biology*, eds. Graff O. & Satchell J.E., pp. 59–71. Amsterdam: North Holland Publishing Co.

TROUGHTON A. (1957) The underground organs of herbage grasses. *Bull.* **44**, *Imp. Bur. of Plant Genetics. Com. Agric. Bur.* Farnham Royal.

TROUGHTON A. (1960) Uptake of ^{32}P by the roots of *Lolium perenne*. *Nature* **188**, 593.

TROUGHTON A. & WHITTINGTON J.W. (1969) Genetic effects on roots. In *Root Growth* ed. Whittington, J.W. 15th Easter School, Nottingham. London: Butterworths.

TRUTER M.R. (1975) Co-ordination chemistry of alkali metal cations. *Ann. Rep. Rothamsted Exptal. Sta.*, 1974, 165–169.

TULL J. (1731). *Horse Hoeing Husbandry*. London.

TYREE M.T. (1970) The symplast concept. A general theory of symplastic transport according to the thermodynamics of irreversible processes. *J. theor. Biol.* **26**, 181–214.

UNESCO (1971) *Final Report of Internatonal Co-ordinating Council of the Progamme on Man and the Biosphere*. Paris.

U.S. Salinity Laboratory Staff (1954) Diagnosis and improvement of saline and alkaline soils. *U.S. Dept. Agric. Handbook* 60.

VAIDYANATHAN L.V., DREW M.C. & NYE P.H. (1968) The measurement and mechanism of ion diffusion in soils. IV. The concentration dependence of diffusion coefficients of potassium in soils at a range of moisture levels and a method for the estimation of the differential diffusion coefficient at any concentration. *J. Soil Sci.* **19**, 94–107.

VAIDYANATHAN L.V. & NYE P.H. (1970). The measurement and mechanism of ion diffusion in soils. VI. The effect of concentration and moisture content on the counter diffusion of soil phosphate against chloride ion. *J. Soil Sci.* **21**, 15–27.

VAIDYANATHAN L.V. & TALIBUDEEN O. (1968) Rate controlling processes in the release of soil phosphate. *J. Soil Sci.* **19**, 342–353.

VAIDYANATHAN L.V. & TALIBUDEEN O. (1970). Rate processes in the desorption of phosphate from soils by ion-exchange resins. *J. Soil Sci.* **21**, 173–183.

VAN KEULEN H. (1975) *Simulation of water use and herbage growth in arid regions*. Centre for Agricultural Publishing and Documentaion, Wageningen, p. 95.

VAN MONSJOU, W. (1975) Food–fertilizer–energy efficiency. *Proc. Fertil. Soc.* **152**, 1–21.

VAN OLPHEN H. (1957) Surface conductance of various ion forms of bentonite in water as the electrical double layer. *J. Phys. Chem.* **61**, 1276–1280.

VASEY E.H. & BARBER S.A. (1963) Effect of placement on the absorption of Rb[86] and P[32] from soil by corn roots. *Soil Sci. Amer. Proc.* **27**, 193–197.

VIETS F.G. (1944) Calcium and other polyvalent cations as accelerators of ion accumulation by excised barley roots. *Plant Physiol.* **19**, 466–480.

VIETS F.G. (1965) The plants, need for and use of nitrogen. In *Soil Nitrogen* eds Bartholomew W.V. & Clark F.E., pp. 543–554. Agronomy Series No. 10.

VIETS F.G. (1972) In *Water deficits and plant growth*, ed. by Kozlowski T.T., pp. 217–239. Vol. 3.

VIRTANEN A.I. & TORNAIEN M. (1940) A factor influencing nitrogen excretion from leguminous root nodules. *Nature (Lond.)* **145**, 25.

VLAMIS J. (1953) Acid soil infertility as related to soil-solution and solid phase effects. *Soil Sci.* **75**, 383–394.

VOSE P.B. (1962) Nutritional response and shoot:root ratio as factors in the competition and yield of genotype of perennial rye-grass. *Ann. Bot.* **26**, 425–437.

WADLEIGH C.H., (1968) Wastes in relation to agriculture and forestry. U.S. Dept. Agric. Washington D.C. Misc. Pub. No. 1065.

WALKER A. (1971) Effects of soil moisture content on availability of soil-applied herbicides to plants. *Pestic. Sci.* **2**, 56–59.

WALKER A. & FEATHERSTONE R.M. (1973) Absorption and translocation of atrazine and linuron by plants with implications concerning linuron selectivity. *J. exp. Bot.* **24**, 450–458.

WALKER J.M. & BARBER S.A. (1961) Ion uptake by living plant roots. *Sci.* **133**, 881–882.

WALKER T.M. (1960) Uptake of ions by plants growing in soil. *Soil Sci.* **59**, 328–332.

WAREING P.F. & COOPER J.P. (eds) (1971) *Potential Crop Production*. London: Heinemann.

WARNCKE D.D. & BARBER S.A. (1972) Diffusion of zinc in soil: I. The influence of soil moisture. *Soil Sci. Soc. Amer. Proc.* **36**, 39–42.

WARNCKE D.D. & BARBER S.A. (1973) Ammonium and nitrate uptake by corn (*Zea mays* L.) as influenced by nitrogen concentration and $NH_4^+:NO_3^-$ ratio. *Agron J.* **65**, 950–953.

WARNCKE D.D. & BARBER S.A. (1974) Root development and nutrient uptake by corn grown in solution culture. *Agron. J.* **66**, 514–516.

WAY J.T. (1850) On the power of soils to absorb manure. *J. Roy. Agric. Soc. Eng.* **11**, 313–379; ibid (1852) **13**, 123–143.

WEATHERLEY P.E. (1963) The pathway of water movement across the root cortex and leaf mesophyll of transpiring plants. In *The water relationships of plants. Brit. Ecol. Soc. Symp.* 3, eds Rutter A.J. & Whitehead F.J., pp. 85–100. London: Blackwell Scientific Publications Ltd.

WEATHERLEY P.E. (1969) Ion movement within the plant and its integration with other physiological processes. In *Ecological aspects of the mineral nutrition of plants. Brit. Ecol. Soc. Symp.* 9. ed. Rorison I.H., pp. 323–340. Oxford: Blackwell Scientific Publications Ltd.

WEAVER J.E. (1962) *Root development of field crops*. New York: McGraw-Hill.

WEIHING R.M. (1935) The comparative root development of regional types of corn. *J. Amer. Soc. Agron.* **27**, 526–537.

WELBANK P.J. (1964) Competition for nitrogen and potassium in *Agropyron repens*. *Ann. Bot. N.S.* **28**, 1–16.

WELBANK P.J., GIBB M.J., TAYLOR P.J. & WILLIAMS E.D. (1974) Root growth of cereal crops. *Ann. Rep. Roth. Exp. Sta.* **1973**, 26–66.

WELBANK P.J. & WILLIAMS E.D. (1968) Root growth of a barley crop estimated by sampling with portable powered soil equipment. *J. appl. Ecol.* **5**, 477–481.

WETSELAAR R. (1962) The fate of nitrogenous fertilizer in a monsoonal climate. *Trans. Int. Soil Conf. N. Zealand*, 588–595.

WHISLER F.D., KLUTE A. & MILLINGTON R.J. (1968) Analysis of steady-state evapotranspiration from a soil column. *Proc. Soil Sci. Amer.* **32**, 167–174.

WHISLER F.D., KLUTE A. & MILLINGTON R.J. (1970) Analysis of radial steady-state solution, and solute flow. *Soil Sci. Soc. Amer. Proc.* **34**, 382–387.

WHITE R.E. (1964) Studies on the phosphate potentials of soils. II. Microbial effects. *Plant Soil*, **20**, 184–193.

WHITEHEAD A.G., Dunning R.A. & COOKE D.A. (1971) Docking disorder and root ectoparasitic nematodes of sugar beet. *Ann. Rep. Roth. Exp. Sta. 1970.* **2**, 219–236.

WHITNEY M. & CAMERON F.K. (1903) The chemistry of soil as related to crop production. *U.S. Dept. Agr. Bur. Soils Bul.* 22.

WIERSUM L.K. (1957) The relationship of the size and structural rigidity of pores to their penetration by roots. *Plant Soil.* **9**, 75–85.

WIERSUM L.K. (1958) Influence of water content of sand on rate of uptake of ^{86}Rb. *Nature* **181**, 106–107.

WIERSUM L.K. (1961) Utilization of soil by the plant root system. *Plant Soil.* **15**, 189–192.

WILD A. (1964) Soluble phosphate in soil and uptake by plants. *Nature* **203**, 326–327.

WILD A. (1972) Nitrate leaching under bare fallow at a site in Northern Nigeria. *J. Soil Sci.* **23**, 315–324.

WILD A., ROWELL D.L. & OGUNFOWORA M.A. (1969) The activity ratio as a measure of the intensity factor in potassium supply to plants. *Soil Sci.* **108**, 432–439.

WILD A., SKARLOU V., CLEMENT C.R. & SNAYDON R.W. (1974) Comparison of potassium uptake by four plant species grown in sand and in flowing solution culture. *J. appl. Ecol.* **11**, 801–812.

WILKINS M.B. (1975) The role of the root cap in root geotropism. *Curr. Adv. Pl. Sci.* **6**, 317–328.

WILKINSON G.E. & KLUTE A. (1962) The temperature effect on the equilibrium energy status of water held by porous media. *Soil Sci. Soc. Amer. Proc.* **26**, 326–329.

WILKINSON H.F., LONERAGAN J.F. & QUIRK J.P. (1968a) The movement of zinc to plant roots. *Soil Sci. Soc. Amer. Proc.* **32**, 831–833.

WILKINSON H.F., LONERAGAN J.F. & QUIRK J.P. (1968b) Calcium supply to plant roots. *Science* **161**, 1245–1246.

WILLIAMS B.G., GREENLAND D.J. & QUIRK J.P. (1966) The adsorption of poly vinyl alcohol by natural soil aggregates. *Aust. J. Soil Res.* **4**, 131–143.

WILLIAMS B.G., GREENLAND D.J. & QUIRK J.P. (1967) The effect of poly vinyl alcohol on the nitrogen area and pore structure of soils. *Aust. J. Soil Res.* **5**, 77–83.

WILLIAMS D.E. (1961) The absorption of potassium as influenced by its concentration in the nutrient medium. *Plant Soil* **15**, 387–399.

WILLIAMS R.F. (1946) The physiology of plant growth with special reference to the concept of net assimilation rate. *Ann. Bot. N.S.* **10**, 41–72.

WILLIAMS T.E. (1969) Root activity of perennial grass sward. In *Root Growth*, ed. Whittington, W.J. pp. 270–279. 15th Easter School, Nottingham. London: Butterworths.

WOLDENDORP J.W. (1963) *The influence of living plants on denitrification.* Mededel Handbouwk. Wageningen, 63, 1–100.

WOOD J.T. & GREENWOOD D.J. (1971) Distribution of carbon dioxide and oxygen in the gas phase of aerobic soils. *J. Soil Sci.* **22**, 281–288.

WOODRUFF C.M. (1955) The energies of replacement of calcium by potassium in soils. *Soil Sci. Soc. Amer. Proc.* **19**, 167–171.

WOODS F.W. (1960) Biological anatagonisms due to phytotoxic root exudates. *Bot. Rev.* **26**, 546–569.

WOODWARD, J.M.D. (1699) Some thoughts and experiments concerning vegetation. *Phil. Trans. R. Soc.* **21**, 382–398.

WOOLHOUSE H.W. (1969) Differences in the properties of the acid phosphatases of plant roots and their significance in the evolution of edaphic ecotypes. In *Ecological Aspects of the Mineral Nutrition of Plants*, ed. Rorison I.H., pp. 357–380. Oxford: Blackwell Scientific Publications Ltd.

WRAY F.J. & TINKER P.B. (1969a) A high resolution scanning device for determining the distribution of weak β activity in a planar surface. *J. Sci. Instr. Series* 2, **2**, 343–346.

WRAY F.J. & TINKER P.B. (1969b) A scanning apparatus for detecting concentration gradients around single plant roots. In *Root Growth* ed. by Whittington W.J., pp. 418–422. 15th Easter School, Nottingham. London: Butterworths.

WRAY F.J. (1971) *Changes in the ionic environment around plant roots.* D. Phil. Thesis, Oxford.

WYN JONES R.G. (1975) Excised roots. In *Ion Transport in Plant Cells and Tissues*, eds. Baker D.A. & Hall J.L., pp. 193–229. Amsterdam: North Holland Publishing Co.

YOUNGS E.G. & GARDNER W.R. (1963) A problem of diffusion in the infinite hollow cylinder. *Soil Sci. Soc. Amer. Proc.* **27**, 475–476.

YOUNIS A.F. & HATATA M.A. (1971) Studies on the effect of certain salts on germination, on growth of root, and on metabolism part I. Effects of chlorides and sulphates of sodium, potassium and magnesium in germination of wheat grains. *Plant Soil* **34**, 183–200.

ZAHNER R. (1969) Water deficits and growth of trees. In *Water deficits and Plant Growth II.* ed. Kozlowski T.T., pp. 191–254. New York: Academic Press.

ZIMMERMANN M. (1969) Translocation of nutrients. In *Physiology of plant growth and development.* ed. by M.B. Wilkins), pp. 383–412. New York: McGraw-Hill.

ZSOLDOS F. (1972) Ion uptake by cold-impaired rice roots. *Plant Soil* **37**, 469–478.

AUTHOR INDEX

SUBJECT INDEX